Cold Science

Science during the Cold War has become a matter of lively interest within the historical research community, attracting the attention of scholars concerned with the history of science, the Cold War, and environmental history. The Arctic—recognized as a frontier of confrontation between the superpowers, and consequently central to the Cold War—has also attracted much attention. This edited collection speaks to this dual interest by providing innovative and authoritative analyses of the history of Arctic science during the Cold War.

Stephen Bocking is a Professor with the Trent School of the Environment at Trent University, Canada.

Daniel Heidt is the Research and Administration Manager at the Centre on Foreign Policy and Federalism, St. Jerome's University, Canada.

Routledge Studies in the History of Science, Technology and Medicine

For the full list of titles in the series, please visit: https://www.routledge.com/Routledge-Studies-in-the-History-of-Science-Technology-and-Medicine/book-series/HISTSCI.

Cold Science

Environmental Knowledge in the North American Arctic during the Cold War

**Edited by Stephen Bocking
and Daniel Heidt**

Routledge
Taylor & Francis Group

LONDON AND NEW YORK

First published 2019
by Routledge
2 Park Square, Milton Park, Abingdon, Oxon OX14 4RN

and by Routledge
52 Vanderbilt Avenue, New York, NY 10017

First issued in paperback 2020

Routledge is an imprint of the Taylor & Francis Group, an informa business

British Library Cataloguing-in-Publication Data
A catalogue record for this book is available from the British Library

Library of Congress Cataloging-in-Publication Data
Names: Bocking, Stephen, 1959-author. | Heidt, Daniel, 1985-author.
Title: Cold science : environmental knowledge in the North American Arctic during the Cold War / Stephen Bocking and Daniel Heidt.
Description: Abingdon, Oxon ; New York, NY: Routledge, 2019. |
Series: Routledge studies in the history of science, technology and medicine; 38 | Includes bibliographical references and index.
Identifiers: LCCN 2018048108| ISBN 9781138043961
(hardback: alk. paper) | ISBN 9781315172736 (ebook : alk. paper)
Subjects: LCSH: Science and state--United States--History--20th century. | Science and state--Canada--History--20th century.
| Cold War--Political aspects. | Arctic regions--Research--History--20th century. Classification: LCC Q127.U6 B5725 2019 | DDC
909/.09130825--dc23 LC record available at https://lccn.loc.gov/2018048108

ISBN 13: 978-0-367-66038-3 (pbk)
ISBN 13: 978-1-138-04396-1 (hbk)

Typeset in Times NR MT Pro
by Cenveo® Publisher Services

Contents

Contributors

Dawn Alexandrea Berry is a Postdoctoral Research Specialist/Contract Lead with SNA International Supporting Initiatives at the Defense POW/MIA Accounting Agency. Dr. Berry's research centers on diplomatic history's intersection with foreign policy, security studies, and business and environmental history, and explores how technological advances and competition for strategic resources affect global geopolitics in moments of economic crisis and war. She completed her doctoral work at the University of Oxford, UK, and has taught at universities in Canada, China, the United Kingdom, and the United States. The work for this article was completed while she was a Postdoctoral Fellow at the Einaudi Center for International Studies, Cornell University, USA.

Stephen Bocking is Professor of Environmental Policy and History in the School of the Environment at Trent University, Peterborough, Ontario, Canada. He teaches courses on science and politics, environmental history, and global political ecology. His research examines the evolution of environmental knowledge in relation to politics. He is co-editor of *Ice Blink: Navigating Northern Environmental History* (2017), editor of *Biodiversity in Canada: Ecology, Ideas, and Action* (2000), and author of *Nature's Experts: Science, Politics, and the Environment* (2004) and *Ecologists and Environmental Politics: A History of Contemporary Ecology* (1997).

Ryan Dean is a PhD candidate in the Department of Political Science at the University of Calgary, Canada. His doctoral dissertation examines the development of Canadian Arctic security policy since 1985. His recent co-edited volumes include *(Re)Conceptualizing Arctic Security: Selected Articles from the Journal of Military and Security Studies* (2017) and *Canada's Northern Strategy under the Harper Conservatives: Key Speeches and Documents on Sovereignty, Security, and Governance, 2006-15* (2016)."

Matthew Farish is an Associate Professor of Geography at the University of Toronto, Canada, where he teaches courses in cultural and historical geography. His research concerns the militarization of the planet by the United States during the middle decades of the twentieth century, the geographical knowledge generated to aid this process, and the North American landscapes created or transformed as a result. He is the author of *The Contours of America's Cold War* (2010).

Daniel Heidt is a Fellow as well as Research and Administration Manager with the Centre on Foreign Policy and Federalism, St. Jerome's University, Canada. His research interests include the Canadian Arctic during the Cold War, Canadian politics, and federalism. Heidt is currently finishing a history of the JAWS program with P. Whitney Lackenbauer.

Victoria Herrmann is President and Managing Director of The Arctic Institute, a non-profit organization dedicated to Arctic security research, where her research focuses on human security of remote communities. She is also a National Geographic Explorer, working with coastal communities in the United States and U.S. territories on climate change adaptation. Herrmann teaches sustainability management at American University, Washington, DC, and science communication at the University Centre of the Westfjords, Iceland. She was previously a Fulbright awardee to Canada and a Gates Scholar at Cambridge University, UK, where she received her PhD in Geography.

Adrian Howkins is a Reader in Environmental History at the University of Bristol, UK. His work focuses on the environmental history of the polar regions, and on parks and protected areas more broadly. He is author of *Frozen Empires: An Environmental History of the Antarctic Peninsula* (Oxford University Press, 2016) and *The Polar Regions: An Environmental History* (Polity Press, 2015). Additionally, he has co-edited *Antarctica and the Humanities* (Palgrave, 2016) and *National Parks Beyond the Nation* (University of Oklahoma Press, 2016). For the past seven years he has been a co-PI on the US National Science Foundation funded McMurdo Dry Valleys Long Term Ecological Research site in Antarctica.

Henrik Knudsen studied the history of ideas and history of science and technology, receiving his PhD from Aarhus University, Denmark in 2005. Currently he holds a position as archivist and senior researcher at the Danish National Archives. His research interests cover science and technology in the global cold war, science diplomacy, and nuclear history.

P. Whitney Lackenbauer is the Canada Research Chair (Tier 1) in the Study of the Canadian North and a Professor in the School for the Study of Canada at Trent University, Peterborough, Ontario, Canada. His books include *China's Arctic Ambitions and What They Mean for Canada* (co-authored 2018); *Canadian Armed Forces Arctic Operations, 1945–2015: Historical and Contemporary Lessons Learned* (co-edited, 2017); *The Canadian Rangers: A Living History, 1942–2012* (2013); *Canada and the Changing Arctic: Sovereignty, Security, and Stewardship* (co-authored 2011); and *Arctic Front: Defending Canada in the Far North* (co-authored 2008). His research focuses on Arctic policy, sovereignty, security, and governance issues; modern Canadian and circumpolar history; military history and contemporary defence policy; and Indigenous-state relations in Canada.

Julia Lajus is an Associate Professor of Department of History, St. Petersburg School of Social Sciences and Humanities of the National Research University Higher School of Economics, St. Petersburg, Russia. Her research interests include environmental and technological history of biological resources, especially in marine and polar areas, and history of field sciences such as fisheries science, oceanography, and geophysics. Julia has published four books in Russian (co-authored and edited), about twenty chapters in monographs published by leading international publishers, and papers in international journals.

Tess Lanzarotta is a Lecturer in the History of Science and Medicine at Yale University, USA, and a SSHRC Postdoctoral Fellow at the Dalla Lana School of Public Health at the University of Toronto, Canada. Her research explores both how biomedicine operates as a source of colonial governance and how it has been transformed into a site for the expression of Indigenous sovereignty. Her current book project is a history of biomedical research and public health in Cold War Alaska.

Mark Nuttall is Professor and Henry Marshall Tory Chair of Anthropology at the University of Alberta, Canada, and Fellow of the Royal Society of Canada. He is also visiting Professor of Climate and Society at the University of Greenland and Greenland Climate Research Centre. He is a co-author (with Klaus Dodds) of *The Scramble for the Poles* (Polity, 2016), author of *Climate, Society and Subsurface Politics in Greenland: Under the Great Ice* (Routledge, 2017), and co-editor of *The Routledge Handbook of the Polar Regions* (2018).

Robert Page has enjoyed a lengthy career as both an academic professor and administrator including publications on Arctic pipelines and permafrost, and a private sector business executive in the energy field, and is currently a corporate director in Canada and the United States. He spent three years in the regulatory hearings on the Mackenzie Valley and the Trans-Alaska Pipelines in both the US and Canada and gave testimony to the US Senate Environment & Public Works Committee. He is a Fellow of the Royal Canadian Geographical Society and the International Emissions Trading Association, Geneva, Switzerland. He had direct dealings with Soviet colleagues.

Peder Roberts is Associate Professor of Modern History at the University of Stavanger, Norway, and a researcher at the Division of History of Science, Technology and Environment at KTH Royal Institute of Technology, Stockholm. He has published extensively on the history of polar research with a particular focus on science, politics, and environmental knowledge. Peder is currently the leader of the European Research Council-funded project Greening the Poles: Science, the Environment, and the Creation of the Modern Arctic and Antarctic.

Rafico Ruiz is a Social Sciences and Humanities Research Council of Canada Banting Postdoctoral Fellow in the Department of Sociology at the University of Alberta, Canada. He holds a PhD in Communication Studies and the History & Theory of Architecture from McGill University, Canada. He studies the relationships between mediation and social space, particularly in the Arctic and Subarctic; the cultural geographies of natural resource engagements; and the philosophical and political stakes of infrastructural and ecological systems. His work appears in a number of journals and edited collections, including the *International Journal of Communication*, the *Journal of Northern Studies*, *Continuum: Journal of Media & Cultural Studies*, and *Communication* +1.

Andrew Stuhl is an Assistant Professor of Environmental Studies at Bucknell University, USA. His current work examines histories of knowledge and environment in modern treaties in Arctic North America as well as the colonial and environmental histories of offshore oil development in the Beaufort Sea. He is author of several articles, book chapters, and essays, including *Unfreezing the Arctic: Science, Colonialism, and the Transformation of Inuit Lands* (University of Chicago Press, 2016).

Lize-Marié van der Watt is a researcher at the Division for History of Science, Technology and Environment, KTH Royal Institute of Technology, Stockholm, Sweden. She studies the histories of polar pasts and polar futures, and her publications focus on environmental history, critical geopolitics, and heritage studies.

Matthew S. Wiseman is a SSHRC Postdoctoral Fellow with the Bill Graham Centre for Contemporary International History and the Department of History at the University of Toronto, Canada. He earned his PhD from Wilfrid Laurier University and the Tri-University Graduate Program in History in 2017. His doctoral dissertation, currently under contract and forthcoming with a major academic press, examines the role of science and technology in the formation and development of Canada's Arctic defence policy during the early Cold War. Matthew's work has appeared in such leading publications as *Canadian Military History* and the *Journal of the Canadian Historical Association*.

Part 1

Introductory perspectives

1 Introduction: Cold War science in the North American Arctic

Stephen Bocking and Daniel Heidt

Signs of change are evident across the Arctic: melting ice and permafrost, strange weather, unfamiliar "southern" species, and other indicators of accelerated warming. A once "pristine" region is now, like the rest of the planet, undergoing uncertain but dramatic transformation. There are other novelties as well. Arctic resources—oil, gas, diamonds, fish—are attracting attention, and a soon-to-be ice-free ocean promises new shipping routes. Arctic and non-Arctic nations alike are asserting their interests in the region, while Indigenous Peoples seek self-determination and a role in shaping its future. A "scramble" for the Arctic is underway.[1] In this scramble, knowledge has become central—as evidenced by initiatives such as the International Polar Year (2007–2008), ongoing mapping of the Arctic seabed, and efforts by the Arctic Council and other institutions to track circumpolar environmental change.

Recent events speak not just to the Arctic's future, but also to its past. At Camp Century in northwest Greenland, built in the early 1950s by the Americans near their air base at Thule, wastes once assumed to be frozen for all time are being exposed as the ice cap melts.[2] Their unexpected appearance illustrates how our experience of change in the Arctic is rooted in history, including the time when the region was a center of Cold War strategic tensions. Other activities today tell similar stories. Scientists tracking global environmental change rely on studies of the Arctic's ice and atmosphere undertaken decades ago by the superpowers and their allies. Interest in Arctic resources mirrors Cold War demands for strategic materials. Surveys of the Arctic seabed build on knowledge originally accumulated to guide ships and submarines. Cold War interventions reshaped Indigenous ways of life, provoking efforts to assert self-determination that continue today.

Scientific investigation was a defining feature of the Cold War Arctic. Once the Second World War reversed the assumption that the region was devoid of strategic value, the United States and Canada built roads, airfields and weather stations from Alaska to Iceland: an effort demanding knowledge of land, waters, and the weather.[3] After the war, bombers and missiles redefined the Arctic as a zone separating the two superpowers: a vulnerable flank demanding surveillance and a capacity for military operations. Science was mobilized to meet these requirements. Yet much of the scientific work that took place in the Arctic had no direct link to strategic priorities. Nations did not merely administer and exploit the Arctic, but redefined the region as their territory: surveying, categorizing and putting in order northern landscapes, resources and peoples.

This book examines "cold science": scientific activity in the Arctic during the Cold War. Our authors consider the evolution of this scientific activity in the context of the region's environmental and human histories, and the histories of Cold War science and Arctic affairs. Our focus is on North America: Canada, the United States, and Greenland. This place and time merits closer study. Above all, they provide an opportunity to examine the implications for scientific activity of a range of political and social developments: strategic tensions, efforts by allies to align with a superpower while asserting their autonomy, demands for natural resources, emerging concerns about the Arctic environment, and Indigenous Peoples' struggles to define their own futures. The inclusion of Greenland highlights the contested definition of North America. Even as Greenland was governed from Denmark, its strategic location made it susceptible to the United States' Cold War imperatives.[4]

The contexts of cold science

Arctic science was embedded within the larger context of the Cold War. The transformation of war from a conflict between neighbors to global confrontation made science a strategic necessity: relevant to watching enemies, inventing weapons, and supporting operations. Aircraft, submarines, satellites and other technologies created a demand to predict and control the environments in which they operated. Novel lines of inquiry and disciplines emerged, including an array of "environmental sciences": meteorology, geology, seismology, and oceanography. These sciences emerged as particularly crucial because information about the Earth—from gravitation to ocean currents to wind, encompassing landscapes, seas, and the atmosphere—became a strategic imperative. Their reach now extended everywhere, even into the upper atmosphere, the ocean depths, and the polar regions.[5] The concept itself of the "global environment" emerged, becoming evident in the International Geophysical Year (1957–58), and was soon embedded in transnational scientific organizations.[6] Scientists undertook experimental interventions, testing strategies of nuclear, chemical and biological warfare, as well as efforts to control the weather.[7] Evolving strategic priorities had a variety of consequences for science. For example, the 1963 Partial Test Ban Treaty made atmospheric science, including fallout monitoring, seem less urgent.[8] On the other hand, the Plowshares Program included studies of "peaceful" nuclear explosions, including Project Chariot—a plan to demonstrate their use by excavating a new harbor in Alaska.[9] New ways of organizing research appeared, including "Big Science": combinations of expertise and technologies applied to strategic objectives beyond the capabilities of individual scientists. A vast research complex, soon designated the "military-industrial-academic complex," linked power and the production of knowledge. As Stuart Leslie noted, the era "virtually redefined what it meant to be a scientist or an engineer," transforming institutions and agendas, reshaping the relations between science and the state.[10] These developments also provoked tensions and debate among scientists, both within disciplines and in the wider scientific community—including both a divergence of the physical and biological environmental sciences, as each responded to distinct disciplinary, social and political contexts, and their intersection in interesting and intellectually productive ways—as seen, for example, in whale science.[11]

The Cold War is usually said to have extended from the late 1940s to the dissolution of the Soviet Union in 1991. Yet as several historians have stressed, the Cold War was not a single episode, and viewing it as one risks obscuring political and social developments that extended beyond the geopolitical strategies of the superpowers. It evolved through a period of deep tensions during the 1950s, culminating in the 1962 Cuban Missile Crisis. This was followed by détente in the 1970s; and then renewed antagonisms in the early 1980s. In the 1970s novel concerns also appeared that transcended the East-West division: the postwar boom gave way to economic stagnation; anxieties about energy emerged, particularly in relation to Middle East conflicts; and the environment became a new focus of politics and policy.[12] Our understanding of Cold War science must accommodate this variegated nature of international and domestic politics, particularly by acknowledging that it responded to priorities beyond those implied by strategic imperatives.

As founding members of the military-industrial-academic complex, government and industry were assumed to share the same goals and expected to collaborate to achieve them. Accordingly, technological research commonly addressed both strategic and economic objectives. Military aviation was adapted for commercial purposes. Industrial interests managed government research operations: serving, for example, as contractors for the Atomic Energy Commission's national laboratories. The use of corporations as prime contractors in defense projects—an arrangement integral to Big Science—was described as the "American Method."[13] The oil industry supported geophysical and geochemical research—an instance of how science in service to Cold War industrial economy pushed the boundaries of resource extraction to the limits of the Earth.[14] Ways of thinking characteristic of the Cold War were also applied more widely. Rational resource planning was itself redefined as a strategic necessity.[15] International economic development, guided by experts, justified interventions in the "Third World."[16] Strategic technologies produced information that could be applied to other purposes: satellite images and computer models transformed civilian meteorology; nuclear technologies from the American Atomic Energy Commission (AEC) reshaped the material culture and practices of civilian medicine and biology (including ecology); field biologists seized on radio collars and remote sensing as Cold War icons of technological modernity.[17] Cold War science was thus marked by contingency and possibility, generating outcomes beyond those imagined by its patrons.

The emergence of environmentalism provoked further novel relations between Cold War science and social concerns. The iconic image of the "Earth rise," captured by astronauts orbiting the moon, demonstrated how surveillance technology could generate alternative perspectives: presenting the planet not as a space to be known and controlled, but as a fragile global environment.[18] Atmospheric scientists took on new issues, with NASA (National Aeronautics and Space Administration), itself a Cold War icon, becoming the largest American funder of climate science.[19] But research newly focused on the environment could still exhibit traces of its strategic origins. Global environmental monitoring programs were based on strategic monitoring systems, and experience with fallout framed studies of environmental contaminants, including persistent toxic chemicals.[20] Cold War systems engineering combined with ecology to form a view of the Earth as a global ecosystem amenable to management: some perceived a carefully regulated spaceship environment as a reassuringly technocratic alternative to an irrational, disorderly Earth.[21]

But amidst these diverse social and institutional imperatives, the continuing importance of individuals must also be noted. A few scientists acting as diplomats helped redefine international relations, by urging a "consensual hegemony" between the United States and its allies.[22] Harry Wexler of the United States Weather Bureau advanced his ambition of "making meteorology global" by applying computers, satellites, tracers and other technologies to the study of weather, the effects of nuclear testing, and the consequences of nuclear war. Lloyd Berkner framed much of the plan for the International Geophysical Year.[23] As well, many scientists receiving strategic patronage were still able to pursue their disciplinary agendas. American physicists used their association with the military to pursue a range of theoretical and empirical investigations. The study of Cold War science must thus consider how individuals (and their disciplines) formed a "mutual orientation" with their patrons, and how this was expressed through practices, instruments, sites and concepts.[24] This implies understanding disciplines as not simply concepts and practices formed by autonomous research communities, but as phenomena constituted out of the negotiation between scientists and their institutional patronage. This negotiation occurred at a variety of scales, often coalescing around certain objects, such as natural features, instruments, or even the planet itself.[25]

That scientists could define much of the terms of this negotiation reflected the belief that they should be "on tap": available when needed, but otherwise able to pursue their own interests. Scientists themselves found agreeable this view of the relation between science and power. This autonomy also had an ideological value, particularly in the United States, where, as expressed in Vannevar Bush's report, *Science, the Endless Frontier,* the notion of "basic" science served as an expression of free inquiry and international cooperation. More generally, science served as a symbol of modernity and of a rational world order. In return, strategic security became defined not as a matter of political debate, but as an objective principle founded on a national consensus, and fulfilled through technical expertise.[26]

Yet the evolving political contexts of the Cold War also had consequences for views of the social roles of scientists. By the late 1960s the American and Canadian militaries were losing interest in supporting research, even as the public was demanding more oversight of scientific activities.[27] Military research came under scrutiny, especially at universities such as MIT and Stanford that had strong ties with the Department of Defense. Science was increasingly commercialized, as industry became an important source of funding. There were also more general shifts in the social roles and status of experts: accustomed to a privileged role in decision-making, with ample discretion (reflecting the trust in expertise characteristic of modernity), scientists and the institutions that employed them were challenged by demands that these decisions be open to public scrutiny and accountability. Science itself became a resource for industry and activists alike, and scientific debates increasingly took place in public, with disagreements between experts put on display. The tools of Cold War strategy were now used not to reinforce structures of power, but to challenge them: computer modeling encouraged debate about the "limits to growth," and ecosystem ecology, originally fostered by the American nuclear establishment, was now invoked by critics of industrial society. The authority of science itself was questioned, as skepticism about experts took hold—instilling a sense of science being "in crisis."[28]

Cold War Arctic science

For the purposes of this volume the significance of the brief review of Cold War science presented above lies in the agenda it suggests for the study of science in the Arctic. As Canadian historian Kenneth Coates once noted, the North has consistently been defined by the "conceptual frameworks and intellectual paradigms" of southerners.[29] So it was during the Cold War: the Arctic became a strategic "laboratory," as science shifted from the era of exploration towards a more systematic, state-centered model.[30] American initiatives included the Arctic Research Laboratory in Alaska (established in 1947, becoming the Naval Arctic Research Laboratory in 1967) and the Snow, Ice and Permafrost Research Establishment (1949), as well as activities in northern Canada and at Thule Air Base in Greenland (including Camp Century). The Danish government established its own research operations in Greenland, studying weather, geology, the ionosphere and other aspects of its physical environment. In Canada the Defence Research Board (DRB) assumed broad responsibility for strategic research, with a special focus on the North. In 1959 the Polar

Continental Shelf Project began operations, supporting oceanographic and related research. The Arctic Institute of North America facilitated cooperation between the United States and Canada across a range of research areas.[31] Almost every aspect of the Arctic physical environment became of interest, from the atmosphere to the Earth's gravity and magnetic fields, and the Arctic Ocean's ice cover. This interest encompassed not only the horizontal but also the vertical: what lay beneath ice, water, permafrost, rock or snow was also of concern. The result was a total science of the environment across a range of fields: from geophysics, to glaciology, to atmospheric chemistry.[32]

Much of this scientific activity was motivated by the need to enable people, aircraft, ships, submarines and missiles to work, communicate and travel across, above and below this territory. The Arctic became a proving ground: a site for adapting southern technology and bodies to northern conditions. Ensuring people would be able to operate in this environment required testing human, mechanical and structural endurance.[33] Efforts such as Operation Muskox in 1946 (a land-borne expedition across several thousand kilometers of northern terrain) evaluated the mobility of land forces and their equipment. The Defence Research Northern Laboratory at Fort Churchill on Hudson Bay pursued studies of clothing, equipment, fuels, lubricants, and nutrition; even insects were studied, in order to learn how to operate amidst blackfly swarms. Studies of ice engineering at Camp Century examined whether permanent structures could be constructed within the Greenland ice cap. The point was to gain the "terrain intelligence" needed to work with, and not against, Arctic conditions.[34]

Arctic science in North America extended beyond what could be justified in strategic terms. When Harald Sverdrup visited the United States and Canada in 1948, he noted that while the military had a strong influence on Arctic research, all scientific work was encouraged.[35] Scientists at the Arctic Research Laboratory tracked whale and bird migrations, surveyed polar bears and lemmings, and pursued ethnographic research with Inuit. The Defence Research Board defined its northern mandate broadly, supporting a variety of projects by university scientists. Greenland too served as a platform for wide-ranging studies of the physical environment. As a wider range of agencies became involved in northern science, they provided opportunities for scientists to frame research questions other than in terms of strategic priorities—redefining the Arctic as a natural "laboratory" for a variety of disciplines. These relations between strategic and non-strategic research reflected how scientific, military, and civilian communities maintained close ties in the Arctic, while sharing in the role of science as a marker of the state's presence.

Arctic science had several distinctive features. Some stemmed from the place itself: cold, ice, and the seasonal contrast between dark and light were the most obvious ways in which this environment differed from temperate regions; other differences, including unusual atmospheric conditions and animals' adaptations to Arctic conditions, became apparent through study. The environment also imposed special logistical requirements for housing and travel, and distinctive challenges for field practices. This place also challenged scientists' authority: knowledge gained in temperate regions often did not apply to the Arctic—as they found when they tried to build on permafrost, or study the effects of radioactive fallout. These features reinforced the region's status as distinctive—an extreme, "hostile" environment, anomalous in relation to "normal" temperate regions. Scientists responded in a variety of ways. They usually saw themselves as visitors, producing knowledge that could mediate the relation between other outsiders and this unfamiliar environment. Airplanes and other technologies distanced scientists from the Arctic environment, while aerial surveys and mapping helped render this terrain legible. The persistent view of the north as a frontier (for exploration, testing oneself against the elements, or resource exploitation) was invoked, as was the ambition of conquering a wilderness that had become an adversary.[36] These attempts were often struggles, and scientists sometimes failed. Their simplified, schematic view of the north also often produced unexpected consequences, particularly for the environment.

Yet for scientists, this environment, while distinctive, did not exist in isolation. The region became an element of global surveillance efforts.[37] Meteorologists sought to understand the Arctic's role in global weather patterns, particularly the formation of polar fronts. International Geophysical Year research exemplified the dual regional and planetary significance of Arctic science: synoptic observations defined the Arctic as a distinctive environment, while embedding the region within global systems.[38] Climate change emerged as an issue soon after the Second World War with observations of warming—a phenomenon especially relevant to ice, and so one that aroused much strategic interest. While this warming was not immediately attributed to human activity, much of what we have since learned about climate change is due to Cold War research, with the Arctic and Antarctic— ironically, the environments least hospitable to humans—becoming observatories of a newly vulnerable global environment. The Greenland and Antarctic ice caps provide an archive of climate change, exhibiting the correlation between warming and carbon dioxide, as well as traces of centuries of industrial pollution, challenging assumptions of a "pristine" Arctic.[39] Analytical and modeling techniques shifted environmental science from description to prediction, noting, for example, the prospect of an ice-free Arctic Ocean.[40] In the late 1950s puzzling concentrations of Strontium-90 in caribou and Inuit demonstrated the role of the Arctic environment in mediating the relations between humans and fallout. Studies of circulation of dust and other substances in the atmosphere reinforced perceptions of the Arctic's vulnerability to the long-range transport of contaminants.[41]

Arctic science became defined by transnational flows of people and ideas—a web of knowledge production and circulation that responded to both the region's environmental challenges and to political and strategic imperatives. The United States' hegemonic relations with its allies gave it a measure of authority to operate beyond its borders, including in Greenland.[42]

North America and the Nordic countries cooperated in strategic science, particularly in fields such as meteorology. Links also formed across the Cold War frontier, aligning with the belief in science as a basis for diplomacy. The "natural affinity" implied by shared environmental challenges spurred interest in technical cooperation between Canada and the Soviet Union: intellectuals such as Harold Innis saw the Soviet experience of applying science to northern development as a model, Lester Pearson (the Canadian minister of external affairs, who became prime minister in the 1960s) visited Moscow in 1955, and the DRB's sea ice specialist, Moira Dunbar, learned Russian to access Soviet sea ice literature—acknowledging the advanced state of their Arctic science.[43]

These international relations also hint of the place of the Arctic in the larger history of the Cold War. Some global aspects of the Cold War were also experienced in the Arctic, but in ways distinctive to the region. By the early 1960s the focus of superpower confrontation had shifted elsewhere, to Berlin, Cuba, the Middle East and other flashpoints, while technological developments made military confrontation in the Arctic seem less likely: no longer was it imagined as a potential battleground. Military activities in the region (including scientific work) declined accordingly. Détente in the early 1970s took on a distinctive character in the Arctic: concerns about the status of polar bears provided an opportunity to test East-West cooperation on an issue with relatively low stakes; a little over a decade later Arctic environmental monitoring gave Soviet leader Mikhail Gorbachev a similar opportunity. Other international developments were also reflected in Arctic science, but in ways specific to the region. Anxieties over energy stimulated interest in Arctic oil and gas, and environmentalists described the Arctic as distinctively fragile. The presence of Indigenous Peoples, and their efforts to assert their own identities (including their own knowledge systems), became after the 1970s a distinctive feature of Arctic science.

Political divisions within North America also influenced scientific activities—reminding us that Cold War tensions were not only between East and West. That "American," "Canadian," or "Danish" Arctic science emerged out of a transnational landscape was the negotiated outcome of institutions and policies that defined research activities as national efforts. Maintaining a presence in the North became a priority for Canada, with not just military but civilian institutions such as the Canadian Wildlife Service and the Geological Survey of Canada building on the sense of nationhood produced by the war, while balancing the American presence in its territory. Similarly, the Polar Continental Shelf Project, while collecting geophysical data for American rocket launches, asserted a Canadian presence in the Arctic. Even the study of radio waves was enlisted, through research that defined a distinctive Canadian ionosphere.[44] In Greenland as well, science became a kind of "political performance": a means of asserting (Danish) national authority over territory, through studies of glaciers, the ionosphere, and other aspects of the physical environment.[45] Boundaries within national territories were also evident in the encounter in Canada between southern authority and Indigenous understandings of place, and in the evolution of self-government in Greenland.[46]

Just as Cold War science did generally, Arctic science exhibited close relations between strategic and corporate interests, but with features specific to this environment. One was a shared interest in learning how to handle its distinctive challenges. Petroleum development and strategic activities both drew on "terrain intelligence," particularly when it came to operating on permafrost and ice.[47] The Arctic's function as a proving ground extended to industrial technologies such as pipelines and drilling rigs. Efforts to adapt aviation and radio communication to local conditions similarly responded to this dual strategic and industrial interest. Western Electric developed radar technology for Distant Early Warning (DEW) Line stations constructed near the Arctic Circle, while transportation contractors, including civilian airlines, moved people, equipment and supplies into position; several of these airlines would go on to play substantial roles in enabling mobility across the Arctic.[48] Strategic imperatives and resource development both demanded knowledge of vertical territories, and so surveys addressing one priority also benefited the other. More generally, strategic and economic projects exhibited a similar view of Arctic nature: a high modernist perspective based on a schematized, depoliticized view of the region, emphasizing its ties to the rest of the world.[49]

Indigenous Peoples experienced the Cold War in a variety of ways. At Great Bear Lake in northern Canada, Dene carried uranium ore as it began its journey to the American atomic bomb (code named the "Manhattan Project"); their work left a legacy of cancer. In the 1950s Inughuit (Inuit of Greenland) were relocated to make way for the Thule air base, DEW Line stations influenced Inuit settlement and hunting, and Project Chariot in Alaska raised fears of radioactive contamination.[50] In the 1980s low-level flight training in Labrador disrupted Innu lives. Toxic waste at abandoned DEW Line stations and other Cold War relics remains an ongoing issue. Indigenous Peoples' relations with science itself have been no less complex. During the early Cold War scientists viewed them as a resource: at the Arctic Aeromedical Laboratory experimental studies of acclimatization sought a racialized basis—an "Eskimo" physiological secret—for tolerating cold. Other studies involved administering radioactive iodine.[51] However, Arctic science soon developed in ways that marginalized Indigenous people: equipped with airplanes, helicopters and field stations, researchers no longer needed to rely on traditional ways of travel and survival. Scientific activities increasingly proceeded at a distance from Indigenous people, even within their traditional territories. The view of the Arctic as an empty strategic space erased local knowledge and experience, while scientists' ties with colonial authority fed Indigenous Peoples' distrust of their actions and motives, epitomizing the strains between views of the Arctic as a place that is home or "away."

These tensions would eventually play a role in northern political developments. By the 1970s, as Indigenous people began to assert rights of self-determination, they were also forming new relations with scientists.

Territorial and subsurface knowledge gained through Cold War surveys was applied to land claims negotiations. Scientists increasingly recognized the value of Indigenous knowledge, and northern governments, particularly in Canada, asserted its central—though still contested—role in environmental management. Today, Indigenous perspectives on contaminants, including radioactive materials and other forms of "fallout" such as persistent organic pollutants, illustrate the potential for other ways of understanding the Arctic environment, including novel combinations of Indigenous and scientific knowledge.[52] After long being ignored, Arctic Indigenous People have begun to be heard. Yet scientific narratives of the impact of global environmental change on the Arctic can still muffle their voices, or hear them only as the pleas of victims. Cold War science must thus also be understood in terms of postcolonial perspectives: the Indigenous presence challenges conventional definitions of knowledge, and highlights the need to acknowledge different ways of knowing the Arctic.[53]

These issues define an agenda for the study of Cold War Arctic science, one that places it within the broader history of Cold War science, while acknowledging its distinctive Arctic and North American contexts. This distinctive history exemplifies the role of place and space in forming knowledge, and the movement of ideas, materials, and power between regions. The Arctic itself has come to be understood less in terms of a strict frontier separating north and south, than as a multiplicity of spaces defined by local places, residents and visitors, by global geopolitical, economic and environmental contexts, and by the place of polar regions in the imagination.[54] Within these spaces science has repeatedly served to link different political and cultural conversations— about military strategy, economic development, environmental protection, or Indigenous rights—on the basis of a shared interest in certain features of the environment, such as permafrost, sea ice, or northern wildlife. In pursuit of this agenda our authors explore both Cold War science and science conducted during the Cold War—that is, scientific activities more immediately tied to strategic priorities, as well as the wider array of research conducted across the Arctic. Together, they map the Cold War's northern edges.

Exploring cold science

Following this introduction, Part 2 begins with several chapters that examine the relations between science and strategy. Mark Nuttall travels to Melville Bay in northwest Greenland, where, drawing from his long experience in this region, he considers its shifting identities: as Indigenous homeland, navigational hazard, strategic priority, and place for science. These identities have also been tied to ways of knowing—particularly when the arrival of strategic science apparently erased Indigenous knowledge from the bay. From Greenland we move across the Davis Strait to Canada, where, as P. Whitney Lackenbauer and Daniel Heidt explain, soon after the Second World War a meteorological network known as JAWS—Joint Arctic Weather Stations—were constructed.

Called into existence by demands for forecasting for agriculture, civil aviation as well as strategic operations, the network exemplified the cooperative ties between the United States and its Cold War allies. The United States Weather Bureau and the Canadian Department of Transport were essential in establishing this network, illustrating the importance of civilian agencies, agendas and cultures in Arctic science, even during the tense early years of the Cold War. In our next chapter, Matthew S. Wiseman examines studies of human adaptation in northern Canada, as described in two films produced in 1948 and 1954. These films provide valuable perspectives on how soldiers trained for cold-weather warfare, while the application of science to operating in the "hostile" environment that they portrayed revealed assumptions about Inuit bodies and adaptation. Shifting attention to Alaska, Victoria Herrmann examines *Life* magazine's coverage of Alaska during the Second World War and the early Cold War. Words and images presented Alaska as, alternatively, a distant strategic space, part of the American homeland, or a site for the deployment of advanced technology in the form of DEW Line stations. Finally, Matthew Farish returns us to Canada, to consider the formation of postwar scientific institutions, including the Arctic Institute of North America and the Defence Research Board. These were the handiwork of a small number of individuals, linked by shared northern experiences, who together forged close relations between academia (particularly McGill University) and the military. Chief among their ambitions were the application of the human and medical sciences to northern challenges, through the concept of "man in the north"— the generic white, male soldier—a synecdoche for southern sojourners able to operate in the north.

In Part 3, our contributors extend our attention beyond the strategic, exploring the diverse imperatives—economic, environmental, and Indigenous—that, beginning in the 1970s, also motivated scientific study of the Cold War Arctic. Robert Page provides a participant's view of the discovery and exploitation of Arctic oil and gas. This discovery provoked study of the technical challenges involved in extracting and shipping them south, particularly that of building pipelines on permafrost. Complex cross-border circumstances, including the similar challenges the Soviets were experiencing with their oil industry, and tensions between Canada and the United States, demonstrated the intersections of technology, economics, and international relations. Stephen Bocking extends this theme, focusing on the transition from strategic to resource research as the Canadian petroleum industry shifted northwards. These research areas shared several features, illustrating how, even as resource extraction and environmental protection became priorities, ways of doing science that had formed during the early Cold War could persist. Next, we turn to an unusual resource: icebergs. As Rafico Ruiz explains, both science and the media helped justify redefining ice as a source of freshwater, amidst an amalgam of commercial and nationalistic priorities and international cooperation. Finally, Andrew Stuhl extends our portrayal of cold science to encompass Indigenous land claims and their relations with scientific and other forms of knowledge. These claims asserted visions based on the region's living resources, and on a view of the Arctic itself as a homeland—a perspective sharply contrasting with the views that dominated much of the Cold War.

In Part 4, we turn to Cold War science as a transnational phenomenon. As Dawn Berry explains, the strategic significance of Greenland, especially to Americans, began to emerge even before the Cold War, through the efforts of airborne explorers. Her account extends our understanding of the origins of the Cold War, while illustrating the continuing influence of scientific, technological, military and political interests. Lize-Marié van der Watt, Peder Roberts and Julia Lajus examine how institutions have shaped the transnational ties that established Arctic research as a circumpolar phenomenon. By comparing the development of these institutions across several countries, they demonstrate the importance of a few well-connected individuals, of concepts of professionalization in Arctic science, and of evolving views of Arctic places. In the next chapter Henrik Knudsen turns our attention to Greenland in the 1960s, and the sometimes contentious relations between Denmark and the United States, particularly concerning the launching of rockets used to probe the ionosphere. His account illustrates the implications for science and politics of the unequal relations between the United States and its allies, particularly amidst unexpected events, including the crash of a B-52 bomber. Moving to Canada, P. Whitney Lackenbauer and Ryan Dean analyze what happened when a Cold War artifact—the Cosmos 954 satellite—crashed in 1978 in the Northwest Territories. The incident provoked elaborate efforts by the United States and Canada to recover the remains, involving soldiers, aircraft and radiation detectors. The episode illustrated how Cold War technologies could produce not just unintended outcomes, but also the means of cleaning them up. Finally, Tess Lanzarotta examines an effort in the 1980s to use medical research as the basis for collaborating across the Cold War divide. Through the Alaska Siberia Medical Research Program the health of Indigenous populations became central to a peace-building initiative. The episode illustrates the interesting forms that East-West scientific relations could take at a time when all assumptions had become open to challenge.

In Part 5, our final chapter by Adrian Howkins explores the lessons to be learned about Arctic science through comparison with developments in Antarctica. As he explains, Cold War science in these regions exhibited some similarities, but also profound differences. Attention to these similarities (cold, ice, remoteness, difficult access) and differences (presence or absence of permanent inhabitants; national territories versus international preserve), can, as Howkins shows, generate insights into the history of scientific activities in both polar regions.

Together, these chapters broaden our understanding of Cold War Arctic science in North America, particularly by extending our analysis to encompass its social and political contexts, such as the relations between the United States and its allies, the roles of economic priorities and social contexts, and the presence and influence of Indigenous People. They also exhibit the interplay of geography and history in Cold War science: that is, it was situated in specific places, and made mobile in response to the diverse interests and priorities that have shaped contemporary science.

Notes

1. Klaus Dodds and Mark Nuttall, *The Scramble for the Poles: The Geopolitics of the Arctic and Antarctic*, (Malden, MA: Polity, 2015).
2. Jeff D. Colgan, "Climate Change and the Politics of Military Bases," *Global Environmental Politics*, 2018, 18(1): 33–51.
3. P. Whitney Lackenbauer and Matthew Farish, "The Cold War on Canadian Soil: Militarizing a Northern Environment," *Environmental History*, 2007, 12: 920–50.
4. Planning of this volume began at a workshop hosted by Trent University in April 2016. Besides our contributors, we would like to thank the other workshop participants, who contributed much to defining the perspective on Cold War history of science that this volume represents: Tina Adcock, Dag Avango, Fikret Berkes, Michael Bravo, Patricia Cochran, Ronald Doel, Petra Dolata, Christopher Furgal, Amanda Graham, Shelagh Grant, Peter Kikkert, Tina Loo, Heather Nicol, Mitchell Patterson, Richard Powell and Sverker Sörlin. We would also like to thank the Social Sciences and Humanities Research Council of Canada (Award number 611-2015-165), Trent University, the Swedish Research Council and the Symons Trust for their support of this workshop.
5. Useful overviews of these developments and their contexts include: J. R. McNeill and Corinna R. Unger, eds., *Environmental Histories of the Cold War*, (New York: Cambridge University Press, 2010); Naomi Oreskes and John Krige, eds., *Science and Technology in the Global Cold War*, (Cambridge, MA: MIT Press, 2014); Roger D. Launius, James Rodger Fleming and David H. DeVorkin, eds., *Globalizing Polar Science: Reconsidering the International Polar and Geophysical Years*, (New York: Palgrave Macmillan, 2010); Jacob Hamblin, *Arming Mother Nature: The Birth of Catastrophic Environmentalism*, (New York: Oxford University Press, 2013); Ronald E. Doel, "Constituting the Postwar Earth Sciences: The Military's Influence on the Environmental Sciences in the USA after 1945," *Social Studies of Science*, 2003, 33(5): 635–666; Simone Turchetti and Peder Roberts, eds., *The Surveillance Imperative: Geosciences during the Cold War and Beyond*, (New York: Palgrave Macmillan, 2014); Matthias Heymann, Henrik Knudsen, Maiken L. Lolck, Henry Nielsen Kristian H. Nielsen and Christopher J. Ries, "Exploring Greenland: Science and Technology in Cold War Settings," *Scientia Canadensis*, 2010, 33(2): 11–42.
6. Susan Barr and Cornelia Lüdecke, eds., *The History of the International Polar Years (IPYs)*, (Heidelberg: Springer, 2010); E. Jerry Jessee, "A Heightened Controversy: Nuclear Weapons Testing, Radioactive Tracers, and the Dynamic Stratosphere," in: James Rodger Fleming and Ann Johnson, eds., *Toxic Airs: Body, Place, Planet in Historical Perspective*, (Pittsburgh: University of Pittsburgh Press, 2014), 152–180; Robert Poole, "What Was Whole about the Whole Earth? Cold War and Scientific Revolution," in: Turchetti and Roberts, eds., *The Surveillance Imperative*, 213–235; Sebastian Vincent Grevsmühl, "Serendipitous Outcomes in Space History: From Space Photography to Environmental Surveillance," in: Turchetti and Roberts, eds., *The Surveillance Imperative*, 171–191; James Rodger Fleming, "Polar and Global Meteorology in the Career of Harry Wexler," in: Roger D. Launius, James Rodger Fleming and David H. DeVorkin, eds., *Globalizing Polar Science: Reconsidering the International Polar and Geophysical Years*, (New York: Palgrave Macmillan, 2010), 225–241.

7. Fleming, "Polar and Global Meteorology"; Hamblin, *Arming Mother Nature*; Donald Avery, *Pathogens for War: Biological Weapons, Canadian Life Scientists, and North American Biodefence*, (Toronto: University of Toronto Press, 2013).
8. Matthias Dörries, "The Politics of Atmospheric Sciences: 'Nuclear Winter' and Global Climate Change," *Osiris*, 2011, 26(1): 198–223.
9. Dan O'Neill, *The Firecracker Boys*, (New York: St. Martin's Griffin, 1994).
10. Stuart W. Leslie, *The Cold War and American Science: The Military-Industrial-Academic Complex at MIT and Stanford*, (New York: Columbia University Press, 1993): 9; on "Big Science" generally as a Cold War phenomenon, see: Jon Agar, *Science in the Twentieth Century and Beyond*, (Malden, MA: Polity, 2012), 330–339.
11. Ronald E. Doel, "Quelle Place pour les Sciences de l'Environnement Physique dans l'Histoire Environnementale?" *Revue d'Histoire Moderne and Contemporaine*, 2009, 56(4): 137–164; Max Ritts and John Shiga, "Military Cetology," *Environmental Humanities*, 2016, 8(2): 196–214.
12. Rüdiger Graf, "Détente Science? Transformations of Knowledge and Expertise in the 1970s," *Centaurus*, 2017, 59: 10–25; Matthias Heymann, "1970s: Turn of an Era in the History of Science?" *Centaurus*, 2017, 59: 1–9.
13. Matthew Farish and P. Whitney Lackenbauer, "Western Electric Turns North: Technicians and the Transformation of the Cold War Arctic," in: Stephen Bocking and Brad Martin, eds., *Ice Blink: Navigating Northern Environmental History*, (Calgary: University of Calgary Press, 2017), 261–292.
14. Ronald E. Doel, "The Earth Sciences and Geophysics," in: John Krige and Dominique Pestre, eds., *Companion to Science in the Twentieth Century* (London and New York: Routledge, 2003).
15. Thomas Robertson, "'This is the American Earth': American Empire, the Cold War, and American Environmentalism," *Diplomatic History*, 2008, 32(4): 561–584.
16. Odd Arne Westad, *The Global Cold War: Third World Interventions and the Making of Our Times*, (Cambridge: Cambridge University Press, 2005).
17. Jessee, "A Heightened Controversy"; Angela N. H. Creager, *Life Atomic: A History of Radioisotopes in Science and Medicine* (Chicago: University of Chicago Press, 2013); Etienne Benson, *Wired Wilderness: Technologies of Tracking and the Making of Modern Wildlife*, (Baltimore: Johns Hopkins University Press, 2010).
18. Robertson, "This is the American Earth"; Sheila Jasanoff, "Heaven and Earth: The Politics of Environmental Images," in: Sheila Jasanoff and Marybeth Martello, eds., *Earthly Politics: Local and Global in Environmental Governance*, (Cambridge: MIT Press, 2004); Jessee, "A Heightened Controversy."
19. Dörries, "Politics of Atmospheric Sciences"; Erik M. Conway, "Bringing NASA Back to Earth: A Search for Relevance During the Cold War," in: Oreskes and Krige, eds., *Science and Technology in the Global Cold War*, 251–272.
20. Toshihiro Higuchi, "Atmospheric Nuclear Weapons Testing and the Debate on Risk Knowledge in Cold War America, 1945–1963," in: McNeill and Unger, eds., *Environmental Histories of the Cold War*, 301–322; Soraya Boudia, "Observing the Environmental Turn through the Global Environment Monitoring System," in: Turchetti and Roberts, eds., *The Surveillance Imperative*, 195–212.
21. Peder Anker, "The Ecological Colonization of Space," *Environmental History*, 2005, 10(2): 239–268; Sabine Höhler, "'Spaceship Earth': Envisioning Human Habitats in the Environmental Age," *GHI Bulletin*, 2008, no. 42: 65–85; Sabine Höhler, "The Environment as a Life Support System: The Case of Biosphere 2," *History and Technology*, 2010, 26(1): 39–58.
22. John Krige, *American Hegemony and the Postwar Reconstruction of Science in Europe*, (Cambridge: MIT Press, 2006).

23. Dörries, "Politics of Atmospheric Sciences"; Fleming, "Polar and Global Meteorology"; Allan A. Needell, "Lloyd Berkner and the International Geophysical Year Proposal in Context: With Some Comments on the Implications for the Comité Spéciale de l'Année Géophysique Internationale, CSAGI, Request for Launching Earth Orbiting Satellites," in: Launius et al., eds., *Globalizing Polar Science*, 205–224.
24. Michael Aaron Dennis, "Postscript: Earthly Matters: On the Cold War and the Earth Science," *Social Studies of Science*, 2003, 33(5): 809–819; Dan Kevles, "Cold War and Hot Physics: Science, Security, and the American State, 1945–56" *Historical Studies in the Physical and Biological Sciences* 20 no. 2 (1990): 239–264; Sverker Sörlin, "Narratives and Counter-Narratives of Climate Change: North Atlantic Glaciology and Meteorology, c.1930–1955," *Journal of Historical Geography*, 2009, 35: 237–255; Naomi Oreskes, "Science in the Origins of the Cold War," in: Oreskes and Krige, eds., *Science and Technology in the Global Cold War*, 11–29; John Krige, "Concluding Remarks," in: Oreskes and Krige, eds., *Science and Technology in the Global Cold War*, 431–441.
25. Needell, "Lloyd Berkner."
26. Dennis, "Postscript: Earthly Matters"; Naomi Oreskes, "Science, Technology, and Ideology," *Centaurus*, 2010, 52: 297–310; C. A. Miller, "Scientific Internationalism in American Foreign Policy: The Case of Meteorology, 1947–1958," in: C. A. Miller and P. N. Edwards, eds., *Changing the Atmosphere: Expert Knowledge and Environmental Governance, Politics, Science, and the Environment*, (Cambridge: MIT Press, 2001).
27. Krige, "Concluding Remarks."
28. Jon Agar, "What Happened in the Sixties?" *British Journal for the History of Science*, 2008, 41(4): 567–600.
29. Kenneth Coates, "The Discovery of the North: Towards a Conceptual Framework for the Study of Northern/Remote Regions," *The Northern Review*, 1994, 12/13: 15.
30. Matthew Farish, "The Lab and the Land: Overcoming the Arctic in Cold War Alaska," *Isis*, 2013, 104: 1–29.
31. Matthew Farish, "Creating Cold War Climates: The Laboratories of American Globalism," in: McNeill and Unger, eds., *Environmental Histories of the Cold War*, 51–84; Ronald E. Doel, Kristine G. Harper and Matthias Heymann, eds., *Exploring Greenland: Cold War Science and Technology on Ice*, (New York: Palgrave Macmillan, 2016); Ronald E. Doel, Robert Marc Friedman, Julia Lajus, Sverker Sörlin and Urban Wråkberg, "Strategic Arctic Science: National Interests in Building Natural Knowledge—Interwar Era through the Cold War," *Journal of Historical Geography*, 2014, 44: 60–80.
32. Peder Roberts, "Scientists and Sea Ice Under Surveillance in the Early Cold War," in: Turchetti and Roberts, eds., *The Surveillance Imperative*, 125–144; Edward Jones-Imhotep, "Communicating the North: Scientific Practice and Canadian Postwar Identity," *Osiris*, 2009, 24: 144–164.
33. Matthew S. Wiseman, "The Development of Cold War Soldiery: Acclimatisation Research and Military Indoctrination in the Canadian Arctic, 1947–1953," *Canadian Military History*, 2015, 24(2): 127–155; Farish, "Creating Cold War Climates"; Ronald E. Doel, Kristine C. Harper, and Matthias Heymann, "Introduction: Exploring Greenland's Secrets: Science, Technology, Diplomacy, and Cold War Planning in Global Contexts," in: Doel, et al., eds., *Exploring Greenland*, 1–22.
34. Wiseman, "Development of Cold War Soldiery"; Farish, "Creating Cold War Climates"; Andrew Iarocci, "Opening the North: Technology and Training at the Fort Churchill Joint Services Experimental Testing Station, 1946–64," *Canadian Army Journal*, 2008, 10(4): 74–95; Peter Kikkert, "The Polaris

Incident: 'Going to the Mat' with the Americans," *Journal of Military and Strategic Studies*, 2009, 11(3): 1–29; Henry Nielsen and Kristian H. Nielsen, "Camp Century—Cold War City Under the Ice," in: *Exploring Greenland*, 195–216.

35. Stian Bones, "Science In-between: Norway, the European Arctic and the Soviet Union," in: Sverker Sörlin, ed., *Science, Geopolitics and Culture in the Polar Region: Norden Beyond Borders*, (Burlington: Ashgate, 2013), 143–169.

36. Liza Piper, "Introduction: The History of Circumpolar Science and Technology," *Scientia Canadensis*, 2010, 33(2): 1–9; Lackenbauer and Farish, "Cold War on Canadian Soil".

37. Néstor Herran, "'Unscare' and Conceal: The United Nations Scientific Committee on the Effects of Atomic Radiation and the Origin of International Radiation Monitoring," in: Turchetti and Roberts, eds., *The Surveillance Imperative*, 69–84.

38. Sverker Sörlin, "Narratives and Counter-Narratives of Climate Change: North Atlantic Glaciology and Meteorology, c.1930–1955," *Journal of Historical Geography*, 2009, 35: 237–255; Doel et al., "Strategic Arctic Science"; Fleming, "Polar and Global Meteorology".

39. Adrian Howkins, *The Polar Regions: An Environmental History*, (Malden, MA: Polity, 2016); Richard B. Alley, *The Two-Mile Time Machine: Ice Cores, Abrupt Climate Change, and our Future*, (Princeton: Princeton University Press, 2000); Janet Martin-Nielsen,"'The Deepest and Most Rewarding Hole Ever Drilled': Ice Cores and the Cold War in Greenland," *Annals of Science*, 2013, 70(1): 47–70.

40. Doel et al., "Strategic Arctic Science."

41. Higuchi, "Atmospheric Nuclear Weapons Testing"; O'Neill, *Firecracker Boys*; Stephen Bocking, "Toxic Surprises: Contaminants and Knowledge in the Northern Environment," in: Bocking and Martin, eds., *Ice Blink*, 421–464.

42. Sverker Sörlin, "Introduction: Polar Extensions—Nordic States and their Polar Strategies," in: Sörlin, ed., *Science, Geopolitics and Culture in the Polar Region*, 1–19; Sverker Sörlin, "Circumpolar Science: Scandinavian Approaches to the Arctic and the North Atlantic, ca. 1920 to 1960," *Science in Context*, 2014, 27(2): 275–305; Jørgen Taagholt and Jens Claus Hansen, *Greenland: Security Perspectives*, Trans. Daniel Lufkin. (Fairbanks, Alaska: Arctic Research Consortium of the United States, 2001), 29; Matthias Heymann, "In Search of Control: Arctic Weather Stations in the Early Cold War," in Doel et al, eds., *Exploring Greenland*, 75–98; Janet Martin-Nielsen, "The Other Cold War: The United States and Greenland's Ice Sheet Environment, 1948–1966," *Journal of Historical Geography* 2012, 38: 69–80.

43. Sörlin, "Polar Extensions"; Bones, "Science In-between"; Roberts, "Scientists and Sea Ice"; Doel et al., "Strategic Arctic Science"; William J. Buxton, "Northern Enlightenment: Innis's 1945 Trip to Russia and its Aftermath," in: William J. Buxton, ed., *Harold Innis and the North: Appraisals and Contestations*, (Montreal: McGill-Queen's University Press, 2013), 246–272.

44. Richard C. Powell, "Science, Sovereignty and Nation: Canada and the Legacy of the International Geophysical Year, 1957–58," *Journal of Historical Geography*, 2008, 34: 618–638; Jones-Imhotep, "Communicating the North."

45. Sörlin, "Polar Extensions"; Lisbeth Lewander, "The Nordic Arctic Periphery: Fragments from Fieldwork," in: Sverker Sörlin, ed., *Science, Geopolitics and Culture in the Polar Region*, 393–409; Kristian H. Nielsen, "Small State Preoccupations: Science and Technology in the Pursuit of Modernization, Security, and Sovereignty in Greenland," in: Doel et al., eds., *Exploring Greenland*, 47–71.

46. Hans M. Carlson, *Home is the Hunter: The James Bay Cree and Their Land*, (Vancouver: UBC Press, 2008); Kirsten Thisted, "Discourses of Indigeneity: Branding Greenland in the Age of Self-Government and Climate Change," in: Sörlin, ed., *Science, Geopolitics and Culture in the Polar Region*, 227–258.

47. Andrew Stuhl, *Unfreezing the Arctic: Science, Colonialism, and the Transformation of Inuit Lands*, (Chicago: University of Chicago Press, 2016).
48. Farish and Lackenbauer, "Western Electric Turns North"; Daniel Heidt and P. Whitney Lackenbauer, "Sovereignty for Hire: Civilian Airlift Contractors and the Distant Early Warning (DEW) Line, 1954–1961," in: P. Whitney Lackenbauer and W. A. March, ed., *De-Icing Required! The Historical Dimension of the Canadian Air Force's Experience in the Arctic*, (Ottawa: Department of National Defence, 2012), 95–112.
49. Dodds and Nuttall, Scramble; Tina Loo and Meg Stanley, "An Environmental History of Progress: Damming the Peace and Columbia Rivers," *Canadian Historical Review*, 2011, 92(3): 399–427; Ronald E. Doel, Urban Wråkberg, and Suzanne Zeller. "Science, Environment, and the New Arctic." *Journal of Historical Geography*, 2014, 44: 2–14.
50. Frank Tester and Peter Kulchyski, *Tammarniit (Mistakes): Inuit Relocation in the Eastern Arctic, 1939–63* (Vancouver: UBC Press, 1994); David Damas, *Arctic Migrants/Arctic Villagers: The Transformation of Inuit Settlement in The Central Arctic* (Montreal: McGill-Queen's University Press, 2002); Peter C. Van Wyck, *The Highway of the Atom*, (Montreal: McGill-Queen's University Press, 2010); Erik Beukel, Frede P. Jensen, and Jens Elo Rytter. *Phasing out the Colonial Status of Greenland, 1945–54: A Historical Study.* (Copenhagen: Museum Tusculanum Press, 2010); Lackenbauer and Farish, "Cold War on Canadian Soil."
51. Matthew S. Wiseman, "Unlocking the 'Eskimo Secret': Defence Science in the Cold War Canadian Arctic, 1947–1954," *Journal of the Canadian Historical Association*, 2015, 26(1): 191–223; Farish, "The Lab and the Land".
52. Bocking, "Toxic Surprises"; D. L. Downie and Terry Fenge, eds., *Northern Lights Against POPs: Combatting Toxic Threats in the Arctic*, (Montreal: McGill-Queen's University Press, 2003); Martina Tyrrell, "Making Sense of Contaminants: A Case Study of Arviat, Nunavut," *Arctic*, 2006, 59(4): 370–380.
53. Michael T. Bravo, "Voices from the Sea Ice: The Reception of Climate Impact Narratives," *Journal of Historical Geography*, 2009, 35: 256–278; Thisted, "Discourses of Indigeneity"; Emilie Cameron, "Securing Indigenous Politics: A Critique of the Vulnerability and Adaptation Approach to the Human Dimensions of Climate Change in the Canadian Arctic," *Global Environmental Change*, 2012, 22(1): 103–114.
54. John McCannon, *A History of the Arctic: Nature, Exploration and Exploitation*, (London: Reaktion, 2012); Doel et al., "Science, Environment, and the New Arctic."

Part 2
Strategic science

2 Ice and the depths of the ocean: probing Greenland's Melville Bay during the Cold War

Mark Nuttall

In early March 1988, I traveled by dog sledge on Melville Bay's sea ice for the first time. I had been there before in the summer and early autumn, traveling by small boat on the open water and staying at hunting and fishing camps along the coast and inner islands. I had also moved regularly across the ice by dog sledge with hunters a little further south in the Upernavik district of Northwest Greenland since the previous December. This time, I was on a long journey with polar bear hunters from Savissivik and Upernavik. We headed far out and away from the coast to places where large icebergs were trapped in the winter ice. It was at these spots that polar bears hunted seals. They would lie in wait, in grooves and furrows in the snow around an iceberg, or on the iceberg itself, for seals to emerge from their breathing holes and haul themselves up onto the ice. These niches are called *qorfiit*, places where polar bears hunt; the bears will seek out furrows or depressions, or enlarge them or make grooves themselves, and settle in. While *qorfiit* refers to these indentations, furrows and grooves, the name also describes how their hollowed out nature resembles a chamber pot (*qorfik* is the singular for chamber pot or a bucket used as a toilet). And so, in approaching furrows in the snow, or when ascending an iceberg to seek polar bears, or to use the iceberg as a vantage point from which to scan the icescape, hunters will look for signs of polar bear urine or excrement in them. They will also put an ear to the ice and listen for scratching sounds a nearby polar bear may be making with its claws (*kukkilattorpaa*), as it makes a resting and hiding place in the snow and on the ice.

During this, my first winter sojourn on Melville Bay, the hunters with whom I traveled taught me how to watch and listen to the sea ice and icebergs carefully, and to pay close attention to the sounds that reveal much about the behavior and movement of ice. At the end of a long day, following polar bear tracks or looking for seals to hunt ourselves, we made camp, set up tents over the sledges, fed the dogs, and made our own meal—usually a pot of boiled seal meat. Against the hiss of the primus stove, we discussed the day's events, the nature of animals, and the hunting plans for the following day. But we also listened to the creaking of the ice and talked about what could be moving in the water below.

One evening, we had a conversation about the things that possibly lurked under the ice and the vessels that must disturb the ice itself as they broke a way through it. The hunters suspected that American submarines were lying in wait deep down in the water for a long time (*kitsumavoq*). They could be resting there, they said, in their own *qorfiit* at the bottom of the sea, and on the look out for Russian ships in the open sea or submarines beneath the ice. But could there also be Soviet submarines waiting down there for American ships and submarines? Regardless of where they came from, though, the hunters wondered what excrement the submarines left in these *qorfiit*. Sometimes, they said, they noticed strange-looking bubbles on the surface of the water, or rising through a headland crack in the ice—very different from the air bubbles indicating the presence of marine mammals or fish. Was this urine or something more sinister that the submarine expels from its body? How would this affect the sea, the ice, seals, whales, and fish? Is this why it was sometimes difficult to find polar bears? The seals often behaved as if they were frightened by something elsewhere moving out there in the water, they remarked. Put your ears close to the ice, I was told, and lis-ten carefully. You can hear something down in the sea that hums and grum-bles (*qatimaappoq*). Submarines and ships also *kukkilattorpaa* the ice. And at times when hunters encountered ice that had loosened and broken away (*kaavoq*), they felt it was not always because of the "way of the ice" (*sikuup ileqqui*) with its own twists and turns, but instead because something like a submarine could have surfaced through it.

Greenlandic waters during the Cold War

This chapter contributes to recent scholarship on the history of Greenland and the geophysical sciences during the Cold War period,[1] but it derives from a larger, ongoing anthropological project on human-environment relations in northern Greenland which has a concern with the shaping and nature of place.[2] My focus is Melville Bay and the adjacent waters. I explore how north Greenland environments, and the non-human entities which compose them, such as ice and water, became objects of scientific enquiry. This research was often inspired by American military interests to ensure the safe and efficient movement of shipping, while at the same time also marking out the region as strategic and geopolitical spaces of surveillance and observation. From the 1950s onwards, north Greenlandic waters were vital sea lanes for vessels establishing and supplying the infrastructure and installations of meteor-ology, radar systems, and defense. However, ice often hindered, obstructed and frustrated the smooth flow of maritime traffic and considerable effort went into tracking, sizing up and measuring ice and icebergs, probing ocean depths, and surveying, mapping and charting the possibilities of transiting safely and efficiently through more fluid, ice-free passages. This increased activity, though, also brought the risk of marine pollution (as well as more ship and aircraft noise) and disruption to animal migrations.

The ethnographic vignette with which I began this chapter is important for setting the context for my discussion, and I shall return towards the end of it to a consideration of how the kinds of things people imagined to be deep within as well as on the surface of the sea (and the unseen presence of such things) enter into contemporary local accounts and narratives of environmental change and resource development, and how these are framed by stories of Cold War happenings and anxieties over toxic legacies and environmental disturbance. For my traveling companions in Melville Bay in 1988, memories of the USNS *Potomac* oil spill, which had occurred just over a decade before in 1977, for instance, were still fresh. Today, people talk about animals as well as waters, ice, and places in the wider coastal landscape that are themselves agitated and unsettled by shipping, seismic surveys, and helicopters, but they often contextualize these concerns with reference to earlier events and experiences.[3] Along with scientific activities that were associated largely with military or security operations, during the last couple of decades of the Cold War period, Melville Bay and areas further north (such as the North Water Polynya) were also sites of interest for oceanographic research to understand water flow, heat budgets and ice formation, and for geological surveys of the seabed that aimed for greater understanding of its extent and nature, but which also revealed something about the region's hydrocarbon potential. Large-scale surveillance programs and research initiatives ensured Northwest Greenland was a busy region—military operations, aerial reconnaissance flights, and scientific expeditions crisscrossed the skies and seas, yet the Indigenous inhabitants were not necessarily informed about such activities; nor were they consulted about their knowledge and understanding of the nature of ice and water, clouds, winds and weather patterns.

The social anthropological fieldwork I first carried out in Northwest Greenland in the late 1980s—not just in Melville Bay, but elsewhere in the hunting and fishing communities along the coast—took place in an environment that was very much part of a wider securitized space of Arctic seas and skies (I also did my research towards what became the end of the Cold War period). It was a space that had been probed, measured, mapped and charted for a century or more, but which was still relatively unknown in terms of ice formation and movement, glaciology, bathymetry, and the shape of the seabed. Melville Bay is in the northeast corner of Baffin Bay, on Northwest Greenland's continental shelf, where it forms a long curve of the coastline. This coast is an archipelago dominated by bedrock— around nineteen large glaciers stretch across it, reaching down to the sea and calving thousands of icebergs each year, many of which drift southward into the western part of Davis Strait (and along the coast of Baffin Island), eventually joining the Labrador Current. Melville Bay's bathymetry remains largely uncharted in many parts, but it is known to be some 300–500 m (984–1640 ft) deep in places [there is a deep trench of 600–700 m (1969–2297 ft) and a deep basin of 1100–1200 m (3609–3937 ft)]. Two types of sea ice can occur in Baffin Bay and Melville Bay: multi-year ice normally enters from the Arctic Ocean as drift ice through Nares Strait and stays on the Canadian side of Baffin Bay, while first-year ice is predominant in Melville Bay and the Greenland side of Baffin Bay. Sea ice in Melville Bay usually forms in late September or early October, tends to be land fast, and forms inside the fjords independently of the ice conditions in Baffin Bay. Even in summer, the water is rarely free of drifting pack ice, and large icebergs fill the bay and dot the horizon.

Its Greenlandic name, *Qimusseriarsuaq* ("the great dog sledging place"), says much about how Indigenous people travel in and across Melville Bay during winter and spring. This is a significant place for hunting polar bears, seals and walrus, and narwhals. Its English name commemorates the second Viscount Melville, Robert Dundas Saunders, who was first lord of Britain's Admiralty from 1812-1827. Part of the coastline and marine areas of Melville Bay have been protected as an International Union for Conservation of Nature (IUCN) Category 1 nature reserve since 1977, but conservation regulations were first put in place in 1946 and current wildlife management and environmental protection procedures are endorsed by a Government of Greenland executive order from 1989, which also allows and governs specific hunting and fishing practices by Indigenous people living in Northwest Greenland. These measures protect polar bear habitat, a summer population of belugas and a breeding population of narwhals. Melville Bay is also an important breeding area for ringed seals and Sabine's gull.

Melville Bay has long been seen as an in-between place, and there was nothing particularly inherent about it that made it key for security purposes during the Cold War (apart from it being close to a possible route for Soviet submarines to enter North American Arctic waters by way of the Arctic Ocean). By this, I mean it had not been identified as a site for military observations or for the setting of weather stations, for instance, but it was a place—part of the Northabout Route—through which US naval shipping had to transit en route to strategic installations. For the non-Indigenous visitors to the region over the centuries, it has been seen predominantly as a dreaded, ice-choked stretch of water. For nineteenth-century whalers and explorers, to cross Melville Bay was to endure an arduous, tortuous, and dangerous journey to reach a safer, more open space, to hunt in the whaling grounds of the North Water Polynya and Lancaster Sound, to seek a way to the Northwest Passage, or to enter stretches of open water leading to the North Pole. For a while in the nineteenth century, discovery expeditions also thought the way to the North Pole was via Smith Sound and Nares Strait and so the ice of Melville Bay was something that mariners needed to enter and get *through*; in narrative accounts of whaling and exploration, the ice frustrates and Melville Bay was not a place in which to linger (it was approached nervously and with fearful anticipation), but many whaling ships were nonetheless stuck and crushed in the ice. It was also seen as a forbidding ice barrier that separated the Thule region and the Inughuit (or Polar Inuit) from West and South Greenland.

Observing and controlling northern seascapes

Oceanographic data from Baffin Bay, Melville Bay, and the seas of the Canadian Arctic Archipelago and northern Greenland had been acquired on various discovery expeditions in the late nineteenth century as well as specific scientific voyages during the early twentieth century.[4] However, American military activities during the immediate post-Second World War period and the early decades of the Cold War demanded a more complete understanding of the properties of ocean, sea ice, and glaciers. A strategically vital part of the North Atlantic during Second World War, Greenland constituted critical sea and air space for the campaigns of the Allied forces, and the United States built air bases in west, south and east Greenland (Greenland Base Command had its headquarters at Narsarsuaq). Knowledge of weather patterns was vital for naval and merchant fleet convoys and for aircraft movements between North America and Europe so Greenland became central to weather forecasting. US forces remained in Greenland after the war, and a new defense agreement was signed by the Danes and the Americans in 1951. The North Atlantic Treaty Organization (NATO) had in effect precipitated this by requesting that the US and Denmark should make arrangements for military facilities in Greenland to be available for use by the armed forces of other NATO parties in defense of the area under the North Atlantic Treaty, which was signed in 1949. The 1951 agreement was, in a sense, an implementation of the treaty and reassurances were given that this would not infringe upon or threaten the sovereignty of the Kingdom of Denmark. Construction of Thule Air Base began in 1951 and was completed in 1953.

Along with other geophysical sciences, oceanography became enrolled in Cold War scientific endeavors to understand the Earth, and so attention also turned to the surveillance of sea ice and to investigating the depths of Arctic seas.[5] From the 1950s, research on the physical oceanography of the wider north Baffin Bay area, the North Water Polynya, the waters stretching to the Nares Strait region and beyond, and the Arctic ice zone, became a priority for the US Navy for reasons of acquiring data for safer navigation and the unhindered movement of icebreakers and supply ships. But it was also important for gaining a greater understanding of coastal geographies, icefields, the volumetric space of northern waters, and the shape and formation of the seabed. It was vital to be able to see what lay ahead on the surface of the sea in terms of charting the location and extent of ice and icebergs, and over towards the horizon, and it was just as critical to know what was within the sea, such as the submerged parts of icebergs. Michael Dennis points out that field sciences such as oceanography "take place in spaces that are neither easily contained or controlled. Nonetheless, the need to contain and control those spaces drives much of the research."[6] In *War and Cinema*, Paul Virilio argues that war is less about victories on battlefields and securing territories, and more about how innovations in technology, particularly technologies of representation could influence and shape the visual perceptions of the battlefield and military action.[7] For Virilio, advances in technology, whether through radar images, maps and other forms of representation, were all key in military and scientific efforts to make a "dense country" knowable and "transparent." As Ronald Doel, Tanya Levin, and Mason Marker describe, when Bruce Heezen and Marie Tharp from Columbia University created the first comprehensive physiographic map of the North Atlantic Ocean basin in 1957, they drew on depth profiles of the seafloor and other information derived from US military-funded oceanographic research carried out during the early years of the Cold War.[8] Mapping and representing the seabed and oceanic troughs and ridges reflected particular kinds of motivations for doing such work in the first place.

During the Cold War, north Greenlandic waters and the Arctic ice zone, including Melville Bay, came to be viewed, like other northern ocean spaces such as the North Atlantic, as "a vast topography for military surveillance."[9] They also became spaces for military transit to supply and secure operations at Thule Air Base. The "dense country" of pack ice, icebergs, glaciers, fjords, inlets and islands, and the murky glacial waters of some stretches of the inner coast, needed to become transparent to planners, strategists and mariners. As essential areas for maritime patrol and transport systems, Baffin Bay and Melville Bay became Cold War geopolitical seascapes[10] during an era when oceans were becoming subject to initiatives to turn them from wild, unruly, vast stretches of water to knowable and controllable spaces. On the east coast of Greenland, for instance, and into the North Atlantic, the Greenland-Iceland-UK (GIUK) gap became an area of ocean surveillance in the 1960s and 1970s through American and British efforts.[11] The GIUK gap was a choke point that had been critical for Allied naval and merchant fleets during Second World War and its strategic importance continued into the Cold War period: controlling choke points such as the GIUK gap was critical for allowing the flow of military and commercial shipping, but also for monitoring Soviet vessels and submarines. NATO allies took strong anti-submarine defense measures (including placing and operating underwater sensor and listening posts) to detect, deter, intercept and prevent movement of the Soviet Union's nuclear ballistic missile submarine force.[12] The assumption was that the fleet had to pass through the GIUK gap to be able to launch missiles at the American eastern seaboard, and some contemporary scholars continue to emphasize the importance of controlling the GIUK gap to secure UK and US defense systems today. As Østreng pointed out over forty years ago, however, advocates of the GIUK gap monitoring and detection system overlooked the capability of Soviet nuclear ballistic missiles to cross wider distances, and he suggested that the Arctic Ocean may have served the Soviets just as well as a missile launching and transit area.[13]

Surveys of Baffin Bay and the waters and ice further north extended this oceanographic knowledge and panoptic reach. Knowledge of the depths and contours of the sea bed of Melville Bay, and of its glaciers and mountain heights, and how Greenland's vast inland ice affected cloud formation and winds, was necessary to make it legible: enabling weather and ice forecasting and the plotting of routes for ships and aircraft. Making Baffin Bay and Melville Bay transparent required a perspective that mapped and modeled the sea so that it became possible to gaze within and investigate its depths, its volume and density. Gísli Pálsson's idea of how the sea is represented by fisheries biologists and managers as a virtual aquarium (important for how they are able to visualize and measure fish stocks and deep sea ecosystems) seems apt in understanding how northern Greenland's waters were shaped and modeled by technologies and practices of observation, surveillance, the measurement of water columns, and the mapping of the sea bed.[14] But looking within the depths and peering into the darkness of the sea was only part of the process of making the Arctic knowable; ice observers also looked down on ice, sea, glaciers and ice sheets from above, and extended their vision through the skies, air and clouds, and up into the atmosphere.

The US Navy sea ice reconnaissance program

By the 1950s, the development of submarine technology meant that Arctic sea ice no longer presented a formidable barrier to some naval operations within ocean depths. On the surface of the water, though, the ice (as well as icebergs and smaller pieces of ice of glacial origin) still impeded shipping and had implications for aircraft logistics and movement in Arctic areas. Ice-forecasting techniques were developed in the late 1940s, mainly by studying aerial photographs taken on flights to and over the North Pole by US Air Weather Service (AWS) Operation Ptarmigan aircraft.[15] The first AWS weather reconnaissance flight over the North Pole was undertaken by a B-29 on 17 March 1947. The construction of air bases, the development of a US-Canada program for establishing weather stations, and increased flights to and over Greenland and Arctic Canada, meant that knowledge of both sea ice and glaciers was essential not just for planning military operations, but for assessing the feasibility of emergency landings on ice surfaces as well. Understanding the distribution of ice, and its nature, variety, thickness and extent was also of vital importance for the operations of the US Military Sea Transportation Service (MSTS), which played a major role in the development of military infrastructure and resupplying the growing network of defense and weather stations in the Arctic.

Beginning in 1950, the MSTS had the overall responsibility for the annual US military sealift in the US and Canadian Arctic and in Greenland.[16] Military building programs, such as those that developed Thule Air Base and Camp Century (the latter, some 240 km (149 miles) east of Thule, and at an elevation of some 2000 m (6562 ft), was a facility complete with a nuclear power plant that was constructed as part of the covert Project Iceworm, which had intended to develop an extensive system of tunnels and place nuclear missile launch sites under the Greenland inland ice), expanded significantly in the 1950s. The need for weather forecasting and for greater surveillance of northern skies to detect Soviet aircraft and possible missiles heading to Europe and North America led to major investment in research and in constructing and maintaining weather and radar stations (some of which were part of the Distant Early Warning (DEW) Line network in the North American Arctic). All this meant American and Canadian icebreakers and other ships were deployed in the operations that built and resupplied DEW Line radar stations, meteorological stations, and airbases, carrying cargo, equipment, and fuel, as well as personnel. The sea lanes were getting busier.

In the early to mid–1950s, more than 100 ships operated in Arctic waters on an average year transporting construction equipment, building materials and men to do the heavy lifting and labor. Icebreakers were assigned to escort MSTS ships, as well as to supply Arctic bases and stations such as Alert and Eureka on Ellesmere Island, or at Resolute on Cornwallis Island. But the ice continued to obstruct, hinder and thwart marine transportation. In 1951, for instance, a large US convoy en route to Thule became trapped in the ice in Baffin Bay. At the time, knowledge of the Baffin Bay ice regime was scant and, as it was later understood, the ice extent was less along parts of the West Greenland coast. Had the Americans known that it was better to steer a course north (and then south again) by taking the Greenland route rather than heading into the middle of Baffin Bay, they felt they could have avoided "the loss of many critical man-hours of construction time" and the cost of "many millions in ship damage associated with convoy movement through the center of the pack."[17] It was clear they needed better information and forecasting techniques to slice a safer and more straightforward passage through the frozen surface of the sea or, if possible, to avoid as much ice as they could. Further exploration of sea lanes in Baffin Bay and Melville Bay was necessary and the MSTS asked the US Navy Hydrographic Office to expand its ice reconnaissance program.

The Hydrographic Office and the US Naval Oceanographic Office produced a series of technical reports on the distribution of ice in Baffin Bay and Davis Strait, and forecasts and long-range ice outlooks for the Eastern Arctic in the 1950s and 1960s. Such data were also of interest to the British government which, in 1953, requested Hydrographic Office ice forecasts for the use of British trawlers fishing in West Greenland's Disko Bay region.[18] The early survey work undertaken by the Hydrographic Office mobilized civilian scientists, meteorologists and oceanographers in aerial ice reconnaissance. This involved considerable risks, which underscored the need for detailed information leading to improved knowledge and understanding of how and when ice formed, and the dangers it posed in the military's operational areas and adjacent regions. By the early 1950s, for example, helicopters (mainly US

Navy Bell HTL-4s) had taken over from fixed-wing aircraft to fly ahead of the icebreakers on reconnaissance trips to peer further into the distance and spot leads in the ice and stretches of open water. But the use of helicopters did not come without peril, as illustrated by the work undertaken by the US Coast Guard cutters *Westwind* and *Eastwind*. In late June 1954, *Westwind* sailed into Melville Bay on its way to Thule. It soon met heavy ice and one of its two helicopters took off to assess the conditions—a malfunction occurred when it was not too far from the ship and it hit the edge of the ice; both the pilot and the ship's executive officer were killed. It was a clear day, and weather conditions were not to blame. That same season, in August, both helicopters from *Westwind*'s sister vessel *Eastwind* also suffered malfunctions. The first was forced to land at sea north of Thule, while the second suddenly lost lift on an ice observation flight as *Eastwind* made its way through heavy ice from Thule Air Base to deliver supplies to the weather station at Alert on northern Ellesmere Island. The second helicopter made an emergency landing on a stretch of ice and was badly damaged (there were no fatalities from these latter incidents, but the Navy grounded its fleet of HTL helicopters following the 1954 operations). With *Eastwind* having no operational helicopters, further ice reconnaissance was impossible and left the crew unable to predict the movement of the ice. Heading towards Alert, the vessel soon became trapped in the ice pack—and the hull, despite being doubly reinforced, was torn open. *Westwind*, which was now at Thule, made a course to the north, reached her sister vessel and towed her back to the harbor at the base. Alert had to be resupplied by aircraft and did not receive equipment and machinery it was expecting that summer. As an aside, following *Eastwind*'s resupply of Alert two years previously in August 1952, the ship was trapped in heavy ice and forced to drift within 420 nautical miles of the North Pole—after this adventure it participated in Operation Muskrat in northern Baffin Bay, which included the launch of rockets from high altitude balloons.

Technical reports, ice manuals and ice forecasts became essential for navigation in Baffin Bay and north Greenlandic waters, but so did information about the shape, contours and angles of the coastline, the nature of glaciers, the movement of icebergs and the very depths of Melville Bay. For example, Henry Kaminski, who worked in the Applied Oceanography Branch of the Hydrographic Office's Oceanography Division, authored a major technical report emphasizing the importance of observing, monitoring and forecasting ice conditions. Kaminski was killed when the US Navy ice reconnaissance aircraft he was traveling in as an ice observer crashed near Paget Point on Ellesmere Island on 16 April 1954, and the accident emphasized the dangers and difficulties of the surveillance of Arctic ice zones. He had played a key role in the development of how the US Navy's ice forecasting service should proceed in its efforts to understand the physical properties of sea ice and its behavior under stress,[19] and his report on the *Distribution of Ice in Baffin Bay and Davis Strait* was published posthumously in 1955. He had also been involved in Operation Blue Jay, the code name for the (initially secret) operation that constructed the air base and strategic command facilities at Thule. While being careful to point out that there were variations in sea ice from year to year, Kaminski's report provided rich and detailed information about the formation, thickness, consistency and break-up of ice, as well as what kind of ice one would expect to find in particular places—coastal waters, for instance, "abound in glacial ice," while inlets and fjord contain icebergs, bergy bits and growlers. It presented normal (or average) sea ice conditions in the Baffin Bay-Davis Strait region, and it was intended to be "a useful guide and reference in determining feasibility of emergency landings on ice, and the planning and execution of operations, particularly when modified to reflect the deviations from normal factors controlling the growth, distribution, and behavior of sea ice."[20] The report conveyed the sense that, with enough information and knowledge, US operations could, despite deviations from the norm, anticipate ice conditions and use this knowledge to plan ship and aircraft operations. This assertion was supported by the report's credentials to be based upon the observations of personnel of the US Navy Hydrographic Office during several hundred aerial ice reconnaissance surveys, as well as recent ship and shore-based ice observations and the historical records from exploratory voyages in the Arctic over the previous century.

The US Naval Oceanographic Office produced long-range ice predictions for the Eastern Arctic for MSTS operations and were based on an aerial reconnaissance in spring, evaluation of oceanographic and climatic data on the growth of sea ice during the previous winter, and a study of historical ice and climatic information. For example, the outlook for mid-May to mid-August 1965 (the report was published in April), gave the following forecast:

> Increasing temperatures and westward drift should result in rapid disintegration during the latter part of June and early July. Concentrations in Melville Bay should be primarily open pack by 10 July with areas of close pack remaining in the Northabout Route until mid-July. Melville Bay should essentially be safe for escorted shipping by 10 July and safe for unescorted shipping by 24 July. However, belts and patches of variable concentrations are expected to remain in this area until approximately 1 August. By mid-August, Melville Bay should be completely ice free with only remnant open and very open pack remaining in central Baffin Bay.[21]

Soundings, depths and submarine passages

In the late 1950s the Hydrographic Office undertook an extensive hydrographic survey in Melville Bay to obtain the precise kind of nautical data required for safe navigation along the Northabout Route. It stayed close to (or within sight of) the coastline from Upernavik to Thule. Civilian scientists were often aboard the survey ships, such as during the voyage of USS *Edisto* in 1959, gathering data which also sometimes fed into Master's and PhD projects. Nathan Fishel's MSc thesis, for example, contains a wonderful account of the methodology and practice of *Edisto*'s hydrographic survey in Melville Bay, giving insight into the soundings and recording of the bay's depths, tidal measurements and observations, the identification of grounded icebergs, photogrammetry, weather information and the formation of sea ice.[22] Working from aerial photographs and triangulation points established by the Danish Geodetic Survey, his thesis describes how the hydrographic work in Melville Bay involved the mobilization of considerable resources and labor to erect (and later dismantle) a series of relay stations to carry out measurements and gather data. Fishel emphasized the number and extent of icebergs, noted the difficulty of identifying whether they were grounded or not, and recommended that ships exercise extreme caution when sailing near them. Mariners and travelers in Melville Bay had often reported on how the quality of the light made it impossible to differentiate between land and ice, or how sudden mists and dense fogs bent the world out of shape, deceiving, confusing and disorientating them.[23] Fishel's thesis contains this insight:

> The author became curiously concerned as to why so many islands previously recorded by mariners did not exist, and similarly why previous chart editions compiled by photogrammetric techniques of the off-shore islands also showed no islands where islands do exist. Aside from the personal element of error, it became apparent during the survey that dirty icebergs, existing at the time (of) the flight filming, could easily be mistaken from aerial photographs for small islands, and small islands could easily be hidden by icebergs. The ice since gone offers each succeeding year an entirely different appearance. In addition, to the same mistakes made by the photogrammetric compilers at the office, the mariners reported hollow arched icebergs for non-existent islands. The refraction and diffusion of light balanced proportionately, the hollowed arch is miraged as a dark island.[24]

He also pointed to the changing coastline under processes of weathering, erosion and glacial surges and how large ice movements "and prominent iceberg creation in this area has undoubtedly altered the isostatic equilibrium."[25]

A Southabout route, tracing a direct line across the open sea of Baffin Bay and nudging the southern edge of Melville Bay was first utilized in 1964 because of severe ice conditions in Melville Bay. Knowledge of ice and icebergs was also improved by data gathered by US submarines patrolling in Melville Bay and elsewhere in Canadian Arctic waters. Alfred McLaren's account, *Silent and Unseen*, describes his experiences of measuring and assessing icebergs in Baffin Bay and Melville Bay in 1960. McLaren served as a diving officer aboard the submarine and his book is full of wonderful descriptions of sailing directly beneath "monstrous icebergs" and of visual encounters with Arctic marine life mediated by technology:

> It was during this first hovering beneath the icepack that many of us discovered the wonders of underwater TV. The monitor in the control room revealed all sorts of fish in our immediate vicinity. We thought we were seeing some cod. Larger dark-gray shapes appeared to be sharks or even small whales. Several of us were sure we had seen a killer whale. Exciting![26]

The submarine was attempting a transit of the Northwest Passage. It headed into northern Baffin Bay and on to Melville Bay because *Seadragon*'s commander, George P. Steele, resolved, in his own words, "'To demonstrate beyond question the ability of a nuclear submarine to enter an area of high ice concentration with safety' before we left Baffin Bay."[27] On 13 August they surfaced some forty nautical miles from Kap York in an area of open water and headed to a stretch of sea close by which lookouts had reported as having more than forty large icebergs. They moved slowly among them to identify the most suitable to make underwater investigations of. Icebergs were photographed, measured, their qualities assessed, and their undersides examined. They learned a great deal about the various ratios of how much of an iceberg was above the water and how much of it was below, and they discovered that "icebergs give off a loud seltzer or hissing noise as they melt."[28] McLaren describes how

> Much discussion followed in the control room on how we might one day apply what we had learned about icebergs and iceberg-infested waters in actual combat situations with our most likely adversary, the Soviet submarine. Certainly, an excellent tactic, upon discovering one, would be to move up as close as possible to the ice canopy overhead to better hide one's presence.[29]

Acquiring knowledge of these High Arctic ocean depths then had a strategic purpose beyond its function for ice forecasts or for charting waters for safer navigation. There were fears that Greenland's ice zone could provide shelter for Soviet submarines. In the late 1950s and into the 1960s, a route into Hudson Bay via the Arctic Ocean close to northern Greenland and south through Nares Strait and Davis Strait was discussed by military strategists in the Soviet Union as a way submarines could approach North America swiftly and undetected and launch missiles at the United States. Even if this was initially intended as propaganda, and if the Soviets did not at that time have the nuclear submarine capability, US and Canadian aerial and ice reconnaissance operations were increased.[30] Voyages such as *Seadragon*'s passage through the Parry Channel were part of a broader initiative to assess the ability to mobilize and deploy submarines in Arctic ice-filled waters, consider the possibilities of icebergs as places of concealment, and identify and mark out possible choke points between Canada and Greenland.

Leaving Kap York and Melville Bay, *Seadragon* set a course for Lancaster Sound and the Northwest Passage. The submarine's orders were to investigate the feasibility of a submarine passage through the Parry Channel and the eastern segment of the Northwest Passage, and specifically a deep-water pass through which a nuclear submarine could travel in the worst conditions in the heart of winter darkness. As well as completing the first submerged transit of the Northwest Passage (during the voyage Steele referred to Edmund Parry's logbook from his 1819 attempt to sail through), *Seadragon* also became the first submarine to pass close to the North Magnetic Pole, then located in Viscount Melville Sound. Entering the Beaufort Sea on 21 August, *Seadragon* headed for the North Pole and surfaced there four days later.[31] During its voyage, *Seadragon* collected profiles of the underside of sea ice in M'Clure Strait (as did another submarine, USS *Sargo* during a cruise in February 1960).[32] In the 1960s, submarines continued to obtain sonar profiles that provided the first ice thickness data in parts of the Arctic such as the Northwest Passage and Davis Strait, but the mapping and identification of ice was also enhanced by the use of side-looking air-borne radar (SLAR) equipment (which, for example, was used on US Coast Guard C-130 aircraft in conjunction with the SS *Manhattan*'s transit of the Northwest Passage in September 1969), and by the development of satellite remote sensing in the 1970s.[33]

Things that bubble and move below the surface of the water

During the Cold War, initiatives to test and maintain missile and bomb sites, monitor Soviet military movements, control airspace and secure sea routes led to the building of infrastructure that often entailed ecological rupture and forms of cultural violence, dispossession and displacement. To say that this was deeply unsettling for many people around the world does not do justice to how they experienced the power and reach of the state through narratives of fear, the creation of scientific-technological landscapes, and expulsion.[34] In the case of northern Greenland, the Inughuit residents of Uummannaq (Dundas) were relocated to the Qaanaaq area in 1953 when the Thule Air Base defense area was expanded to include a weather station at the site of the community's settlement.[35] Cold War environments were environments of violence, in the sense of them being potential battlefields, or from where violence would be enacted (in the case of missile launch sites), or caught up in and affected by accidents and contamination (such as the crash of a B-52 bomber carrying four nuclear weapons on the sea ice 12 km (7.5 miles) from Thule in January 1968).

Indigenous peoples are noticeably absent from reports on ice, water, and the wider geographies of northern Greenland; ice reconnaissance and shipping took place in an Arctic space that appeared empty of people. Knowledge of sea ice and weather was based initially on aerial, ship and land-based observations and, later on, these methods of surveillance were combined with new meteorological techniques and practices, including radar and remote sensing. Surveys and observations, though, did not take note of Indigenous cryospheric knowledge and experience of changes in drift ice and fast ice extent, freeze-up and break-up patterns, glacier fronts and icebergs, and calving processes from glaciers, and how this facilitates both the formation and stability of the fast ice cover. In my work in northern Greenland over the years, I have been struck by how precise and specific Indigenous knowledge of sea ice and icebergs is for travel and navigation—for example, knowing when one must go windward of an iceberg, when it is safe to go between them, or how to identify a wave that is rising because a glacier has calved an iceberg, or because an iceberg has shifted its center of gravity, or how to recognize whether a sudden gust of wind has come from a mountain or from along a fjord. Nor did later scientific interest in the 1960s and in the 1970s in the North Water and in the geological history of the Nares Strait region between Greenland and Canada take local understandings into consideration, despite there being a rich Indigenous vocabulary that hints at extensive knowledge of the cryosphere, of the movement of ice streams from the Greenland inland ice, or that describes how icebergs have created troughs, ridges, passages, and furrows in the seafloor.[36] By and large, for the people who lived within the boundaries of Melville Bay or who traveled there to hunt and fish, the effects of Cold War activities were somewhat surreptitious. The effects and legacies, however, are apparent in the way people talk today about animals and places being agitated.[37] This sense of agitation is said to derive in part from recent seismic surveys and oil exploration vessels, as well as pronounced changes to the Baffin Bay sea ice regime, but memories of an oil spill over forty years ago linger.

On the morning of 5 August 1977, *Westwind* was escorting the USNS *Potomac*, a petroleum tanker, through the Melville Bay ice in an intermittent but dense fog. *Potomac* was carrying a cargo of arctic-blend diesel fuel (called bunker C oil) to Thule Air Base, but an iceberg holed one of its fuel tanks and 107,000 gallons of oil spilled into Melville Bay. The US response was swift, *Westwind* towed *Potomac* to Thule, representatives of the US Coast Guard National Strike Force and the military sea lift command arrived at the air base on a coast guard C-130 on 8 August, and set out on *Westwind* for the scene of the accident. The C-130 was then deployed to locate and map the extent of the slick. A clean-up operation ensued. A joint National Oceanographic and Atmospheric Administration and Ministry for Greenland report concluded that the spill significantly increased the pollution in Melville Bay but likely had no lasting ecological effect.[38]

The report is notable in that the authors considered the *Potomac* incident an unprecedented opportunity to carry out a number of studies to assess and evaluate an oil spill in pristine Arctic waters. The accident happened in calm seas with light winds, which prevented the slick from dispersing for a few days. Interestingly, it took several hours for the Danes to be alerted, but when they received the news they saw the spill as presenting a way to gather some baseline data given interest at the time in oil exploration in Greenlandic waters, as the report emphasizes:

> At 0010 GMT on August 6, the Ministry for Greenland received a message that 390 tons of oil had been spilled by the USNS POTOMAC in Melville Bay, Greenland. The Ministry recognized that this opportunity to study an oil spill in Arctic water was of particular interest because of the ongoing oil exploration off West Greenland and the importance of hunting marine mammals and birds in this area by native hunters. Moreover, they wanted the opportunity to exercise and test recent contingency plans for oil spill research. The Ministry, therefore, redirected the research vessel ADOLF JENSEN from its scheduled cruise to a special cruise into Melville Bay to study the oil spill.[39]

However, it also downplayed the possibility of significant or lingering effects of the oil spill on wildlife and local hunting and fishing livelihoods:

> Birds and seals were not abundant in Melville Bay during the August field study period. A few flocks of auks and kittiwakes were observed as well as a few solitary gulls and fulmars. None were observed to be influenced by the oil. Fifty-five seals were observed, 43 of which were ringed seals. Twenty-five of the seals were spotted well south of the spill site. Only 4 seals were seen in the oiled area; however, no unusual behavior was observed. Starting in September, 5 weeks after the spill, the first of 29 reports of oiled sealskins surfaced. All of these reports came from native eskimo seal hunters. Eighteen of the 29 skins were delivered to Denmark for analyses. Ten of the 18 were surficialy contaminated by petroleum hydrocarbons, while the remaining 8 were clean. Seven of the contaminated skins may have contained oil from the USNS POTOMAC. No damage to the skin underneath oiled hair was observed during histological examinations.[40]

The report concluded that the fuel spill was an "unfortunate accident" but that "excellent cooperation" between operational and scientific personnel from the US, Greenland and Denmark "not only led to a better understanding of the fate of oil in the cold marine environment but also its impact on Arctic marine ecology."[41] No testimony from the "native eskimo seal hunters" appears in the report, but the hunters with whom I traveled—and the people with whom I lived—during my first sojourn in Northwest Greenland in the late 1980s talked a lot about how they remembered seals being dirty and their skins spoiled by oil around the time of the *Potomac* incident (and

I have continued to hear these stories on my recent visits to the Upernavik district). *Qalaliavoq* is a word I learned to describe something bubbling under the surface of the water, indicating something alive, moving, spreading; and oil was described in this sense—there was a feeling at the time that oil was still there, as were submarines moving silently in the murkier depths of the sea (and stories of periscopes raised above the surface of the water gave further credence to what people said they knew was happening, but which they were not informed about by the Greenlandic and Danish authorities). Just as a hunter lies in wait in his kayak for a passing seal (*qamavoq*), so, people said, submarines also lie in wait (*qamavaa*) in *qorfiit* etched out of the sediment on the sea floor, and this may explain why hooks, lines and nets sometimes get caught, snagged, entangled and lost (*nassippoq*).

Conclusions

While recent scholarship has revealed much about scientific interest in Greenland during the Cold War, Melville Bay's importance as a Cold War site of passage has received less attention. Science and techniques of observation sought to make it knowable (largely for the purposes of sea transportation and improved weather forecasting), but it remained a somewhat marginal site in terms of strategic importance, an ice-filled, if dangerous space, that military and coast guard vessels had to pass through to get to other, more vitally-important places. Melville Bay could hardly be ignored, however—the difficulties encountered in getting through and across it en route to Thule Air Base and other installations such as radar and weather stations made it a rather conspicuous space on the map of the High Arctic. It was not a region that was elided from the military map, but it remained, as it had been for a century or more in the European and North American imaginations, a liminal space. Today, though, Melville Bay has an increasingly prominent place in regional and global discussions concerned with climate change and environmental sensitivity in northern Greenland (the *Potomac* oil spill is of considerable relevance for contemporary understanding of the region's place in discussions of development and conservation, and people in the Upernavik district and in Savissivik have talked to me about it when they have expressed their concerns over oil development or possible oil spills following from increased shipping). The entire sea ice regime in Northwest Greenland has undergone significant changes over the past few decades. The marine ecosystem supports the livelihoods of communities in the region, so changes to sea ice have had far-reaching effects for hunting, fishing, and traveling and for local economies. At the same time, international energy and mining companies are increasingly interested in the possibility of developing mines as well as exploring for oil in offshore waters (extractive industry development continues to be a stated aim of the Greenlandic self-rule government, despite recent downturns in global prices for oil and some minerals). The planning for extractive industries involves political, volumetric and stratigraphic procedures and practices that measure, map, define and demarcate north Greenlandic environments such as Melville Bay as resource spaces. In response to this, an environmentalist intervention seeks to designate large areas of Northwest Greenland as exceptional ecosystems in need of protection.

The World Wide Fund for Nature (WWF), for instance, has defined northern Greenland as part of its Last Ice Area initiative, while the Inuit Circumpolar Council has reconfigured the North Water Polynya as Pikialasorsuaq ("the great upwelling"), and both organizations seek to garner international support for conservation initiatives for these regions. Like many other parts of the Arctic, the north Greenland environment has become many different things subject to a diverse range of interests. It is a place in which animals such as narwhals, polar bears and other marine animals are hunted by Indigenous residents, but which are also enmeshed in a complexity of social relations with them; it has been a site for exploration, discovery, adventure, and commercial whaling; it is a site for fieldwork by international teams of scientists concerned with understanding climate change, sea ice thickness, and ocean-cryosphere interactions; it is a place where marine biologists seek to gather data on marine mammals and fish for the purposes of informing management regimes; it is a site in which conservationists seek to protect the northern ice; and it is a site for the exploration and development of hydrocarbons and minerals, and as such it is important for Greenlandic ambitions for greater autonomy and state formation. Yet, as Melville Bay and the wider north Greenland region assume a greater significance, our understanding of these parts of the High Arctic can only be enriched by examining the history of military and scientific interest in them during the Cold War.

Notes

1. R.E. Doel, K.C. Harper and M. Heymann, eds. *Exploring Greenland: Cold War Science and Technology on Ice* (London: Palgrave, 2016); J. Martin-Nielsen "'The Deepest and Most Rewarding Hole Ever Drilled': Ice Cores and the Cold War in Greenland," *Annals of Science*, 2013, 70(1): 47–70.
2. M. Nuttall, *Climate, Society and Subsurface Politics in Greenland: Under the Great Ice* (London and New York: Routledge, 2017).
3. M. Nuttall, "Places of Memory, Anticipation and Agitation in Northwest Greenland," eds. K.L. Pratt and S.A. Heyes *Language, Memory and Landscape: Experiences from the Boreal Forest to the Tundra* (Calgary: University of Calgary Press, in press).
4. R.D. Muench, "The Physical Oceanography of the Northern Baffin Bay Region," The Baffin Bay-North Water Project Scientific Report No. 1. (Calgary: Arctic Institute of North America, 1971).
5. M.A. Dennis, "Earthly Matters: On the Cold War and the Earth Sciences," *Social Studies of Science*, 2003, 33(5): 809–819; S. Robinson, "Stormy Seas: Anglo-American Negotiations on Ocean Surveillance," eds. S. Turchetti and P. Roberts *The Surveillance Imperative: Geosciences During the Cold War and Beyond*, (New York: Palgrave Macmillan, 2014); P. Roberts, "Scientists and

Sea Ice Under Surveillance in the Early Cold War," eds. S. Turchetti and P. Roberts, *The Surveillance Imperative: Geosciences during the Cold War and Beyond* (New York: Palgrave Macmillan, 2014).

6. M.A. Dennis, "Earthly Matters," 809.
7. P. Virilio, *War and Cinema: the Logistics of Perception* (New York: Verso, 1989).
8. R.E. Doel, T.J. Levin and M.K. Marker, "Extending Modern Cartography to the Ocean Depths: Military Patronage, Cold War Priorities, and the Heezen-Tharp Mapping Project, 1952–59," *Journal of Historical Geography*, 2006, 32(3): 605–626.
9. F. MacDonald, "The Last Outpost of Empire: Rockall and the Cold War," *Journal of Historical Geography* 2006, 32(3): 627–647.
10. MacDonald, "Last Outpost."
11. Robinson, "Stormy Seas."
12. L.M. Alexander and J.R. Morgan, "Choke Points of the World Ocean: A Geographic and Military Assessment," *Ocean Yearbook*, 1998, 7(1): 340–355.
13. W. Østreng, "The Strategic Balance and the Arctic Ocean: Soviet Options," *Cooperation and Conflict*, 1977, 12(1): 41–62.
14. G. Pálsson, "The Virtual Aquarium: Commodity Fiction and Cod Fishing," *Ecological Economics*, 1998, 24(2–3): 275–288.
15. C.C. Bates, T.F. Gaskell and R.B. Rice, *Geophysics in the Affairs of Man: A Personalized History of Exploration Geophysics and Its Allied Sciences of Seismology and Oceanography* (Oxford: Pergamon Press, 1982).
16. A.J. Tait, "The Operational Concept for a Sea Ice Reconnaissance and Forecasting Program during Arctic Operations," in *Arctic Sea Ice.* (Washington DC: National Academy of Sciences-National Research Council, 1958).
17. A.J. Tait, "Operational Concept."
18. C.C. Bates, T.F. Gaskell and R.B. Rice, *Geophysics in the Affairs of Man: A Personalized History of Exploration Geophysics and its Allied Sciences of Seismology and Oceanography.* (Oxford: Pergamon Press, 1982).
19. C.C. Bates, H. Kaminski and A.R. Mooney, "Development of the U.S. Navy's Ice Forecasting Service, 1947–53, and its Geological Implications," *Transactions of the New York Academy of Sciences*, 1954, 14(4): Series II: 162–232.
20. H.S. Kaminski, *Distribution of Ice in Baffin Bay and Davis Strait*, (Technical Report. Washington DC: US Navy Hydrographic Office, 1955).
21. US Naval Oceanographic Office, *Long Range Ice Outlook: Eastern Arctic (1965)* Oceanographic Prediction Division, US Naval Oceanographic Office, Washington DC, 7–8.
22. N. Fishel, "The Technical Aspects of the 1959 Melville Bay, West Greenland 'Two Range Raydist' Hydrographic Season," (Unpublished MSc. Thesis, Ohio State University, 1960).
23. M. Nuttall, *Climate, Society and Subsurface Politics in Greenland: Under the Great Ice*, (London and New York: Routledge, 2017).
24. Fishel, "Technical Aspects," 48.
25. Fishel, "Technical Aspects," 48.
26. A. S. McLaren, *Silent and Unseen: On Patrol in Three Cold War Attack Submarines*, (Annapolis, Maryland: Naval Institute Press, 2015), 83.
27. McLaren, *Silent and Unseen*, 84.
28. McLaren, *Silent and Unseen*, 86.
29. McLaren, *Silent and Unseen*, 86.
30. W. Østreng, "Danish Security Policy: The Role of the Arctic, the Environment and Arctic Navigation," ed, W. Østreng *National Security and International Environmental Cooperation in the Arctic: the Case of the Northern Sea Route.* (Dordrecht: Springer, 1999).

31. It was the third submarine to do so; USS *Nautilus* being the first to surface at the pole in August 1958, and USS *Skate* doing the same in March 1959. See also: G. P. Steele, *Seadragon: Northwest under the Ice*, (New York: E.P. Dutton, 1962).
32. See A. S. McLaren, P. Wadhams and R. Weintraub, "The Sea Ice Topography of M'Clure Strait in Winter and Summer of 1960 from Submarine Profiles," *Arctic*, 1984, 37(2): 110–120, for an account of the analysis of these profiles when they were declassified in the early 1980s.
33. A. Lajeunesse and B. Carruthers, "The Ice Has Ears," *Canadian Naval Review*, 2013, 9(3): 4–9; McLaren, *Silent and Unseen*; McLaren et al., "Sea Ice Topography"; P. Wadhams, A.S. McLaren and R. Weintraub, "Ice Distribution Thickness in February from Submarine Sonar Profiles," *Journal of Geophysical Research: Oceans*, 1985, 90(C1): 1069–1077; J.D. Johnson and L.D. Farmer, "Use of Side-looking Air-borne Radar for Sea Ice Identification," *Journal of Geophysical Research*, 1971, 76: 2138–2155.
34. H. Kwon, *The Other Cold War*, (New York: Columbia University Press, 2010); V. Kuletz, *The Tainted Desert: Environmental Ruin in the American West*, (London and New York: Routledge, 1998); D. Vine, *Island of Shame: The Secret History of the U.S. Military Base on Diego Garcia* (Princeton and New York: Princeton University Press, 2011).
35. J. Brøsted and M. Fægteborg, *Thule-Fangerfolk og Militæranlæg*, (Copenhagen: Akademisk Forlag, 1987).
36. M. Dunbar, "The Geographical Position of the North Water," *Arctic*, 1969, 22(4): 438–441.
37. Nuttall, "Places of Memory."
38. P.L. Grose, J.S. Mattson and H. Petersen, *USNS Potomac Oil Spill*, (Washington DC: US Department of Commerce, National Oceanic and Atmospheric Administration, and Copenhagen: Ministry for Greenland, 1977).
39. Grose et al., *Potomac Oil Spill*, 16.
40. Grose et al., *Potomac Oil Spill*, 133.
41. Grose et al., *Potomac Oil Spill*, 134.

3 Leadership, cultures, the Cold War and the establishment of Arctic scientific stations: situating the Joint Arctic Weather Stations (JAWS)

P. Whitney Lackenbauer and Daniel Heidt

In some cases, the superpower rivalry undoubtedly pushed security-driven scientific inquiry to the forefront. As other chapters in this volume attest, the American military devoted considerable resources to studying Greenland, constructing massive airbases, staging experiments, and operating weather stations.[1] In Canada, the United States Air Force and the Royal Canadian Air Force busily mapped much of the Arctic. The American military also provided nearly all of the logistical resources to construct five weather stations located at Resolute, Eureka, Mould Bay, Isachsen and Alert between 1947 and 1950. The United States Weather Bureau (USWB) and the Canadian Department of Transport (DOT) then each shouldered roughly half of the personnel and financial obligations to run these Joint Arctic Weather Stations (JAWS).

The received history of JAWS conceives the program as an American continental defence initiative foisted upon the Canadians who succumbed to American pressure and concealed what was essentially a military program under "civilian cover."[2] Given technological advances in long-range strategic bombing during the war, the interwar idea of the Arctic becoming the world's 'new Mediterranean' no longer seemed far-fetched either commercially or militarily.[3] Would the region become North America's Achilles' heel? In this view, the reliable passage of western interceptors and bomber aircraft throughout the Arctic necessitated the production and integration of synoptic meteorological data into defence forecasts.[4] The JAWS network, planned according to southern as well global requirements, projected a pioneering and permanent scientific presence into Canada's High Arctic.

Our re-assessment of the archival record yields a different, or at least more nuanced, picture. The JAWS program must be understood within this emerging Cold War context, and early bilateral negotiations over the construction of weather stations in Canada's High Arctic embodied Canada's postwar anxieties of dealing with a superpower interested in the northern approaches to North America. As we have described elsewhere, several

Canadian cabinet ministers viewed these stations within this broad security framework. Fearing a repetition of the largely unsupervised American onslaught into the Canadian Arctic that followed the United States' entry into the Second World War, and lacking adequate northern icebreaking and air transport capabilities and well as a limited pool of experts capable of undertaking major northern research programs itself, the Canadian cabinet was keen to avoid a renewed American military buildup in the North. Continental security, many officials feared, would be used to trump Canadian control on the ground.[5]

The JAWS program ultimately proved these fears to be misplaced. While concerns about American disregard for Canadian sovereignty were raised by some trying to explain occasional interpersonal misunderstandings at the stations, the history of the JAWS program would ultimately become a beacon of bilateral cooperation. Moreover, the USWB—which sought the construction of the stations and ultimately provided most of the human and budgetary resources to turn southern visions into Northern scientific outposts—envisioned the program as a civil endeavor and it was this latter view, shared by the Canadian Department of Transport, that ultimately informed the program's conception, construction, and operational stages. In fact, one Canadian official who tried to impose military-style discipline upon station personnel ultimately had to depart before the station's operations completely devolved.

Setting the context

The USWB (a civilian agency) spearheaded the JAWS program as a component of a post-war effort to gather sufficient meteorological observations to produce accurate long-term weather forecasts. As Kristine Harper notes, Dr. Francis W. Reichelderfer, the Chief of the U.S. Weather Bureau and one of the first American disciples of the Bergen School of Meteorology, defended his civilian department's traditional hegemony on weather matters from military incursions. Although he was willing to court military patronage for major projects that served his research agenda, his story serves as an important reminder about the need to carefully parse overlapping lines between civilian projects, military funding and logistical support, as well as conflicting political messaging offering justifications for particular courses of action.[6]

When supporting a budgetary proposal for the project to the US House of Representatives, Reichelderfer admitted that "it is very essential from a defense point of view to have full coverage of reports of weather likely to have a bearing on our theatre of operations." Yet his statements emphasized the civilian economic and industrial benefits of the proposed program, which he believed would start with five to six American-built Arctic stations and would stimulate "other countries to do their share by establishing stations under their own flags in their own parts of the Arctic." The economic benefits of the proposed program could exceed a billion dollars each year. An example he gave related to drying raisins, which could be protected from rainfall if farmers had access to reliable 36-hour forecasts. Thus linking the Arctic to farming in California, Reichelderfer highlighted how building a network of polar weather stations would improve predictive capacity. "Without the information from the Arctic," Reichelderfer concluded, "we are lacking some of the data necessary to do weather forecasting in a more quantitative and scientific manner."[7] Similarly, Charles Hubbard, who had played a pivotal role in establishing the Crystal stations in the eastern Canadian Arctic and Greenland for the US Army Air Force during the war and then conceived and on-boarded Reichelderfer to the JAWS cause, and who would subsequently serve as a civilian project officer leading the Arctic weather station initiative,[8] offered balanced testimony before the committee. For an estimated $200,000 per Arctic station, he emphasized the civilian benefits of long-range forecasting for American life, from farming, to construction, to transportation, to merchandising.[9]

Hubbard and the USWB recognized that emerging continental security concerns opened a window of opportunity to secure military support for civilian initiatives—including JAWS. His detailed January 1946 report furnishing detailed specifications for buildings, transportation requirements, operational timetables, and personnel depended on the Army and Navy for transportation and supplies, making their "full approval" essential to actually implementing the civilian program. Hubbard's primary objectives that spring were reconnaissance flights and exploration, establishing a base in the western Arctic (on Banks or Melville Island), and setting up a fuel cache and aviation facilities at Thule, Greenland. He pushed the US government to approach Canada and Denmark for approvals as soon as possible—and initially recommended that the U.S. retain responsibility for the entire project. The Canadians would insist on participating for "national prestige," he anticipated, and would require sovereignty guarantees to allay their concerns. Nevertheless, he sought to confine Canada's contribution to a few personnel or bush pilots, given that the U.S. had possessed the practical capabilities to build and operate the stations and, in his view, would accrue the greatest benefit from them.[10]

After the Canadian cabinet approved the Joint Arctic Weather Stations program as a bi-national undertaking on 28 January 1947,[11] technical experts from both countries met in Ottawa late the following month to determine how they would build and sustain the stations. The Americans accepted Canada's proposed plan for nine stations, but insisted that available transport and supply would dictate progress. Similarly, the Americans presumed that technical experts could adjust the precise locations of the stations based on operating problems and reconnaissance data gleaned from the field.[12] Reichelderfer recommended Winter Harbour on Melville Island as the location for the main base, and refused to consider other possibilities at that time. Hubbard proposed to establish a satellite station at Eureka

Sound on the west coast of Ellesmere Island as soon as possible, and both countries agreed to assemble their civilian contingents for Eureka by mid-April—less than two months away.[13] The officials also agreed to general guidelines on personnel and infrastructure. Canada would contribute the officer-in-charge (OIC) and half the personnel at each station, as well as an RCMP representative at the main station. All permanent installations at the stations and adjacent airstrips would remain Canadian property, thus allaying possible sovereignty concerns. The United States would provide the other half of the personnel, construct "temporary" buildings, and cover the bulk of the costs, including meteorological equipment, transportation, fuel, and supplies. The executive officer, as the senior American at each station, would oversee American staff subject to the Canadian OIC's policies and would report to the U.S. Weather Bureau on technical matters.[14]

With these understandings in place, the U.S. and Canadian weather services launched the JAWS program. Eager to still establish the first stations that year, Hubbard worked with the US military to quickly position supplies assembled during the preceding year, as well as coordinate with the Canadians, before sending the first team North. Slidre Bay, along the eastern shore of Eureka Sound, was the first station to be established. On 7–8 April, US Strategic Air Command flew the station crew and four loads of cargo to Slidre Bay. The temperature was –42°F and the wind blowing when Canadian Jud Courtney and his colleagues arrived and set up temporary shelter for a midday meal. No one was excited at the prospect of sleeping in small, inadequate tents, so Courtney's plan to build a Jamesway hut "met with instant and unanimous approval." By 7:30 pm, the staff had erected and heated the building, prepared hot meals, and set up the radio equipment.[15] The next morning began an endless cycle of airlift operations that strained the small station staff over the following week. All told, the spring 1947 airlift delivered 110 tons of food, fuel, and consumable stores—enough to sustain the station crew for more than a year.[16] Because initial equipment shortages prevented the staff from beginning their full-scale weather observation program, they invested their energies in construction. "The novelty of pioneering in the set up of a new station" wore off after a few months, so Courtney decided to "tighten up on organization and supervise the various projects a little more stringently to avoid half measures" which would undoubtedly create problems later. "In all cases personnel showed good sense and a spirit of cooperation rarely encountered elsewhere," the station officer-in-charge applauded, indicating that his leadership style was effective and appropriate.[17]

Task Force 68 ventures North

The planning team, however, was resolved to improve on its experiences earlier in the year setting up Eureka. The group sent to Ellesmere Island under Courtney's leadership had been hastily assembled. Although the Canadian meteorological service still had to "scrounge" to find suitably trained personnel, Canadian James Donald Cleghorn's team would be fully staffed and equipped.[18] On 7 May 1947, the American Chief of Naval Operations stood up Task Force 68 or, as it was popularly known, Operation Nanook II. Its basic mission for that summer was to provide logistical support to the USWB in establishing a main station at Winter Harbour, installing an automatic weather reporting station along Lancaster Sound, and resupplying the existing weather stations at Thule and Slidre Fiord (Eureka).[19] Actual planning began when high ranking US Army officers, naval officers with previous Arctic and Antarctic experience, and Weather Bureau officials convened in Washington about a week later. Cleghorn, who had been picked as the officer-in-charge for the main station to be build that summer, also attended. Before the war, he was an ornithologist and the associate curator at the Redpath Museum at McGill University in Montreal. Originally commissioned in the Black Watch (Royal Highland Regiment) of Canada,[20] he had wartime experience commanding Canadian and American troops in isolated conditions as the former base commander at Camp Churchill. He had also been responsible for maintaining the moving force during Exercise Musk Ox in 1946, managing discipline, accommodation, and rations.[21] In early 1947, Lieutenant Colonel Cleghorn secured his release from the military so that he could head up the station planned for Melville Island where, in addition to his other duties, he expected to study the local fauna and collect bird and mammal specimens for the National Museum of Canada.[22]

Sixteen men would serve with Cleghorn at the main station: seven other Canadian staff, eight Americans, and an RCMP constable.[23] They would bring ample equipment and supplies for the new station (as well as permanent buildings for Eureka), including tractors, heavy airstrip graders, power generators, pre-fabricated housing, fuel oil, clothing, food and emergency rations, as well as meteorological and other scientific equipment—enough materiel for the station to operate for at least two years without further resupply.[24] Logistical planners invested tremendous effort in marking the shipments in various colors and numbers to show their destination and classification.[25]

On 1 July 1947, Captain Robert S. Quackenbush, Jr., who had been the executive officer of the Navy's Antarctic operation *High Jump* the previous winter, took command of Task Force 68 and its three ships. When the vessels came into range of Winter Harbour, aerial reconnaissance revealed that the route to Melville Island remained choked with impassable ice. Faced with limited options, Hubbard decided to lead a scientific party to survey conditions at Resolute Bay. Planners had not considered the ships failing to reach Melville Island, but after two unsuccessful attempts to reach Winter Habour (and ship damage), Hubbard accepted that he needed to move to an alternative site.[26] By late August, Hubbard formally requested the Canadian government to approve the relocation of the planned hub station to Resolute. The Canadian government, however, remained unconvinced, passing a message through the US Weather Bureau instructing the task force to survey sites on Beechey Island and the southwest coast of Devon Island. Hubbard reported back that Resolute was the most feasible.[27] In the end, the two countries' weather services formally approved the new location at Resolute Bay on 29 August.

Two days later, crews began hastily unloading the ships and pounding stakes into the ground. With winter looming and ice threatening to force out the ships, eight LCMs (fifty-foot landing barges) shuttled cargo from the *Wyandot*, anchored about a mile offshore, to the beach around-the-clock. Two crews working twelve-hour shifts discharged 3,500 tons of cargo, including 200,000 gallons of gasoline, in less than a week.[28] It was chaotic. Although logisticians had carefully classified the cargo in Boston before the ships headed north, the disorganized state of the offloaded crates at Resolute mirrored the problems encountered at Eureka earlier in the year. "I am sure it will sound absurd," Cleghorn recalled, "but we worked throughout the entire period of cargo discharge without a manifest showing exactly what was loaded, and therefore, what we could expect to find at the discharge and supply points." On the ground, the station personnel discovered that many of the boxes did not contain the items they were supposed to, while some contained no markings at all. Fresh fruit and vegetables came ashore before heated warehouse space was built, and 90 percent of this food was destroyed by frost.[29]

The state of the war surplus equipment that arrived also bred frustrations, forcing work crews to compete for limited resources to complete beaching operations, station construction, and the preparation of an airstrip. As soon as heavy equipment arrived onshore, crews quickly uncrated and fitted it together—and detected problems. "The mechanical difficulties encountered at Resolute Bay were not attributable to either the latitude or weather conditions," Major Taylor observed. The US Engineers were convinced that the Army had sloughed off old equipment on the JAWS project that had been written off. Except for the graders, shovels, scrapers, and "rooter," all of the equipment sent northward was used—and well worn. It took mechanics until 13 September to get all of the tractors into serviceable condition, and the four cargo trucks shipped north were in continuous demand but broke down repeatedly.[30] Because the US Army engineers were preoccupied with this machinery, unskilled naval personnel from the two ships were conscripted to set up the prefabricated buildings "at the very last minute, and under severe handicaps." As a result, the structures were "not too well done"[31]—with unfortunate implications for the personnel that winter.

None of the ships were set up to overwinter in the Arctic, and their tasking ordered them to return south when weather or ice conditions rendered further operations "unprofitable."[32] When a calm, cold night produced a film of slush over the entire bay by the morning of 11 September, the Navy became desperate to get away before being blocked by ice floes. Two days later, the Canadian and American weather station personnel moved ashore to occupy their newly built quarters. That afternoon, Hubbard, Quackenbush, and Cleghorn presided over a flag raising ceremony with speeches and tributes to the achievements, and the naval task force steamed out of Resolute Bay just after midnight.[33] "At 18:45 hours the two ships ... passed from view behind Cape Hotham, leaving behind seventeen civilians of the weather station staff and forty officers and men of US Army Engineers," Cleghorn noted.[34]

Initial operations at Resolute Bay

The hasty departure left problems in its wake. Insufficient time and warehouse space prevented the crews from storing all the supplies indoors. Blizzards soon covered the outdoor caches with deep, hard snow-drifts, which concealed their location and their contents. Furthermore, faulty packaging failed to protect some food and supplies from sea water, rain and snow. The station consequently faced shortages of many items during its first winter of operation. Trying to take an inventory of the satellite stores that fall proved "an unhappy experience as well as a waste of time," given that the staff had no list of what the supply dumps contained. "We decided that it would be unwise to open the boxes to see what they held," Cleghorn explained, "for fear that the drifting snow would leak in and spoil the contents."[35]

Amidst ongoing construction, station staff tried to establish a routine. On 5 September, Canadian meteorologist R.W. Rae had set up USWB maximum and minimum thermometers in an improvised screen and kept a daily record of temperature extremes, as well as brief daily weather notes later compiled into a monthly report.[36] The RCMP post was set up on 16 September, and the next day Cleghorn drafted and discussed local station rules and regulations with a full meeting of the station staff. Camp fatigue parties washed dishes, disposed of garbage, drew water, and filled latrines and oil stoves. An army truck drove away the first inquisitive polar bear that approached the station. When it returned a few days later and broke into food boxes, Cleghorn was forced to shoot it. With the freshwater ice now nine inches thick, the entire station staff cut and stacked 3,000 blocks at the station to carry them through the winter.[37]

On 1 October, Cleghorn offered a positive appraisal of the local situation. The airstrip was 6,400 feet long, full electrical power had been turned on in all of the buildings, and Rae planned to start taking regular synoptic observations about 10 October and pilot balloon observations a few days later. "We have no personnel problems," he told Andrew Thomson, the controller in Toronto. "Everybody is pulling his own weight and relations between the Americans and ourselves are on a very cordial basis. There have been minor misunderstandings and some differences of opinion on both sides, but these are to be expected in any normal operation of this kind. I blame the prolonged strain and overwork, rather than any personal animosity for any small outburst of temperament in the past."[38]

Less than two weeks later, Cleghorn reported the first signs of personnel trouble. Although the staff made steady progress "in all branches of our work," Cleghorn reported that the station personnel, both American and Canadian, did not "like to be tied down by a set of rules and regulations." The American executive officer and chief mechanic objected strongly to the "no smoking" rule in the powerhouse and garage, insisting that a man had a right to smoke at his place of work. Cleghorn rescinded his order for "the sake of peace," but he was concerned about this violation of the fire prevention plan—even though the men insisted that they would be careful. When the OIC drew up a set of rules and presented them at a staff meeting, "they were accepted in silence, although they were simple and pretty local in character, dealing with such matters as mealtimes, conservation of fresh water, use of vehicles and warnings to personnel about such things as crossing unsafe ice, or undertaking lone hikes into the surrounding country." Cleghorn removed them from the station notice board when he heard whispers that this constituted undue strictness and "regimentation." The former Army officer remained conscious of his military background, and "tried very hard to live that part of his past down" given that the station represented a very different "time and place" than his previous postings. He tried to dismiss the "touchiness" as a carryover of the difficult voyage and the "unhappy confusion" before the ships departed, and reassured his superiors that "we have shaped up a new course, and everyone is trying very hard to readjust and put up a good showing."[39]

A few months later, Cleghorn painted a more pessimistic picture. Even before they had set sail from Boston, the former military officer had been skeptical about the group's lack of experience and the lack of instruction they had received on duties and conditions of Arctic service. His fears were confirmed on the ground. "From the start it was painfully evident that the majority were more interested in the higher salaries offered than in any other prospect," he complained. "With one, or perhaps two, exceptions, no one was the least bit inspired by the prospect of going north to do important work." No one displayed "the slightest trace of the 'expedition spirit,'" and they were unwilling to do anything for which they had not been hired specifically. Cleghorn was disgusted by "the uncooperative and even hostile attitude shown when I had to round them up to do essential work connected with the handling of supplies, or camp and household duties, and a similar attitude when I asked for strict adherence to station rules and regulations." Loose organization, inadequate equipment and supplies, and long work days heightened confusion, frustration, and ultimately resentment. Unhappiness reigned, Cleghorn concluded, because the personnel recruited for the station "had not realized, and were quite unprepared, to face the isolation from the rest of humanity, from customary relations and usual scenes that life at an Arctic post involves."[40]

By this point, Cleghorn's perspective on the US contingent had shifted profoundly. He had considerable background experience dealing with

American civilian and service personnel, and suggested that he had "always been able to get along with them at all times." This group was "entirely different" from any he had previously encountered:

> In my opinion they resented being placed under Canadian command, and having to abide by a set of uncompromising station rules, regulations and the laws and ordinances of the Territories. I remember hearing one of them say that they were afraid of losing their identity, and no amount of reasoning on my part of the international aspect and co-operative effort of the project seemed to make the slightest difference to this manner of thinking. I showed no discrimination whatsoever in my dealings with them, and international relationship remained on a high level, but it was this sort of thing that kept me at a high nervous pitch and convinced me that my task would have been far happier and less complicated had the expedition been entirely Canadian from beginning to end.[41]

Cleghorn recognized that "American participation, material aid, and good will" were essential to the program. He also believed that the Americans were arrogant, suffered from a host of administrative and technical problems that plagued planning and execution, and rejected Canadian authority. The confusing lines of command and control—with Navy, Army, and Weather Station supervisors all controlling specific aspects of the project, and often failing to solicit a Canadian opinion—made the situation impossible in the former military officer's eyes.

RCMP Constable Harry Aimé, a detached observer, concluded that Cleghorn embodied the real problem. He did not fit in with the station team, openly criticized those around him, and was a "busy-body" who injected himself in others' affairs. "A dreamer, he was likely to be found reading about polar exploration, including Scott's trek to the South Pole," the police officer recounted in his memoirs. Cleghorn "fantasized that if Resolute Bay had to be abandoned, we would all trek to Dundas Harbour." Given the lack of RCMP facilities at Dundas, the constable noted, this was "a most unrealistic idea." But Cleghorn seemed aloof from the realities of life at a small, civilian station. "Almost every day there were tensions and, if none arose, Cleghorn would create some. He just wasn't suited for northern isolation."[42]

Tragic incidents compounded morale problems at Resolute that fall. First, a polar bear severely mauled Edwin (Ted) Gibbon, a Canadian radio operator, within the camp area in the blustery, grey morning of 24 October. He had gone outside to notify Cleghorn, who was sleeping in the dormitory on the other side of the mess hall, of an incoming flight. "As I walked towards the mess hall, ... out of the corner of my eye, I saw the bear charging toward me on all fours" about four feet away, Gibbon later recounted. He tried to dart behind a sled, but the bear cut him off with a blow to the head. In a semi-conscious state, Gibbon recalled looking up at the bear's face. "He had his paws around the back of my neck and seemed to be trying to break it." As they wrestled, Gibbon shoved his arm into the bear's mouth to prevent it from biting him. The cook heard his screams and rallied Cleghorn out of bed, who promptly grabbed the station rifle and shot the bear. "Lucky for me the American army doctor was still at the camp," Gibbon explained. "He sewed me up. He said he didn't bother counting the stitches, there were so many. He found that there were teeth and claw marks all over my head and neck and arms." Two days later, the victim was evacuated out on a flight with the departing army engineers and recuperated in a hospital in Montreal before returning home to Ontario.[43] Cleghorn concluded that this situation "showed how totally unprepared [the station personnel] were to receive the full impact of wilderness living." Although he had warned everyone about polar bears wandering near camp, no one had believed an attack possible.[44] The cook, traumatized by the event and haunted by "visions of how a big bear would sink its teeth into his long thin neck," took a handful of sleeping pills and slipped into a four hour "coma" before two of the boys managed to shake him out of it. It was several days before he resumed cooking for the crew.[45]

The second major accident occurred on 7 December when Lorne Manion, a young Canadian Met Tech (meteorological technician) from Saskatchewan, was electrocuted in his room. Climbing into a double bunk bed late one evening after his work shift, Aimé recounted how Manion "gripped the steel frame and at the same time rubbed his moist back against an open ceiling furnace duct." The bed was in contact with an open electrical outlet, causing the duct to short from the opposite side of the building, sending an electrical shock of 220 volts through Manion and killing him instantly. Aimé held a coroner's inquest, which confirmed beyond a doubt that inadequate electrical supplies and careless installation of wiring and fixtures had caused Manion's death. "No one could be blamed for the tragedy—the staff were not electricians—nor were they aware of the dangers," he noted. Morale sank and distrust grew amongst station staff, who believed that the entire program was built on similar, dangerous foundations. Cleghorn did little to reassure them, his credibility now eroded beyond repair. "Cleghorn began his evidence [before the coroner's inquest] by relating the books he was reading at the time," Aimé recalled. "I instructed him that only the facts were required, not his opinions or assumptions. The staff was ecstatic. For once, Cleghorn was not in the driver's seat."[46]

By this point, Cleghorn had failed as a leader. "I had hoped that once the ships sailed for the United States and we were left on our own resources we would become one united group, but such was not to be," he confessed. "Small cliques were forming, there were hurt feelings and misunderstandings all around until I could see there was nothing more I could do to restore their confidence, since the leadership I offered was not acceptable to them in any form, and was even resented." Despite his previous experience in managing a remote base, Cleghorn had had enough of Resolute. He spoke with Constable Aimé, who strongly advised him to resign. On 12 December, he sent a message to Andrew Thomson, his supervisor in Toronto. "Imperative that I report to you in person regarding entire situation here," Cleghorn communicated. "Urgency demands travelling aircraft due here twentieth and total weight seven hundred pounds." Thomson was "shocked" to receive the message, which gave no clear reason why Cleghorn felt it "imperative" to abandon his post. Cleghorn's forwarding letter simply stated that he had to head south "to clear up some of the misunderstandings, improve our methods and aid in the welfare of those serving in the north." The reference to 700 lbs, however, implied that the officer-in-charge had taken all of his personal belongings with him and did not intend to return to Resolute. Cleghorn's reply, which documented his experiences that fall, insisted that he faced a "very threatening" situation at the station. "I am sure that had I remained for another month, violence, in some form or another, would have broken out, since I had absorbed all the nonsense and abuse that I was prepared to take under the circumstances and I decided that it was going to stop then and there," he explained. "Having made this decision, I had two courses left open to me, and I chose the rational one."[47] On 23 December, Cleghorn boarded a B-17 aircraft and headed for Goose Bay, Labrador.

Safely ensconced in the comforts of southern Canada by mid-January 1948, Cleghorn authored a dizzying array of recommendations to improve the weather station program. He criticized everything from the 30/06 hard point ammunition supplied to station, which would not have stopped a polar bear unless cleanly shot through the heart, to the overly lavish American procurement system. He did not think that pillowcases, electric grills, and fresh turkey were essential to the well-being of station staff. "It is true that a little luxury now and then is a morale builder," Cleghorn conceded, "but for a long pull under northern conditions, adequate bedding and good wholesome fare and lots of it is more to the point." Creature comforts became expectations, leading individuals to complain "when the bed sheets provided are unbleached cotton instead of linen."[48] Cleghorn also called on Canadian and American officials to more carefully select and screen applicants serving at isolated Arctic stations, listing the traits that he considered most desirable: youthfulness, physical strength, the desire for adventure, good common sense, a willingness to accept responsibility, self-sufficiency, and an extroverted personality. "He should be made fully aware of all the phases of Arctic life, its isolation, climatic conditions, its joys as well as its dangers," Cleghorn emphasized. "He should be told there are, theoretically speaking, two Arctics—the high Arctic of the weather stations, and the one which he has heard and read, ... the Arctic of the trading posts, the natives and their hunting camps." The latter already had "its community life, interest, and colour," while the station experience "is what you make it."[49] Cleghorn considered himself to be the "experienced Canadian" well-suited to assume liaison responsibilities with the US Weather Bureau and ensure the program followed a more cooperative and efficient course, but there was no doubt that he had retreated from a difficult situation at Resolute.

Cleghorn's superiors, undoubtedly disappointed with his performance, did not avail themselves of his offer. The Deputy Commissioner of the NWT, R.A. Gibson, met with the RCMP commissioner on 6 January 1948 and learned that "he had been advised confidentially by his representative at Cornwallis Island that Colonel Cleghorn ... was unable to measure up to his responsibilities [as senior officer], whereas the Americans sent an outstanding officer who is a natural leader. Colonel Cleghorn consulted the policeman to see what he should do and the policeman advised him to seek a recall." When Cleghorn passed through Ottawa a few days later, he paid a visit to J.G. Wright in the NWT office and offered his perspective. "He said the weather station staff had been under strict naval supervision on the way north & were 'fed up,'" Wright recounted. When they learned that "an ex-military man" would be in charge of the station, "they immediately resented him. There was a lot of insubordination against his camp rules," and after Manion was accidentally electrocuted "all blamed [Cleghorn] for not seeing to it that the wiring of the U.S. Army engineers had been properly done!" Perceiving his position to be "impossible," Cleghorn asked to be recalled.[50] Senior officials obliged. "He turned out to be a disaster," Robert McTaggart-Cowan recalled. Cleghorn had to be replaced or "he would have endangered the project."[51]

The weather station at Resolute survived Cleghorn's abrupt departure. R.W. (Bill) Rae, the resident meteorologist and senior Canadian, assumed the OIC role at the station for the next two years.[52] Where Cleghorn had failed, Rae succeeded. His report on living conditions at Resolute, reprinted in the *Christian Science Monitor* in early May 1948, indicated a vastly improved situation. The station staff kept busy and interpersonal relations were generally free of friction. "There is always plenty of work to be done and the amount of spare time left over for hobbies or recreation is relatively small," Rae noted. Given space limitations, table tennis proved an ideal form of indoor recreation. "The entire set is homemade except for the Ping-Pong balls, which I begged from the United States-Danish weather station at Thule," Rae described. "We are presently in the midst of a hectic handicap tournament for the table tennis championship of Cornwallis Island. The winner not only receives five chocolate bars, but what is more important, is excused from helping with the dishes for two days."[53] The improved situation at Resolute in the winter of 1948 mirrored that at Thule and showed the importance of solid leadership.[54] Creating the right conditions for simple, predictable routines proved a key recipe for success at an isolated outpost.

The Eureka station also enjoyed a harmonious environment in late 1947 and early 1948. The staff settled into the new wooden prefabricated accommodation building, which represented a vast improvement over the temporary huts. With the station on a full operational basis by early October, Courtney had turned his attention to preparations for the dark period. He kept everyone busy between regular scheduled weather and radio operations so that days would pass quickly. He assembled his staff to remind them to stay alert—animals could surprise them in the dark. The crew installed lighting in all buildings as the daylight waned, completed an inventory of all supplies, and caught up on narrative reports of weather observations. That station's winter and spring water supply would come from icebergs across the sound, which would be accessed during full moon periods each month.[55] Courtney valued the physical exercises involved in simple activities like this, given the lethargy that set in during the darkness.

Conclusions

The activities under Courtney and Rae that winter marked the shift from *creating* the Eureka and Resolute stations to actually *inhabiting* them on a permanent basis. The pioneering crews established the footprint and proved the concept. Courtney and others, in suggesting a litany of improved buildings, scientific equipment, supplies, and clothing, were laying the groundwork for a persistent presence that would transform the stations into permanent hubs fit to facilitate continuous scientific observations.

The JAWS program was the product of requirements for meteorological information in the immediate post-war era. The advent of trans-Atlantic commercial flights, primitive computer-based forecasting and the emerging need for Allied militaries to operate in the Arctic necessitated synoptic surface and upper air High Arctic meteorological data. The USWB and Canadian Department of Transport seized the moment and cultivated a joint program that served this largely civilian combination of interests. While both countries' militaries provided logistical and material support for the program, their attention was generally directed elsewhere. All of the stations initially grappled with surplus equipment that was sub-standard or ill-suited to Arctic conditions. In future years, the American military repeatedly pushed for the Canadian Forces to relieve it of resupply burdens and rebuffed USWB requests for additional funding when its own budgets were strained by other commitments.

The human dimensions of leadership—from personality to style, to accepting the unique physical environment in which the stations were situated—proved instrumental to creating and sustaining functional stations. Jud Courtney demonstrated how a leader with appropriate traits and temperament could overcome adversity and achieve success. Conscientious, agreeable, and sensitive to the needs of his men, he proved well suited for an isolated and confined environment. He ensured that the personnel at Eureka did not succumb to boredom and found ways to motivate them without resorting to his "command" authority. This modelled Hubbard's philosophy (as his wife described it). "The reasons Charlie had for wishing always to keep the stations on a civilian basis were that his experience with the Crystal bases led him to believe that only small, hand-picked groups could operate with maximum efficiency in the arctic," she explained. "Though this has been practically speaking a great difficulty,

Charlie's stations are a whole lot better than the large, cumbersome, groups the armed forces are likely to have operating in the arctic."[56]

The early experiences of station crews also yielded important lessons, highlighting the need for careful planning to supply isolated stations that could not be visited easily—and the need to improvise locally when equipment did not arrive or was in poor condition. Station personnel quickly identified the physical and psychological stresses of adapting to often unpredictable Arctic conditions. Environmental realities directly influenced the form and pace of development. Although station staff living in permanent buildings did not face the same physical challenges as explorers traveling long distances on the land and living in ships, tents, or snowhouses, hazards remained: from polar bears to hastily constructed buildings to extreme weather. Furthermore, isolation from the rest of the world remained a stark reality, however much modern communication and transportation provided unprecedented connections to the south.

The military provided vital construction and logistical support to JAWS, but USWB and DOT staffs affirmed the network's civil character. The involvement of these civil organizations was not a case of "civilian cover" for militarization of the Arctic, as some historians allege.[57] Indeed, the case of Lieutenant-Colonel (retired) Cleghorn, who Hubbard had specifically requested to oversee the establishment of the main weather station based upon his wartime background working at a joint facility, proved that a military mindset did not suit the JAWS environment. His failures demonstrated the necessity of flexible leadership at remote outposts. Although Thomson later suggested that Cleghorn "didn't have the northern qualifications" that the meteorological service believed he had,[58] a more appropriate assessment might be that he did not have the *right* qualifications or traits to successfully lead a *civilian* weather station.

There is little support for Cleghorn's assertions that bilateral friction permeated relationships between Canadian and American personnel on the ground at Resolute. Instead, bilateral relationships between personnel proved to be overwhelmingly cordial at the stations throughout the JAWS program. "Since [the Resolute] station is staffed jointly by Canadians and Americans, it represents an interesting example of a practical application of the good-neighbour policy," a reporter noted in May 1948. "Both groups had to adapt themselves somewhat to the other's point of view, for the procedures in the various phases of the station operating program are a combination of both Canadian and American practices." All told, Canadian R.W. Rae reassured the program directors down south that "the degree of cooperation between the two groups has been excellent."[59] The senior weather bureau officials in Washington and Ottawa got the message. McTaggart-Cowan explained in a 1983 interview that "between Dr. Reichelderfer ... and John Patterson and then Andrew Thomson and then myself, we kept on top of the little bits of friction" like the "trials and tribulations" surrounding Cleghorn. "It was a marvelous example of good partnership in international cooperation."[60]

Oral histories and the archival record support McTaggart-Cowan's observation of strong bilateral cooperation. Historian Gordon Smith aptly assessed the JAWS program "as one of the most important and successful examples of U.S.-Canadian joint endeavor in northern regions," offering "a striking illustration of successful international cooperation and collaboration."[61] For the next quarter century, Canadian and American personnel worked side by side at the stations and, with only rare exceptions, developed a sense of shared purpose and *esprit de corps* that transcended national lines. (Conflicts, when they arose, were usually the result of interpersonal or professional differences, not nationality.) This allowed the joint stations to become a quiet fixture of the Cold War Canadian Arctic through to 1972, when the last Americans departed the stations and Canada assumed full responsibility. It had proven to be a tremendous success. Geographer William C. Wonders lauded that "the Joint Arctic Weather Stations programme was imaginative, venturesome and expensive at the time it was launched," and "proved to be one of the most valuable investments" that the Canadian government made in the High Arctic. "It more than lived up to its expectations in its meteorological and climatological returns" and, as "anchor points" for exploration and development in the High Arctic, the stations enabled "a far-flung programme of even wider scientific value to be implemented"[62] —thus fulfilling Hubbard's expectations that building weather stations in the remotest parts of Canada's Arctic Archipelago would "provide habitations, channels, communications, and transportation which will make it possible for us to penetrate the Arctic for other purposes."[63]

Notes

1. For some of the most recent scholarship on this topic, consult: Janet Martin-Nielsen, *Eismitte in the Scientific Imagination: Knowledge and Politics at the Center of Greenland* (Springer, 2013); Matthias Heymann, "In Search of Control: Arctic Weather Stations in the Early Cold War," in *Exploring Greenland: Cold War Science and Technology on Ice*, ed. Ronald Doel, Kristine Harper, Matthias Heymann (New York: Palgrave Macmillan, 2016), 75–98.
2. See, for example, Shelagh Grant, *Sovereignty or Security? Government Policy in the Canadian North 1936–1950*, (Vancouver: UBC Press, 1988); David Bercuson, "Continental Defense and Arctic Sovereignty, 1945–50: Solving the Canadian Dilemma," in *The Cold War and Defense*, Keith Neilson and Ronald G. Haycock eds. (New York: Praeger Publishers, 1990), 156; David Bercuson, "Advertising for Prestige: Publicity in Canada-US Arctic Defence Cooperation, 1946–48," in *Canadian Arctic Sovereignty and Security: Historical Perspectives*, ed. P. Whitney Lackenbauer (Calgary: Centre for Military and Strategic Studies, 2011), 111–20; Adam Lajeunesse, "The True North As Long As It's Free: The Canadian Policy Deficit 1945–1985," unpublished M.A. thesis (University of Calgary, 2007) 18, 24; and Heymann, "In Search of Control," 92.
3. Kenneth Eyre, "Custos Borealis: The Military in the Canadian North," unpublished Ph.D. dissertation (University of London - King's College, 1980).

4. David Beatty, "The Canadian-United States Permanent Joint Board on Defence," unpublished Ph.D. dissertation (Michigan State University, 1969), 117.
5. Daniel Heidt, "Clenched in the JAWS of America? Canadian Sovereignty and the Joint Arctic Weather Stations, 1946–1972," in *Canadian Arctic Sovereignty and Security*, 145–169; Whitney Lackenbauer and Peter Kikkert, "Setting an Arctic Course: Task Force 80 and Canadian Control in the Arctic, 1948," *The Northern Mariner*, 2011, 21(4): 327–58.
6. Kristine Harper, *Weather by the Numbers: The Genesis of Modern Meteorology*, (Cambridge, Mass: MIT Press, 2008). On Reichelderfer, see Bill Davidson, "On Guard at the Pole: Our Sub-Zero Heroes," *Collier's Magazine*, 21 June 1952, 70; and Jerome Namias, *Francis W. Reichelderfer, 1895–1983: A Biographical Memoir* (Washington: National Academy of Sciences, 1991), http://www.nasonline.org/publications/biographical-memoirs/memoir-pdfs/reichelderfer-francis.pdf (accessed September 18 2018).
7. *Hearings Before the Committee on Agriculture, House of Representatives*, Seventy-Ninth Congress, Second Session on H.R. 4611 (S.765), 22 January 1946, 1–13, National Archives and Records Administration (NARA), RG XPOLA, entry 17, Charles Hubbard Papers, box 1, file Miscellaneous.
8. On his early vision for the weather station program, see Lt. Col. Charles J. Hubbard, "The Arctic Isn't So Tough," *Saturday Evening Post*, 217/9 (26 August 1944), 12. On Hubbard's role, see his extensive papers at NARA, RG XPOLA, Entry 17.
9. *Hearings Before the Committee on Agriculture*, 22–28. Hubbard quoted a supporting letter from the Secretary of the Navy suggesting that the proposed stations were "primarily intended to aid in the development of civil and commercial air transportation and, if enacted, would have no direct bearing upon the steps which may be taken by the military services in the interests of national defense." *Hearings Before the Committee on Agriculture*, 26.
10. "A Report on Recommended U.S. Weather Bureau Arctic Operations for the period of April 1946 to July 1, 1947, in compliance with Public Law 296, 79th Congress," NARA, RG XPOLA, entry 17, box 5, file "Analysis of Possible Arctic Ops, 1946." See also Shelagh Grant, "Weather Stations, Airfields, and Research in the High Arctic, 1939–1959—An American Perspective," presented at the Canadian Historical Association Annual Meeting on 27 May 1990, and subsequently revised as an unpublished two-part article on "American Defence of the Arctic, 1939–1960."
11. "Northern and Arctic Projects," 28 January 1948, LAC, RG 2/18, vol. 57, file A-25-5. Ironically, cabinet proposed a more ambitious program than the Americans had contemplated.
12. "Copy of substance of United States reply to Canadian Note No. 16 of February, 1947," Libraries and Archives Canada (LAC), RG 22, vol. 732, file SE-4-1-83.
13. Joint Meeting of United States and Canadian Technical Experts to Discuss the Establishment of Arctic Weather Stations, 25–26 February 1947, LAC RG 22 Vol 732 File SE-4-1-83. See also LAC, RG 25, vol. 8177, file 6700-40, pt.1. The terms of this agreement were summarized by Mr. St. Laurent in a note to Mr. Atherton on March 18. See LAC, RG 25, vol. 3347, file 9061-A-40C, pt.2. On previous visits to Winter Harbour and a detailed description of the area, see Moira Dunbar and Keith R. Greenaway, *Arctic Canada from the Air*, (Ottawa: Defence Research Board, 1957), 237–38. On Rowley, see John MacDonald, "Graham Westbrook Rowley," *Arctic*, 2004, 57(2): 223–24.
14. Joint Meeting of United States and Canadian Technical Experts to Discuss the Establishment of Arctic Weather Stations, 25–26 February 1947, LAC RG 22 Vol 732 File SE-4-1-83.

15. Courtney Report Eureka Sound 1947–1948, 11–13, NARA, RG 27, entry 7, box 1, file Reports, Eureka.
16. Courtney Report Eureka Sound, 15 April 1947; Report on the Inspection of Eureka Sound and Thule Weather Stations during April 1947, LAC, RG 93, box 26, file 11-10-11, pt. 2. The aircraft operated on wheels because the skis bounced on the hard snow surface while the wheels tended to cut down to the hard ice. Approximately 50 per cent of the total tonnage was carried by a C-82, which proved especially well adapted because of the ease of loading and discharging cargo from it.
17. Meteorological Division-Department of Transport-Canada and U.S Weather Bureau-Department of Commerce-United States, *A Review of the Establishment and Operation of the JAWS at Eureka Sound, Resolute, Isachsen, Mould Bay and Alert and a Summary of the Scientific Activities at the Stations, 1946-1951* (Ottawa, 1951), 9–10; Courtney Report Eureka Sound 1947–1948, 20. Courtney found the Quonset hut completely unsuited to Arctic conditions—particularly when trying to erect it in cold and windy weather (p.22).
18. H.L. Keenleyside to Undersecretary of State for External Affairs, 1 May 1947, LAC, RG 25, vol. 3841, file 9061-A-40c; Memorandum for Under-Secretary of External Affairs, 1 May 1947, LAC, RG 22, vol. 732, file SE-4-1-83. On CMS staff shortages following the war, see Canadian Meteorological Service, Oral History Project, Patrick D. McTaggart-Cowan, interviewed by D.W. Phillips, 5 October 1983, 22, transcript available at https://opensky.ucar.edu/islandora/object/archives%3A7619 (accessed September 18 2018).
19. Chief of Naval Operations to Commander in Chief, U.S. Atlantic Fleet, CNO conf. ser. 072P33, 7 May 1947, NARA, RG 27, entry 5, box 12, file Report of Task Force 68, 1947. Secondary tasks including training personnel; testing ships and materiel in Arctic conditions; making observations of geographical, navigational, and aviation interest; and collecting detailed hydrographic, meteorological, and electro-magnetic propagation data, as well as other scientific experiments in line with the "limited scope of this operation." The USN also provided logistic support and lift to the US Army Engineers tasked with building an airstrip at Winter Harbour.
20. "Capt. J.D. Cleghorn," *Montreal Gazette*, 30 September 1942, in "James Donald Cleghorn," McGill University War Records, McGill University Archives, 0000-481.01.2.e0431.
21. "Cold Weather Trials: Papers on Exercise 'Muskox'," terms of reference, 9 October 1945, DHH 746.033(D1); Lieut.-Colonel G.W. Rowley, "Exercise Muskox," *The Geographical Journal* 1947, 109(4/6), 176; Kevin Thrasher, "Exercise Musk Ox: Lost Opportunities" (MA thesis, Carleton University, 1998), 21. During Musk Ox, two men under Cleghorn's command died in a fire in the Officer's Mess caused by fuel spillage, casting "a somber mood over the camp." Thereafter, "Cleghorn organized frequent drills to keep the men on their toes." Thrasher, "Musk Ox," 39–40.
22. A.L. Rand to R.A. Gibson, 29 April 1947, LAC, RG 85, vol. 1013, f.17742. He was sworn in as a game officer under the provisions of the North West Game Act on 11 June 1947. It is notable that Andrew Thomson, in his communications with other officials, explicitly "omitted any reference to Colonel" when describing Cleghorn "in view of the civilian nature of the operation of the Winter Harbour Station." Thomson to R.A. Gibson, 6 September 1947, LAC, RG 85, vol. 1013, f.17742.
23. H.L. Keenleyside to Undersecretary of State for External Affairs, 1 May 1947, LAC, RG 25, vol. 3841, file 9061-A-40c; Memorandum for Under-Secretary of External Affairs, 1 May 1947, LAC, RG 22, vol. 732, file SE-4-1-83.

24. DoT-USWB, *Review of the Establishment and Operation of the JAWS*, 10.
25. Major Andrew Taylor, *Report on the Engineering Aspects of the Operations of US Task Force 68 in the Canadian Arctic—1947*, 24 Nov 1947, 71, NARA, RG 27, entry 5, Formerly Security Classified Subject Files 1942-63, box 8, f. Navy Reports-Confidential.
26. Taylor, *Report on the Engineering Aspects*, 66.
27. Cleghorn Report, appendix A, LAC, RG 93, box 26, f.11-10-11 pt.3.
28. Taylor, *Report on the Engineering Aspects*, 68.
29. Cleghorn report, 8.
30. Taylor, *Report on the Engineering Aspects*, 75.
31. Cleghorn Report.
32. Chief of Naval Operations to Commander in Chief, U.S. Atlantic Fleet, CNO conf. ser. 072P33, 7 May 1947, NARA, RG 27, entry 5, box 12, file Report of Task Force 68, 1947.
33. They were prepared to leave on Friday the 13th, but they did not leave until early the next morning because of "naval superstition." Harry Hampton Aimé, *Overalls, Red Serge, and Robes: Life and Adventure in the Canadian North: An Autobiography*, (Hampton Press, 2004), 141.
34. Cleghorn Report, appendix A.
35. Cleghorn Report, 8.
36. R.W. Rae to Controller, Air Services, Meteorological Division, 1 October 1947, LAC, RG 93, box 26, file 11-10-11, pt. 2.
37. Cleghorn Diary, 17 September, LAC, RG 93, box 26, f.11-10-11 pt.3
38. J.D Cleghorn, Officer in Charge to Andrew Thomson, 1 October 1947, LAC, RG 25, vol. 3841, file 9061-A-40. On Rae's activities, see also Rae to Controller, 1 October 1947.
39. J.D. Cleghorn to Controller, Air Services, Meteorological Division, 12 October 1947, LAC, RG 93, box 26, file 11-10-11, pt. 2.
40. Cleghorn Report, 9–10.
41. Cleghorn Report, 6.
42. Aimé, *Overalls, Red Serge, and Robes*, 146. Cleghorn held Aimé in high esteem, extolling that the RCMP constable "has been a great help to me. He is steady and gets on well with everybody." J.D. Cleghorn to R.A. Gibson, 21 November 1947, LAC, RG 85, vol. 1013, file 17742.
43. "Gave Bear Arm to Chew to Save Neck Twisting," *Calgary Albertan*, 6 November 1947. See also various other clippings on LAC, RG 85, vol.1022, file 18554. See also Will C. Knutsen, *Arctic Sun on my Path* (Guilford, CT: Lyons Press, 2005), 256, which erroneously states that Gibbon survived but never regained consciousness.
44. Cleghorn Report, 10. For other background on his supposed "warning" to station personnel, see Cleghorn to Gibson, 21 November 1947.
45. Aimé, *Overalls, Red Serge, and Robes*, 146.
46. Aimé, *Overalls, Red Serge, and Robes*, 147; Cleghorn Report, 10. For the official RCMP inquiry and report, see LAC, RG 85, v.1021, f. 18517.
47. Andrew Thomson to J.D. Cleghorn, 10 February 1948, and reply, 13 February 1948, LAC, RG 93, box 26, file 11-10-11 pt.3; Aimé, *Overalls, Red Serge, and Robes*, 147; Cleghorn Report, 147.
48. Cleghorn Report, 7.
49. Cleghorn Report, 11.
50. R.A. Gibson to J.G. Wright, 6 January 1948, and marginalia by Wright, LAC, RG 85, vol. 1013, file 17742.
51. Canadian Meteorological Service, Oral History Project, Patrick D. McTaggart-Cowan, interviewed by D.W. Phillips, 5 October 1983, 22, transcript available at https://opensky.ucar.edu/islandora/object/archives%3A7619.

52. Cleghorn diary, 22 December 1947.
53. "Weather Watching in the Arctic," *Christian Science Monitor*, 7 May 1948. The personnel felt elated when a shipment of 20 packages of 15 phonograph records arrived, promising evening entertainment of varied music. Their "eager anticipation turned to keen disappointment" when they unpacked the boxes to find that each one contained the same 15 records.
54. See R. Magruder Dobie, "Why Husbands Go North," *Los Angeles Times*, 9 May 1948.
55. Courtney Report Eureka Sound 1947–1948, 34–35.
56. Mrs. Harriet Bissell Hubbard to Mr. Davidson, 1952, NARA, RG XPOLA, Entry 17, Charles Hubbard Papers, Box 1, File Correspondence of Mrs. Charles Hubbard.
57. See, for example, Grant, *Sovereignty or Security*, 174, 215–16, 238; Lajeunesse, "The True North as Long as its Free," 18, 24; and Bercuson, "Advertising for Prestige," 111, which asserts that "the stations would be operated by the United States Weather Bureau, a civilian agency, but they were clearly military in intent."
58. Aimé, *Overalls, Red Serge, and Robes*, 155.
59. "Weather Watching in the Arctic," *Christian Science Monitor*, 7 May 1948.
60. Canadian Meteorological Service, Oral History Project, Patrick D. McTaggart-Cowan, interviewed by D.W. Phillips, 5 October 1983, 22–23, transcript available at https://opensky.ucar.edu/islandora/object/archives%3A7619 (accessed September 18 2018).
61. Smith, "Weather stations in the Canadian North," *Journal of Military and Strategic Studies*, Spring 2009, Vol. 11, Issue 3: 63.
62. William C. Wonders, "The Joint Arctic Weather Stations (JAWS) in the Queen Elizabeth Islands," in *Essays on Meteorology and Climatology in Honour of Richmond W. Longley*, ed. K.D. Hage and E.R. Reinelt (Edmonton: Department of Geography, University of Alberta, 1978), 415. For a recent examination of research in the High Arctic, see Richard C. Powell, *Studying Arctic Fields: Cultures, Practices, and Environmental Sciences* (Montreal and Kingston: McGill-Queen's Press, 2017).
63. *Hearings Before the Committee on Agriculture, House of Representatives*, Seventy-Ninth Congress, Second Session on H.R. 4611 (S.765), 22 January 1946, NARA, RG XPOLA, Entry 17, Charles Hubbard Papers, Box 1, File Miscellaneous.

4 Frontier footage: science and colonial attitudes on film in Northern Canada, 1948–1954

Matthew S. Wiseman

Between 1948 and 1954, the National Film Board of Canada (NFB) produced two films depicting Canadian military personnel training for cold-weather warfare in the subarctic region of Hudson Bay near Churchill, Manitoba. The Defence Research Board (DRB), a branch of the Department of National Defence and Canada's first peacetime military science organization, cooperated with the Army Directorate of Military Training to sponsor the production of the first film, *Going North* (1948).[1] In the midst of escalating Cold War concerns over Arctic security, the DRB and the Army produced *Going North* as the first in a series of winter training films for soldiers, civil servants, and scientists heading north. The film depicted winter weather, northern topography, and other environmental challenges through the experiences of five soldiers training on the subarctic barrens and in the sparsely forested areas of Churchill. Released six years later, the film *Vigil in the North* (1954) also depicted soldiers training to perform military duties in the North, although, unlike the first, the NFB produced this second film as Cold War propaganda for public audiences.[2] As historical products of the early Cold War, both films provide a glimpse into southern perceptions of northern Canada and are valuable for understanding the role of science in reinforcing hegemonic attitudes and state goals toward modernization and development in the Indigenous North.

A comparative analysis of each film suggests military and government representatives utilized science to respond to Cold War anxieties and construct cultural identities for the North. Despite a six-year production gap, the two films shared a distinctly similar narrative: science represented "civilization" and provided an ethnographic measuring stick, dichotomizing North-South and the peoples who represented both cultural spaces. From this colonial point of view, the ability to wield scientific knowledge distinguished southern newcomers from the Indigenous Peoples they encountered.[3] As newcomers to the North attempted to understand how Indigenous Peoples lived and functioned in the environmental extremes of northern Canada, the demonstration of intellectual superiority reinforced the perception of southern dominance. Science gave newcomers the authority *they* needed to cultivate a distinct and secure space for the southern body in

Canada's northern environment, thus acting as a cultural marker for southern visitors to the North during the earliest years of the Cold War.

To the senior officials who sponsored the films discussed here, northern Canada represented both a geostrategic entity and a set of ideas. Considering the elasticity of southern perceptions toward northern Canada, Samantha Arnold has suggested that historians interpret the North as a space where newcomers explored and expressed nordicity as a challenge to Indigenous culture.[4] Joan Sangster recently picked up on this theme when she defined postwar northern contact zones as "places of cultural interaction characterized by asymmetrical power relations."[5] Analyzing southern perceptions of northern Indigeneity allows us to comprehend the complex colonial relationships that developed among northern Indigenous Peoples and southern visitors to the North. Visual records are particularly useful for understanding northern contact zones, as Peter Geller reminds us in his research on photography in the North.[6] Geller's approach to "reading" images suggests the visual appropriation of the North for mass audiences was important to claiming territorial and cultural sovereignty over northern Canada. Reconstructing the historical context behind the "production, circulation, and reception" of visual records helps explain how southern Canadians perceived and approached the North as a space of Indigeneity and environmental challenges.[7]

Understanding southern ideas about the physical and mental capacities of Inuit in relation to the natural cold environment of northern Canada is also useful for unpacking the colonial undertones of each film discussed here. In 2013, Patrizia Gentile and Jane Nicholas suggested that historians of Canada "consider the contested body as a fundamental category of analysis."[8] Echoing Geller's approach to visual analysis, Gentile and Nicholas argued historians must "read" bodies and representations of body to understand and explore the sociocultural issues that underpin human relationships. In the military context of the early Cold War, human performance in the cold received heightened scientific attention. Canadian and American scientists studied military and civilian bodies in preparation for Arctic warfare.[9] Indigenous life in the North served as a visual and cultural marker against which southern soldiers and scientists could test the capabilities of newcomers to the North. Although separated by six years, both *Going North* and *Vigil in the North* reflected the role of science in creating, supporting, and perpetuating these ideas during the early postwar period.

Going North (1948)

Shortly after being discharged from the Canadian Army in 1945, soon-to-be Arctic filmmaker Doug Wilkinson took up a position with the National Film Board to supervise the filming of Musk Ox, a non-tactical military exercise that began at Churchill during the frigid cold winter of 1946.[10] Born in Toronto in 1919, Wilkinson traveled extensively in the eastern Canadian Arctic after graduating from high school in 1937.[11] He wrote and directed more than forty documentary films during his career, shot mostly in the Far North between the late 1940s and the late 1960s. His films documented Inuit life on the land and the challenges endured by the settled communities he encountered. He also wrote about his experiences living among Indigenous Peoples, often expressing empathy for the social changes forced upon northern residents.[12] In January 1947, Wilkinson returned to Churchill and traveled farther north to direct the footage of *Going North*. He flew two hundred miles north of Churchill to Eskimo Point (Arviat) on the northwest shore of Hudson Bay in the southeast corner of modern-day Nunavut, where he shot scenes of Inuit for inclusion in the film.[13]

The Canadian Army and the Defence Research Board had initially considered releasing Arctic training films in three phases. *Going North* was a result of the first phase, which aimed to produce a film illustrating how to overcome the difficulties of Arctic climate and terrain.[14] The producers designed the film as a tool to educate government officials, military personnel, and scientists for work and semi-permanent occupation in the North. The second phase proposed the production of a series of short films on methods to counteract the effects of the Arctic climate, while the third proposed the production of films on Arctic tactical doctrine. Although military and defense representatives considered Arctic training important to prepare soldiers and civil servants for northern work, *Going North* seems to have been the sole release of the proposed three-phase production plan.

The Defence Research Board was particularly concerned with environmental challenges considered unique to the high latitude, cold-weather climate of northern Canada.[15] During the early Cold War, officials in both the United States and the United Kingdom impressed upon their Canadian counterparts the need to prepare for the possibility of military conflict in cold regions. As the research arm of the Canadian services, the DRB responded by leveraging Canada's northern geography into political power. The Canadian North represented an ideal cold-weather training ground, and the DRB pursued the northern sciences as part of Canada's unique contribution to the western security alliance.[16] Few scientists employed by the Canadian government had much Arctic experience, so the DRB decided to accommodate as many of its personnel as possible on military operations to and from the North.[17] While senior officials showed little desire to station large forces across the northern fringes of the country, Ottawa financed northern research to support the winter warfare training program of the armed forces. Similar to other government agencies interested in the North, the DRB attempted to overcome the northern environment by using technology to normalize and modernize state activities.[18] As reflected in *Going North*, federal funding spurred an influx of northern scientific military research during the early postwar years.

Produced by the National Film Board's Robert Anderson and James Beveridge, *Going North* was screened for the first time in Ottawa in July 1948.[19] The training film introduced viewers to problems encountered when living and working in the North. The film contrasted the subarctic forests to the Arctic barrens, discussed Indigenous "methods of coping" in the North, emphasized the importance of teamwork, and showed the benefits of special clothing and equipment designed for the military.[20] In addition to the film's footage and narration, title sequences and intertitle screens provided brief captions about social and cultural attitudes depicted in the film. The initial title sequence described the Arctic with a romantic lure, as a location full of untapped potential:

> This is the first of a series of winter training films. It is designed as an introduction to the climate, topography and living conditions of the Northland—the least known area of Canada. More than four-fifths of Canada has an arctic or sub-arctic climate, with long winters and short summers. These conditions create certain difficulties of living and travel, which this film necessarily stresses. For centuries men have lived in these regions, solving the problems by knowing the conditions and rules. The experiences of these Canadians—the Indian, the Eskimo and the white man—can help us to live comfortably in the north, exploit its riches, and defend our country.

The title sequences, much like the title *Going North* itself, guided the experience of the viewer. Following Canada's geography, the film's narrative moved north through the subarctic forests to the Arctic barrens. Along the way, the viewer learned about the challenges and possibilities of defense and development in the North.

As a NFB documentary-style dramatization, *Going North* combined re-enactment with real-time footage to portray five male soldiers operating on the snowy barrens in the harsh and cold winter environment of northern Canada.[21] The soldiers portrayed in the film utilized military kit and training skills to operate efficiently and effectively. Portions of the footage concentrated on Herby, a new recruit whose inexperience in the North intentionally served to educate the viewer about the "essentials" of cold-weather military work. In one particular scene, Herby learned about the dangers of overheating and sweating while working outdoors. He also learned how to sleep and cook in a tent, dry clothes at night, handle metal objects in the cold, and navigate using limited landmarks and a magnetic compass that often malfunctioned.[22]

The film depicted a set of complex social interactions. Indicative of a functional civil-military relationship, the film showed cordial and supportive exchanges between military personnel and Indigenous Peoples.[23] Scenes depicted Inuit navigating over various types of northern terrain, and the film showed soldiers replicating Indigenous methods for travel and dress. In this context, the sharing of knowledge reflected a dynamic colonial relationship that challenged the notion of a preconceived power structure.[24] Southern visitors to the North relied on local knowledge to understand and function in a northern setting, replicating the experiences of earlier explorers and scientists who were indebted to the wisdom and kindness of Indigenous residents in the North.

Yet despite the visual portrayal of cooperation, and the clear evidence showing the soldiers' reliance on local Indigenous knowledge, the film's narrator underscored the necessity of western science to northern development:

> Men like the trader and the missionary know the value of the Eskimo philosophy.... The Royal Canadian Mounted Police follow the Eskimo methods of dressing and living off the land. These men enjoy this life and understand it. Many of them on completion of a tour of duty look forward to a return to the North. But when men go north in large numbers, the Eskimo's simple methods no longer work; there aren't enough caribou for meat and clothes; there aren't enough dog teams to pull in the supplies.

Through the application of science and engineering, the Canadian military, as suggested by the film, intended to develop techniques for clothing, shelter, food, and transport to enable large numbers to live and move in northern Canada. The film portrayed military personnel overcoming the elements in preparation for northern defense, but the wider message reflected southern goals for long-term development in the North. Equipped with western knowledge and resources, the military utilized science to thrive in a space largely considered inhospitable in the southern imagination. This portrayal of scientific ingenuity devalued Indigenous knowledge and reinforced southern ideas about social, physical, and mental control in the North.

Unlike the soldiers portrayed in the film, Inuit represented a static people. The Inuit depicted in the film were not actors, nor were they portrayed as re-enacting daily life.[25] The film showed real-time footage of Inuit traveling over snow, constructing shelter, and dressing in parkas and mukluks. "The Arctic barren land is the home of the Eskimo, he never leaves it," claimed the narrator. "His environment provides all his simple requirements and dictates the whole manner of his living." The film also showed an Inuit family cooking, eating, and sleeping together. The footage juxtaposed the Inuit family living in a snow house with the group of soldiers living together in a military tent. While the Inuit family was content, the soldiers enjoyed the benefits of modern amenities: heaters, lamps, cooked meals, sleeping bags, specially designed clothing, and mobility—all necessary components of "civilized life" in the North, according to the film's narrator. For reasons of defense and development, the Canadian military wanted to traverse the North on land, at sea, and in the air.[26] Local knowledge was useful but limited, according to the film, an idea that emphasized the apparent necessity of southern technology.

Inuit featured as a valuable but subordinate educator in a narrative focused on modern science and northern development. Selected references to Indigenous life reinforced the perception that military personnel, armed with the knowledge and benefits of science and engineering, could live and work more comfortably in the North than Indigenous residents: "The great subarctic forests were once the hunting ground of native Indian tribes," proclaimed the narrator. "Their way of life evolved out of their needs and was adapted to the natural conditions around them. Today, many old skills are forgotten and they live precariously on a level often below that of their ancestors." From this perspective, northern residents appeared fixed at a subsistence level, unable to modernize without the support of southern science and engineering. Indeed, both the footage and narration of the film conformed to widespread views about contact and the degradation of northern Indigenous life.[27]

For the producers of *Going North*, the development of winter clothing provided a particularly useful topic to demonstrate cultural *difference* through science. Myra Rutherdale has argued that northern Canada is a significant location to study bodily perceptions and the hybridized use of clothing in the postwar period: "Newcomers frequently referred to traditional northern clothing as 'Eskimo clothing' or used terms like "white man's clothing" to dichotomize race through attire. In that sense clothing assumed explicitly racialized meanings, and was a visible marker of cultural change."[28] The science depicted in *Going North* simultaneously supports and contradicts this view. Although the military created its own forms of Arctic clothing, Inuit attire was essential to the research and development process. Knowledge of fundamental scientific principles acted as a marker of cultural difference, however. As portrayed in Figure 4.1, the film espoused science to make deliberate assertions about Indigenous intelligence. "The Eskimo may not know it," claimed the narrator while this image displayed on the screen, "but his dress is scientifically efficient in its use of still air insulation." The assumption

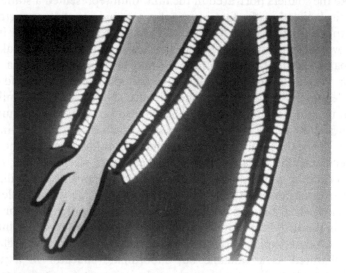

Figure 4.1 The science of insulated clothing: still air is trapped between two separate layers, effectively keeping the wearer warmer than if one layer of clothing is worn. While the Canadian military appropriated this knowledge from Inuit to develop specialized winter clothing, the film suggested that Inuit did not understand the scientific concept of insulation.

Source: *Going North*, National Film Board of Canada, 1948.

that Inuit could neither understand nor explain the science of insulated clothing served to distinguish southern Canadians from northern residents, thus reinforcing, through the demonstration of scientific knowledge, a cultural hierarchy and southern dominance over the Indigenous North.

As a state-sponsored training film, *Going North* projected the "high modernist" goals of the Canadian state in an ambitious yet innocuous manner.[29] Its production style reflected an approach widely associated with the NFB during the early Cold War period, of combining fiction with documentary. Drawing on conventions of both filmmaking techniques, the NFB produced scripted dramatizations that involved performers playing directed roles. Films of this type maintained their documentary significance and claim to social reality while assuming a nationalist tone. Under the paternal guidance of Mackenzie King's Liberal government, Canada experienced an era of modest social reform and NFB productions generally avoided openly didactic coverage of social issues. "Consensus was the trademark," in the words of Canadian film historian Peter Morris. "Opposition tended to be viewed as radicalism and radicalism as one domestic manifestation of the communist conspiracy."[30]

Going North fit this mold, depicting a clear view of the Canadian North at the dawn of the Cold War. The film portrayed Indigenous Peoples and the natural environment as objects of study useful to northern newcomers. Science enabled southerners to function and fulfill the strategic priorities of Canadian defense in the North, or so was the intended message. While much had changed by the release of *Vigil in the North* six years later in 1954, with respect to northern security and development, a comparative analysis of each film suggests a strong degree of continuity in the expression of early postwar colonial attitudes about positioning the southern body in the environmentally challenging Indigenous North.

Vigil in the North (1954)

Produced as part of the wartime *Canada Carries On* (CCO) series, *Vigil in the North* was a short Cold War propaganda film released to Canadian theaters.[31] Designed as a visual demonstration of Canada's Arctic warfare capability, the film depicted military activities at Fort Churchill and particularly those connected with science and defense. *Canada Carries On* was a series of short films produced by the National Film Board between 1940 and 1959, which originated during the Second World War as a means to boost morale on the Canadian home front.[32] The Wartime Information Board funded the series during the war, but the NFB had to secure other sources of funding to continue to produce the theatrical shorts after 1945. Issued monthly and designed for public audiences, films in the *CCO* series focused primarily on topics of national interest, as considered by the NFB. The series "intended to instill in Canadians a knowledge of and pride in themselves as well as a sense of their promise and importance in the world," according to film studies scholar William Goetz.[33]

The *Canada Carries On* series first documented the military and strategic value of the Canadian North in 1944 with the release of *Look to the North*, a short film conceived following the Japanese invasion of the Aleutian Islands.[34] Set within the context of the wartime military buildup in northern Canada and the United States, the film considered how such large-scale construction projects as the Alaska Highway and the Northwest Staging Route could facilitate northern resource extraction and economic development in the postwar period. *Vigil in the North* portrayed a similar view of northern Canada. Although focused exclusively on military activities at Fort Churchill, the film represented science as vital to security and development in the North, an idea that took on a heightened significance with the evolving postwar threat.

Vigil in the North was a product of the early Cold War, created to depict a clear message of Canadian pride and safety. The twelve-minute film began with an ominous aerial view of the open, snow-covered subarctic tundra near Fort Churchill and the icy shore of Hudson Bay showing in the distance. "This is winter in two-fifths of Canada," the narrator said while the aerial view panned on the screen, "a crazy quilt of frozen lakes, frozen muskeg, rocks, and snow. To extend Canada's frontier, Canadians are learning to live in this country; to defend that frontier, Canadians are learning to fight in this country." The film showed soldiers training to overcome the twin enemies of "fear and fatigue," which the narrator described as the initial response of man to the Arctic environment. Once again, science provided the answer and served as a marker for the capabilities of southern men in the North.

Indigenous Peoples were not the primary focus of *Vigil in the North*, but the science depicted in the film reflected southern ideas about Inuit being adapted to northern cold. One scene showed a comparative cold experiment conducted on an Inuk man and a white soldier inside the DRB's research establishment at Fort Churchill, Defence Research Northern Laboratory. The scene showed a DRB scientist recording the physiological response of the two test subjects to an experiment designed to measure pain threshold in the hand. As pictured in Figures 4.2 and 4.3, each man individually placed the middle finger of his right hand into a device while lying flat on his back. The scientist presumably controlled the degree of cold produced inside the device.[35] Although the film did not offer a full explanation of the experiment, the narrator's comments suggested science could help in understanding cold tolerance in the Inuk body:

> The North is hard on men. How hard, the scientists of the northern labs are trying to find out. What happens to the blood vessels of the fingers in extreme cold? Do a man's hands eventually become used to below-zero temperatures, or will the soldier always suffer, no matter how long he has been in the Arctic? The reaction of Eskimos' blood vessels to cold may provide answers, for Eskimos withstand cold far better than white men.

A "pain intensity scale" ranging from "none" to "excruciating" measured the cold threshold of each test subject, and the film clearly depicted pain on

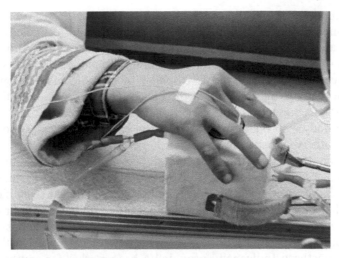

Figure 4.2 The right hand of an Inuk man, with his middle finger placed inside of a device used to measure cold threshold.

Source: *Vigil in the North*, National Film Board of Canada, 1954.

the face of the white soldier while the Inuk man showed no apparent level of discomfort. These images reinforced the hypothesis of the experiment, namely that Inuit had a physiological adaptation to Canada's cold northern climate. Viewers of the film saw a visual representation of Inuit as hardy and primitive, a characterization that categorically fit wider southern ideas about northern Indigenous Peoples being fundamentally *different* or *other*.[36]

The concept of embodiment is useful for understanding the science-environment relationship portrayed in *Vigil in the North*. "The environments and technologies with which we live, play, and work lead us to develop

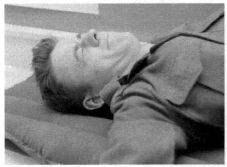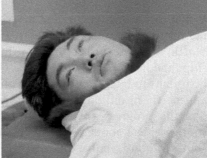

Figure 4.3 Left: White soldier shown experiencing pain from cold experiment conducted on his right middle finger. Right: Inuk man shown experiencing no apparent pain from the same experiment.

Source: *Vigil in the North*, National Film Board of Canada, 1954.

specific modes of bodily perception," as historian Joy Parr has explained.[37] "This tuned reciprocity among body, environment, and technology has historically allowed humans to feel at home, competent, and safe." With regard to the early Cold War Canadian North, the film championed science as a means to intervene and accelerate the development of embodied adaptations. Given adequate funding, the sponsors of the film believed, according to the narrative message represented to the viewer, that scientific research could equip the southern body with the mental and physical capacities necessary to embody life in the otherwise harsh climate of northern Canada. In other words, the film argued that science was the key to developing cold-weather functionality in northern newcomers. Achieving the same level of cold tolerance as Inuit was not the end goal, however. The film suggested soldiers could use scientific knowledge and technological advancements to not only gain but also improve upon the perceived climatic adaptations of Inuit, thus eliminating the possibility of Indigenous Peoples physically outperforming newcomers in the North.[38]

While northern Indigeneity provided a base marker for southern capabilities, science and engineering epitomized the potential of modernization and development. With proper training and operational research, soldiers from southern Canada, as suggested by the film, could overcome the twin enemies of "fear and fatigue" in the North. In addition to the cold threshold experiment, the film also showed scientists measuring oxygen consumption and related bodily responses of soldiers, both inside the laboratory and outdoors in the natural cold. Testing the effects of cold on the human body exemplified the possibilities of modernity. If science could "help inure soldiers to the environmental rigors of northern Canada as they acquired the necessary new methods of travel and survival," the North represented an ideal space for modernization and development, bound only by the limits of western science.

Conclusions

While historians have relied extensively on written documents to examine relations between Indigenous Peoples and southern newcomers in the North, visual records are also valuable for interpreting colonial relationships and northern contact zones. *Going North* and *Vigil in the North* allow historians to interrogate the intersections and consequences of colonialism and northern military research, and especially southern attitudes toward scientific progress, environment, and the human body. Each film reflected southern ideas about northern Canada and the role of science in defining the southern body relative to the natural cold environment. Understanding the portrayal of cold-weather science and civil-military relations in the North is also important to discussions about the appropriation of local Indigenous knowledge during the early Cold War. The Canadian armed forces pursued the Arctic sciences as means to maintain Canada's claim to territorial sovereignty and promote the northward expansion of a southern population.

In the process, scientific knowledge about how to live and work in the North was essential to the realization of state goals for northern modernization and development.

Together, *Going North* and *Vigil in the North* provide insight into the strength and relative longevity of dominant colonial postwar attitudes. Indeed, the narrative structure of each film indicates a strong degree of continuity in southern perceptions toward the Indigenous North during the six-year period discussed here. The perception of the Inuk body in juxtaposition to the southern body is particularly informative. That modern science and engineering could not only replicate but also improve upon traditional Indigenous methods for living and working in the North is indicative of the deep colonialist belief in the power and prestige of western science. *Going North* represented the Arctic as a harsh space to be tamed by science, exploited for riches, and defended by specially trained men. The military applied science and engineering to improve upon Inuit clothing, shelter, cooking, and movement. While the film emphasized the importance of local knowledge to life in the North, the narrative portrayed traditional Inuit methods as outdated and inferior to the scientifically superior kit, living, and transport techniques developed by the military. *Vigil in the North* portrayed the Arctic as an equally challenging environment, ripe for scientific development and postwar modernization. The representation of Indigenous Peoples received comparatively less coverage, underscoring the apparent success of southern science and engineering in the North by depicting the advancements of military research and development at Fort Churchill. While Inuit still served as a valuable cultural marker for southern capabilities in the North, the film emphasized the achievements and practicality of laboratory research and military science to northern progress.

As visual representations of colonial attitudes toward northern development and Inuit life, both films can ultimately help historians understand the role of science in shaping and supporting southern perceptions about northern Indigeneity and the challenging environmental conditions of the North. While military personnel, government representatives, and scientists relied on local Indigenous knowledge to live and work in northern Canada during the period covered here, each film focused on the perceived superiority of western scientific knowledge. The North functioned as a space to demonstrate southern dominance over nature. Military success, as portrayed on screen, gave viewers the presumption that science made southern life in the North possible, comfortable, and sustainable. Science reinforced the current and future capabilities of northern newcomers, stretching the bounds of modernization and development. The demonstration of western science simultaneously devalued local Indigenous knowledge, thus further representing the North as a hegemonic space for the definition of non-Indigenous culture. This portrayal of science and northern Canada, detached from the full context of northern contact and disseminated to government and public audiences through film, provides a strong indicator of the prevailing ideas that shaped and perpetuated southern attitudes toward both the Indigenous North and the cold environment of northern Canada during the early Cold War.

Notes

1. Robert Anderson and James Beveridge, *Going North*, archival film 106B 0148 023 (Montréal: National Film Board of Canada, 1948). For an NFB description and reference to the film, see Donald W. Bidd, ed., *The NFB Film Guide: The Productions of the National Film Board of Canada from 1939 to 1989* (Montréal: National Film Board of Canada, 1991), 214.
2. Nicholas Balla, *Vigil in the North*, archival film 106B 0154 088 (Montréal: National Film Board of Canada, 1954). For a written description of the film, see Bidd, *The NFB Film Guide*, 508.
3. As utilized here, colonialism refers to a sociopolitical relationship characterized by domination of one group over another group or entity, meaning southern Canadians colonized Indigenous Peoples in the North as well as the natural North itself.
4. Samantha Arnold, "'The Men of the North' Redux: Nanook and Canadian National Unity," *American Review of Canadian Studies,* 2010, 40(4): 452–463. For a brief yet detailed discussion of the role of the North in Canadian historical literature, see Janice Cavell, "The Second Frontier: The North in English-Canadian Historical Writing," *Canadian Historical Review*, 2002, 83(3): 364–389. In this article, Cavell challenges the notion that Canadians have always believed in their Arctic heritage stretching back to the pre-Confederation years. See also Sherrill Grace, *Canada and the Idea of North* (Montreal: McGill-Queen's University Press, 2001).
5. Joan Sangster, *The Iconic North: Cultural Constructions of Aboriginal Life in Postwar Canada* (Vancouver: University of British Columbia Press, 2016), 3; also see Sangster, "Irene Baird's 'North and South' in *"The Climate of Power,"* Journal of the Canadian Historical Association*, 2012, 23(1): 283–318.
6. Peter Geller, *Northern Exposures: Photographing and Filming the Canadian North, 1920–45,* (Vancouver: University of British Columbia Press, 2004), 5–6.
7. Geller, *Northern Exposures*, 8.
8. Patrizia Gentile and Jane Nicholas, eds., *Contesting Bodies and Nation in Canadian History,* (Toronto: University of Toronto Press, 2013), 5.
9. On body research in the United States, see the extensive work of historical geographer Matthew Farish, including: "Frontier Engineering: From the Globe to the Body in the Cold War Arctic," *The Canadian Geographer*, 2006 50(2): 177–196; *The Contours of America's Cold War* (Minneapolis, MN: University of Minnesota Press, 2010); and "The Lab and the Land: Overcoming the Arctic in Cold War Alaska," *Isis*, 2013, 104(1): 1–29. On Canadian research, see Matthew S. Wiseman, "Unlocking the 'Eskimo Secret': Defence Science in the Cold War Canadian Arctic, 1947–1954," *Journal of the Canadian Historical Association*, 2015, 26(1): 191–233.
10. Doug Wilkinson, *Land of the Long Day* (Toronto: Clarke, Irwin, 1955), 18. Musk Ox was a three-month exercise to test equipment, air supply, and training techniques in the Canadian North. For details on the exercise, see Kevin Mendel Thrasher, *Exercise Musk Ox: Lost Opportunities* (Ottawa: Carleton University, MA Thesis, 1998); P. Whitney Lackenbauer and Peter Kikkert, *Documents on Canadian Arctic Sovereignty and Security, Lessons in Northern Operations: Canadian Army Documents, 1945–56* (Calgary, AB: Centre for Military, Security and Strategic Studies, University of Calgary, 2016).

11. For biographical information about Wilkinson, see Jim Bell, "Author-filmmaker Doug Wilkinson dies March 12 – Douglas Earle "Qimmiq" Wilkinson: 1919–2008" (obituary), 28 March 2008, http://www. nunatsiaqonline.ca/archives/2008/803/80328/news/nunavut/80328_1055.html (accessed July 17 2017). See also Patrick Baird's book review of *Land of the Long Day*, P.D. Baird, *"Land of the Long Day*, by Doug Wilkinson," *Arctic*, 1955, 8(4): 252–253.
12. Wilkinson, *Land of the Long Day*; see also *Sons of the Arctic,* (Toronto: Clarke, Irwin, 1965).
13. A traditional hamlet, Arviat received its Inuktitut name on 1 June 1989. The English name of Eskimo Point dates to 1921, when the Hudson Bay Company set up a trading post on location. For details on the history of Arviat, see Nunavut Tourism, "Arviat," http://nunavuttourism.com/regions-communities/arviat (accessed May 15 2017).
14. Library and Archives Canada (LAC), Ottawa, Ontario, Canada, Record Group (RG)24 F–1, 3234, file DRBS 3–750–43–2, "Minutes of the 5/47 Meeting of the Arctic Research Advisory," 12 December 1947, 5.
15. For details on the DRB's Arctic research during this period, see Captain D.J. Goodspeed, *A History of the Defence Research Board of Canada* (Ottawa: Queen's Printer, 1958), 175–188; Shelagh D. Grant, *Sovereignty or Security? Government Policy in the Canadian North 1936–1950* (Vancouver: University of British Columbia Press, 1988), 218–220.
16. Edward Jones-Imhotep, "Nature, Technology, and Nation," *The Journal of Canadian Studies*, 2004, 38(3): 5–36.
17. National Archives and Records Administration (NARA), College Park, Maryland, United States, RG319, Box 856, "Defence Research Board, Department of National Defence, Canada: Second Semi-Annual Report, 1 October 1947 – 31 March 1948," Report No DR 5, Ottawa, 5 June 1948, 97.
18. See, for instance, Daniel Heidt, "Clenched in the JAWS of America? Canadian Sovereignty and the Joint Arctic Weather Stations, 1946–1972," in *Canadian Arctic Sovereignty and Security: Historical Perspectives*, ed., P. Whitney Lackenbauer, Occasional Paper Number 4 (Calgary, AB: Centre for Military and Strategic Studies, 2011), 145–169; Ronald E. Doel et al., "Strategic Arctic Science: National Interests in Building Natural Knowledge – Interwar Era Through the Cold War," *Journal of Historical Geography*, 2014, 44: 60–80.
19. While NFB records list Robert Anderson as producer and James Beveridge as executive producer (see Bidd, *The NFB Film Guide*, 214), archival records of the Defence Research Board housed at NARA state that H. Randall produced the film. For the DRB records, see NARA, RG319, Box 856, "Defence Research Board, Department of National Defence, Canada: Third Semi-Annual Report, 1 April 1948 – 30 September 1948," Report No DR 8, Ottawa, 10 December 1948, 151.
20. ed. Bidd, *The NFB Film Guide*, 214.
21. The soldiers depicted in *Going North* might have been actors but neither the footage, nor narration, nor credits of the film made this distinction clear.
22. The high levels of magnetism produced by the North Magnetic Pole occasionally affected military communication and navigation instruments such as the magnetic compass.
23. For an analysis of civil-military relations and the Indigenous North, see P. Whitney Lackenbauer, *The Canadian Rangers: A Living History*, (Vancouver: University of British Columbia Press, 2013).
24. For a detailed discussion of colonialism and the production of scientific knowledge, see Stephen Bocking, "Indigenous Knowledge and the History of Science, Race, and Colonial Authority in Northern Canada," *Rethinking the*

Great White North: Race, Nature, and the Historical Geographies of Whiteness in Canada, eds. Andrew Baldwin, Laura Cameron, and Audrey Kobayashi (Vancouver: University of British Columbia Press, 2011), 39–61.

25. The word *Inuit* is plural, meaning the people; the word *Inuk* is a singular reference to one person. For information on terminology about the Canadian military and the Indigenous North, see Franklyn Griffiths, Rob Huebert, and P. Whitney Lackenbauer, *Canada and the Changing Arctic: Sovereignty, Security, and Stewardship* (Waterloo, ON: Wilfrid Laurier University Press, 2011), 71–72; Lackenbauer, *The Canadian Rangers*, 24–26.

26. See Andrew Iarocci, "Opening the North: Technology and Training at the Fort Churchill Joint Services Experimental Testing Station, 1946–64," *Canadian Army Journal*, 2008, 10(4): 74–95.

27. On the manifestation of southern attitudes toward Indigenous Peoples in this period, see Ian Mosby, "Administering Colonial Science: Nutrition Research and Human Biomedical Experimentation in Aboriginal Communities and Residential Schools, 1942–1952," *Histoire sociale/Social history*, May 2013, 46(91): 145–172.

28. Myra Rutherdale, "Packing and Unpacking: Northern Women Negotiate Fashion in Colonial Encounters during the Twentieth Century," in *Contesting Bodies and Nation in Canadian History*, eds. Gentile and Nicholas, 117–133; quote on page 118.

29. For a critical reappraisal of high modernism in Canadian historical literature, see Tina Loo, "High Modernism, Conflict, and the Nature of Change in Canada: A Look at *Seeing Like a State*," *Canadian Historical Review*, 2016, 97(1): 34–58.

30. Peter Morris, "After Grierson: The National Film Board 1945–1953," *Journal of Canadian Studies*, 1981, 16(1): 3–12; quote on p. 4.

31. Bidd, ed. *The NFB Film Guide*, 508; Canadian War Museum, Hartland Molson Library Collection, *Defence Research Board Newsletter* 1, no. 5 (May 1955).

32. Jack C. Ellis and Betsy A. McLane, *A New History of Documentary Film* (London: Continuum International Publishing Group, 2005), 122–123.

33. William Goetz, "The Canadian Wartime Documentary: *Canada Carries On* and *The World in Action*," *Cinema Journal* 16, no. 2 (Spring 1977): 59–80; quote on p. 62.

34. James Beveridge, *Look to the North*, archival film 106B 0144 176 (Montréal: National Film Board of Canada, 1944). For a written description of the film, see ed. Bidd. *The NFB Film Guide*, 297.

35. Neither the footage nor narration of the film described the exact workings of the device.

36. See references in note 9 for information about biomedical research and racialized experimentation in Canada and the United States during the early Cold War.

37. Joy Parr, *Sensing Changes: Technologies, Environments, and the Everyday, 1953–2003*, (Vancouver: University of British Columbia Press, 2010), 4.

38. This is not to suggest that outperforming Indigenous Peoples was the goal of the scientific experiments shown in the film. The primary aim of the research was the provision of scientific and technical assistance to soldiers performing military duties in the North during winter, when cold conditions and challenging terrain posed a constant detriment to effectiveness and efficiency.

5   Portraying America's last frontier: Alaska in the media during the Second World War and the Cold War

Victoria Herrmann

Alaska has been and remains an imagined piece of America's national narrative. The majority of Americans have never experienced their country's Arctic space. Yet imageries of the North in popular media abound. From accounts of grueling nineteenth century expeditions, to online stories of melting sea ice today, southern publications have provided American audiences a pallet with which to color the Arctic's supposed tabula rasa.

The journey of these narratives from conception to publication moves through conceptual geographies of personal experience, societal contexts, and value systems. The production, or reproduction, of Alaska in news articles and photographs inevitably becomes distorted as it travels from the Arctic Circle to media offices in New York City and Chicago. Much like the Arctic's snowy landscape, Alaska becomes a geography upon which to reflect the anxieties, aspirations, and arguments at play in the national American character. Such reflections are easily recognizable in Arctic news stories today. American audiences identify photographs of polar bears on icebergs as symbols of global climate change and rising sea levels. Stories of Russian military exercises in the Arctic become emblematic of its military actions in Ukraine and the nationalistic anxieties of a new Cold War. These contemporary journalistic constructions are rooted in a long history of remaking Alaska to be consistent with southern themes.

Representations of Alaska in news production peaked in the mid-twentieth century, when the American narrative encountered the territory in a way unlike that previously experienced. The Second World War's Aleutian Islands campaign and the opening of the Cold War's northernmost theater not only brought Alaska into the national military narrative, but into the wartime media's narrative. From the 1940s to the 1960s, news articles about Alaska embodied the rhetorical tone and political contours of America at war.

Using articles from *Life Magazine,* this chapter analyzes how Alaska was contextualized by the media within a space of American nationalism and militarism. In this analysis, the narrative of each event will be illustrated through the esthetic elements of a particular article or set of articles. To complement the macro-features of *Life*'s Alaska visual narratives, textual discourse strands or fragments will be used to map out what particular "truths" about Alaska as an imagined geography were established by *Life*, what consistencies surfaced, and what tensions arose. Reporting of the Second World War and Cold War operations constructed Alaska as both a distant space of wartime strategic importance and a domestic space of cultural familiarity so that Alaska appeared both within and outside the imagined geographies of America. After introducing its methodology and theoretical frameworks, the chapter provides a visual discourse analysis of accounts of the Aleutian Islands campaign and the construction of the Distant Early Warning (DEW) Line. In doing so, it offers a critical analysis of the important but often overlooked power of visual narratives in the construction of Alaska as a politicized space of strategic and cultural nationalism in the American imagination in the decades before statehood in 1959.

Conceptual framework

The analysis to follow explores the relationship between photography and history within the context of media studies. The use of visual expressions of a *zeitgeist* or societal ideology to form a collective memory and understanding of historical events has a long history in Western cultures, where visual manifestations of material reality are often privileged over other ways of understanding. To unpack the American Arctic in the visual narrative of *Life Magazine* requires both a theoretical perspective on the power of the image and a conceptual framework for analyzing how a distant geography is constructed through reproduction and discourse. For scholars of photojournalism like Bonnie Brennen and Hanno Hardt, "the potential of images quickly exceeds its assigned role of illustrating or reflecting the mood of a society whose gullibility or willingness to be deceived only added to its influence".[1] This notion that images transcend their assigned role of mere reflection to become agents of constructing perceptions and realities is a central tenant in the literature on communication and mass media studies. Discourse theory asserts that society's ways of visualizing do not neutrally reflect our world, identities, and social relations, but instead play an active role in creating and changing them.[2] Reality can be scripted, twisted, and framed by imagery to show (or obscure) what is known and accepted.

The decision to focus on *Life Magazine* as the visual source material for this analytical work (drawn from the Google Books digitized archive) is primarily based on its status as one of the most-read photojournalism magazines of the mid-twentieth century. *Life* gave extensive coverage to the Second World War, including a number of special issues meant to be "permanent manual[s] for readers of the news of this world war." The conflict appeared in nearly every issue from the introduction of Adolf Hitler and Benito Mussolini in a December 7, 1936 article (which ominously warned, "today, more than any other man alive, Adolf Hitler is the fulcrum on which peace or war for Europe teeters.") until the United States dropped two atomic bombs on Hiroshima and Nagasaki (appearing in the September 17, 1945 issue). Alaska received considerable coverage, with 18 articles that gave a nuanced visual view into an American soldier's life on the last frontier's front. *Life*'s coverage of Alaska during the Second World War provided a collage of military assets, infrastructure development, everyday life, and military science. Features as varied as the building of the Alaska Highway and soldiers and War Department girls having fun at McKinley Park Hotel during the holiday season were all placed in the context of the climate and geography of the Arctic and sub-Arctic. The Army Engineers' efforts to construct the highway were described as "punching through the wilderness" while the soldiers celebrating Christmas at the McKinley Park Hotel, the "most modern hotel in Alaska," were portrayed as the loneliest soldiers in the country. Stationed in Alaska, "home is far away and in the course of time they learn to curse the bleak beauty of the countryside." Alaska's place in the *Life*'s wartime narrative is indicative of these tensions between nationalism and foreignness, securitization and domestication.

The first stepping stones to the American mainland: the Aleutian Islands campaign

Alaska's first big feature in *Life* came on January 19, 1942, six months before Japan invaded the westernmost stretches of the territory. "ALASKA: U.S. FRONTIER WAITS FOR WAR" was a striking eight-page feature accompanied by 26 images (*Life*, January 1942). Those figures, including photographs and maps, constructed Alaska as a space of both securitization and domestication, tied together by American nationalism. Taking these two themes as the media messages, the figures can be grouped accordingly. The first group consists of two maps and two photographs—one of Sitka's Navy air base and the other of a private bomb shelter in Juneau—that contrast Alaska as a strategic military space. The second group of 22 images visualizes everyday life in Alaska. Through photographs of restaurants, main streets and beauty salons as a familiar and legible space of Americanness, Alaska is reproduced as a domestic place within American nationhood.

The main map on the title spread, occupying nearly three-quarters of the page, depicted Alaska as the central point with air routes connecting the territory with capitals around the world. Thick black lines indicate the routes and distances to New York, London, Berlin, Moscow, Chungking, Manila, Tokyo, Honolulu, and San Francisco. The caption made the map's point clear: "the strategic importance of Alaska is pointedly demonstrated in this polar azimuthal map." The caption went on to note the hour and mileage distances between Alaska and important points on the map. The second map superimposed Alaska onto the continental United States to show its immense size. The visual message of Alaska's central placement, its size, and its close proximity to other geographies (Tokyo, Moscow, Berlin) all moved the idea of Alaska away from notions of remote frozen tundra to instead emphasize its strategic significance in the war narrative.

The inclusion of a map was also noteworthy. Although maps comprised a relatively small portion of news reporting and picture magazines, journalism provided the public with most of its information about places and geographies. Here, the editors of *Life* signaled the importance of geography in its reporting of war in Alaska. As a picture magazine, *Life* understood the dramatic impact of maps, and used them here and in the examples to follow to accentuate the strategic standing of Alaska. These two maps re-picture the territory as a cartographic space of vital importance to America's war effort, "a jumping-off place for attack against Japan or a shield against attack from Japan." Visually and textually, "the Aleutian Islands point[ed] like a dagger at Japan but by seizing them Japan could turn the dagger against America. Whether Alaska is fitted for offense or defense is one of the many military secrets that have settled like a shroud over the Territory."

This article, with its latent theme of strategic importance, took on foundational importance after a small Japanese force occupied the islands of Attu and Kiska in June 1942. In a seven-image spread, *Life* introduced the American public to what would soon become an oft-referenced warfront. The article's opening lines read:

> "Of all the world's far-flung theaters of war, the most murky and menacing last week was the subarctic arena of the Aleutian Islands. There, in cloud-hung wastes of perilous sea and volcanic rock, lurked the enemy. The Japs had moved in quietly under the fog a fortnight ago, scarcely noted amid the spectacular thunders of Midway".[3]

Continuing the January issue's construction of Alaska as a securitized space, the message was clear: "Attu's only importance is strategic." Presented as one of the world's most inaccessible islands, Attu—the westernmost of the Aleutians—was taken by the 301st Independent Infantry Battalion from the Japanese Northern Army six months after the United States officially entered the Second World War. In this issue, the geography of Attu was presented as an animation, complete with U.S. planes raining bombs on a Japanese task force in the waters surrounding the islands. The visual of planes overhead, dropping bombs and splashing into enormous clouds of ash and water below echoed the visuals of the waters around Europe, Hawai'i, and other Pacific islands. The stark images that followed depicted impossibly jagged and unwelcoming islands along the North Pacific Ocean. One image caption read "the air of a bleak Aleutian group was taken through a rare break in the mists that constantly overhand the islands. Fog, rain, gales are unceasing." This, the ominous and harsh frontier, was the geography of war: a zone of militarization and securitization. While the aesthetics of violence were absent, the maps of menacing terrain, of military vehicles dominating the otherwise overwhelming natural landscape speaks to the securitization of Alaska as a geographical narrative presented to readers. It is important to note, however, the presence of violent war visuals in the rest of the magazine.

Positioned between features on "Murder Without Passion Is Nazi Method with Russian Civilians" and "Lightly Armed Japanese Filter Up Through Burma," the war in the Aleutians fit into a jigsaw of visual narratives of the Second World War—photographs of dead bodies, marching men, and exotic places never before explored in this detail.

Alaska was distinctive as a wartime space in the duality with which *Life* constructed it as a front. The lens of war reporting often focused on military movements and actions of conflict, and thus required photographers and journalists to manage readers' encounters with distant but potentially traumatic events.[4] But in Alaska, the effects of the war were as much civilian as military because the conflict became the single biggest factor shaping Alaska's settlement patterns. Recognition of Alaska's strategic military value led to heavy investments in military infrastructure: airfields and bases, roads, docks, and improved communications networks. Military personnel also flooded in, with many remaining after they left the service. These developments not only rapidly increased Alaska's population; they also transformed its demographics. The 1930 US Census marked the first time whites were the majority in Alaska at 50.6 percent. By 1960, nearly four in five Alaskans identified themselves as white.

As Alaska became a space within, rather than outside of, America, *Life* presented it as both a wartime front and a homeland—a combination that the United States had not experienced since the War of 1812. This duality was important not simply because Alaska was moving closer to becoming a formal part of America's union, but also because Japanese bombing and attacks in the Aleutians raised concerns that Japan would invade North America via Alaska. These shifts meant that Alaska had to be militarily and emotionally connected to America's heartland. Making life in Alaska intelligible, visible, and ultimately relatable through cultural associations spoke to one of photojournalism's core tenants. As noted by communications scholar Minla Linn Shields, "photographs are powerful tools for communication because these moments not only represent facts; they also have an ability to speak to viewers on a relatable and emotional level." Much like the construction by European travelers, photographers, and artistic and theater directors of Egypt as a space of visibility that is both distant and domestic, so did *Life*'s photojournalism construct America's northernmost territory. Everyday occurrences familiar to Americans in the lower 48—hair styling and dining out—were used as visual strategies to produce Alaska as an imaginative geography within Americans' horizons of intelligibility and visibility.

While maps inserted Alaska into the American imagination as a strategic wartime geography, the photographs of the 19 January 1942 issue, "ALASKA: U.S. FRONTIER WAITS FOR WAR," framed Alaska as a domestic and familiar space. Titled "IT IS NOT ALL FROZEN WASTE," followed by "PLANES MAKE TRAVEL EASY" and "SALMON & GOLD ARE ITS RICHES," these decidedly non-military images—though still strategic—complicated Alaska's wartime narrative. Here, Alaska was simultaneously made exotic and domestic as the photographs brought places into a tamer, more familiar space within American nationhood. The images conveyed both a narrative of similarity with the American experience and a Wild West narrative of life on the periphery. Photographs of town streets and gold rush prospectors returning from the Yukon constructed small towns in Alaska to "still look like frontier towns," with many more men than women, and fetishized curiosities like district commissioners who could "grant divorce by tearing up marriage licenses." Other photographs drew parallels with continental America. A photograph of women in a beauty parlor described them as dressing "as stylishly as women of the 'outside'". They read the fashion magazines avidly. Grocery store visuals showed similar brands as in New York or Chicago, and couples were seen eating at booth tables that could have been in Memphis. Both the Wild West and the familiar American visual narratives brought Alaska into American nationalism and its national identity, rendering Alaska as an intelligible place and an intimate connection otherwise not available to *Life*'s readership.

Much like the 19 January feature, the images of war and military action in *Life*'s June 1942 Aleutian Islands campaign feature were juxtaposed against photographs of daily life in Attu once the attacks had begun. The photographs, taken in 1938 by archeologist Alan G. May while surveying the Aleutians for the Smithsonian Institution, showed neat houses lined up on Attu's main street, complete with a spotless church. Some oddities were still highlighted in the narrative, such as the school without a schoolteacher and a rugged-looking elected chief who also served as the doctor and priest. Still, the visuals of everyday buildings, objects, and practices of American life made the Aleutians familiar and accessible to readers in ways that military planes and soldiers could not. Instead of making Alaska into a distant warzone, these images made island life American: beneath the mist and fog there were the same trappings of any small U.S. town. Until the U.S. Navy drove the Japanese out of Attu's snowy peaks in the summer of 1943, *Life* continued to cover the Aleutian campaign with enthusiasm and images. The articles continued to depict the Aleutians through maps, models, and photographs of the 930-mile chain of volcanic rocks that stretch westward toward Asia. Throughout these images and texts, the Aleutians continued to occupy their strategic importance and place in the war narrative. In a seven-photograph spread in the 28 September 1942 issue of *Life*, a scale model by Norman Bel Geddes showed the islands "in the position they occupy in the mind's eye of the calculating Japanese Supreme War Council". From here, the islands' "importance cannot be underestimated by any American." The other features on the Aleutians focused on the war effort, the soldiers' advance towards Tokyo, and the fear-inducing reality that "as long as they are there, North America is in peril". Despite this visual narrative of Alaska as a war front, however, the status of Alaska as part of the American strategic and cultural narrative would resurface in later features.

Fighting the coldest war: distant early warning, distancing, and technology

In the 1950s Alaska reappeared in *Life*. Its portrayal of the construction of the Distant Early Warning (DEW) Line in the 1950s reimagined the visual narrative of Alaska as a geography of security, military, and nationalism through the introduction of a new aesthetic theme: technology. Between 1954 and 1957 over 25,000 people were employed to build the roads, towers, antennas, airfields and buildings of the DEW Line, in some of the most isolated environments in North America.[5]

Twelve articles in *Life* featured the DEW Line in Alaska. This reintroduction of Alaska into the magazine came with two thematic changes. First, the DEW Line replaced the photographs of soldiers and traditional military endeavors with new technology. Seaplanes and aerial shots of the Aleutian Island chain were replaced with futuristic domes and advanced scientific equipment. Second, the space of Alaska portrayed expanded from a small slice of its southwest to the wider Arctic landscape. The increasing use of the term "Arctic" and Alaska's connection with the region was heightened in the articles as the DEW Line extended eastward to Greenland, and the territory was reframed within a scientific and international lens. These thematic changes significantly altered the visuals and message presented to readers. In the DEW Line's debut single-page article on 22 August 1955, the map shifted east. Rather than focusing on the Aleutian Islands and their proximity to major centers like Tokyo and Hawai'i, Alaska was the easternmost tip of a dotted line stretching across the North American Arctic. The central focus was not Alaska but the Arctic region, with a headline reading "NEW RADAR SKY-WATCH TO GUARD ARCTIC FRONTIER." The geographic language shifted from Alaska, which only appeared once in the article, to "the North Polar regions," "the Arctic" and "the northern rim of the continent." And instead of militaristic discourse of retaking islands, bombings, and stealth operations, the narrative was one of technology. The feature focused on Bell System and Western Electric's work in building the Line: development, design, engineering, procurement, transportation, construction, installation, testing, and training of operating personnel. The photograph accompanying the painted map showed "DEW Line radar station in the Arctic": the now recognizable photograph of a dark dome on stilts over an otherwise tranquil snowscape of whites and blues. There are no humans present, just the overpowering immensity of technology.

The second installment on the DEW Line, published 30 April 1956, expanded to a five-page spread of photographs where "LIFE presents an up-to-date report on its progress." Further into its full construction, "SPROUTING DOMES ON DEW LINE: U.S. builds a radar warning net in the Arctic" focuses on the building of these domes with "electronic equipment set up to detect approaching aircraft." Reflective white domes dominated the feature's cover image, dwarfing five men dressed for cold walking beneath. In the next photograph, the dome at night is aglow with light. Small accompanying maps illustrated a proposed $68-million extension of the system across the Bering Sea from Point Barrow to Attu Island. The feature, however, did not dwell on the map, instead focusing on the station's technology as well as the men who built it. Photographs of rotating search mechanisms, radar antennas, and workers jackhammering solid permafrost to prepare a tower's foundation completed the article. Gone were the photographs connecting Alaska and the Arctic to the American heartland. The photographs of men here were particular to the system: breakfast tables of workers at the construction camp mess hall and Western Electric technicians feeding foxes peanuts. This detection network now dominated how Americans came to know the Arctic. The "igloo-like radar stations that search the sky" became as familiar as the planes and maps of the Aleutian Islands campaign had been a decade earlier. The images broadened the imagined space of Alaska within America to include science and technology: a narrative of innovation introduced into a remote, exotic, and militarized geography.

Technology not only presented Alaska as a place of strategic importance to America's Cold War, it linked northern settlements and 1950s American mass culture. Dramatic advances in science and technological innovation reshaped American culture and transformed how Americans came to know and live in the world. The development of atomic and hydrogen bombs as well as intercontinental ballistic missiles created a threat shared by Americans stationed at DEW domes in Alaska and by those in cities across America. But more than that, new technologies—color television sets, transcontinental telephone cables, vaccines and contraceptive pills and jet-propelled aircraft—came to define everyday life in the 1950s. The DEW Line may have been more advanced than TV sets and Hollywood movies, but technological and scientific progress was intimately familiar to 1950s American families.

Understanding visual narratives of cold science in the 1950s: *life*, statehood, and the American public

The Aleutian Islands Campaign and the Distant Early Warning Line illuminated the tensions between, and the bridging of, strategic wartime militarism and domestic cultural nationalism in the visual narrative of Alaska during the 1940s and 1950s. Through themes of security, technology, and science, discourses that placed Alaska within the American nation and that defined it as a strategic space were evident in *Life*'s photojournalism. While the reciprocal relations between the military, security science and technology during the 1950s have been amply studied, there has been less investigation of the photojournalism that conveyed these relations to the American public. Scholars examining the communication of science through visual media often focus on the production of materials by American government agencies or by other official institutions such as the National Academy of Sciences.[6] This work contributes to understanding Alaska's political, scientific, and military history but overlooks how continental Americans came to know and understand more distant places such as Alaska through print and visual media. Understanding how Alaska was constructed for public consumption by the leading magazine of the day is not only relevant to understanding public communication during the Cold War—it is important to understanding how the American public perceived Alaska as it fought for statehood. On 3 January 1959, Alaska became the 49th state to enter the union. It has been widely noted that Alaska's desire for statehood was encouraged by the attention it received as a strategic outpost during both the Second World War and the Cold War. Indeed, according to scholars like Laurel Hummel, the evolution of Alaska as a "militarized landscape," born from the Aleutian Island Campaign and the Cold War, was essential to convincing American politicians that statehood was vital.[7]

Like their representatives in Washington, American citizens became convinced of the benefits of statehood. In 1958, a Gallup Poll showed that 73 percent of Americans were in favor of Alaska becoming a state, a statistic quoted by former Governor Ernest Gruening in a May 26, 1958 letter to the editor of *Life*, arguing, "if the Congress is as responsive to public sentiment as an elective body is supposed to be, it ought not hesitate to admit Alaska as the 49th State."[8] Indeed, throughout this period a number of letters to the editor focused on the two moments analyzed in this chapter—the Aleutian Islands Campaign, and construction of the DEW Line. These demonstrated the positive relationship readers had with Alaska during the mid-20th century, reflecting the visual discourse established in *Life*. In at least two "Pictures to the Editor" in 1942 and 1944, readers sent in photographs of the Signal Corps in a plane in the Aleutians (entitled "Hollywood in the Aleutians") and a photograph of a U.S. soldier leaning into a williwaw (a storm distinctive to the Aleutians) entitled "Aleutian Breeze". But it was not just the military visual discourse that readers responded to in *Life*. The connection of Alaska to American life and the pursuit of science is also seen in letters to the editor. In a 1945 issue a CPL. Paul B. Lowney posted a photograph of a church in the Aleutian Islands, noting that "Save for barren surroundings of desolate, rolling tundra, it could well be located in any small U.S. city"—illustrating a connection to American life beyond the military.[9]

Taken together, the photojournalism tropes of militarization created by the Second World War, updated in the Cold War and infused with the cultural commonalities of domestic life, technology innovation, and science not only constructed the visual narrative of Alaska in *Life*, but also the visual narrative reflected back to its editors by *Life*'s readers. This testifies to the complex process by which geography came to be known to the American public in the mid-20th century. Iconic moments like the Aleutian Islands campaign did not exist in isolation of either the social milieu of America or events before, during, and after. Rather, there was an intimate interplay between the images of Alaska and the Arctic in *Life* and other photojournalism narratives in the magazine. The relations between visual images and texts that would have been apparent to *Life* magazine subscribers illustrate the interactions of strategic and cultural nationalism and of science and security during the Cold War.

Notes

1. B. Brennen and H. Hardt, *Picturing the Past: Media, History, and Photography*, (Champaign: University of Illinois Press, 1999), 3–4.
2. G. Rose, "On the Need to Ask How, Exactly, is Geography 'Visual'"? *Antipode*, 2003, 35(2): 212–221; G.Ó. Tuathail, "Review Essay: Dissident IR and the Identity Politics Narrative: A Sympathetically Skeptical Perspective," *Political Geography*, 1996, 15 (6–7): 647–653; John Berger, *Another Way of Telling*, (New York: Random House, 1995).
3. "War in Aleutians," *Life*, June 29, 1942, 33.
4. Stuart Allan, "The Culture of Distance: Online Reporting of the Iraq War," in: Stuart Allan and Barbie Zelizer, eds., *Reporting War: Journalism in Wartime*, (London: Routledge, 2004), 347–65.
5. P. Whitney, Lackenbauer, Matthew Farish, and Jennifer Arthur-Lackenbauer, *The Distant Early Warning (DEW) Line: A Bibliography and Documentary Resource List,* (Calgary: Arctic Institute of North America, 2005).
6. F. L. Korsmo, "Shaping Up Planet Earth: The International Geophysical Year (1957–1958) and Communicating Science Through Print and Film Media," *Science Communication*, 2004, 26(2): 162–187.
7. L. J. Hummel, "The US Military as Geographical Agent: The Case of Cold War Alaska," *Geographical Review*, 2005, 95 (1): 47–72.
8. *Life*, May 26, 1958, 20.
9. *Life*, January 15, 1945.

6   Making "Man in the Arctic": academic and military entanglements, 1944–49

Matthew Farish

In the brief period between the final months of Second World War and the deepening of the Cold War at the end of the 1940s, several new agencies, often working in concert, began to steer the study of northern North America and the broader perception of the region in much of Canada and the United States.[1] Focused on two of these agencies, the bi-national Arctic Institute of North America (AINA) and Canada's Defence Research Board (DRB), and building on the work of several historians, this chapter charts some connections between a modest number of individuals who exerted significant influence over approaches to the postwar Arctic. My emphasis is less on specific initiatives than the institutions that framed and funded them, drawing people together, and sidelining or ignoring other voices and agendas—including, most notably, those of Indigenous northerners.

The mid-century extension of state power into northern Canada and Alaska should not be treated as entirely distinct from previous forms of possession and administration. But as Stephen Bocking has shown, the transition from a late-colonial mode of adventurous fieldwork to an era of aircraft transportation and large frequent military exercises had significant consequences.[2] The purported drama of individual or small-group encounters with Arctic geography was subsumed into a vision of the vast sweep of land and water from Alaska to Greenland: an "official" north, unquestionably the domain of nation-states, conceived, studied, and administered from the distant south.[3] This northern vision had many dimensions, some removed from overt military concerns.[4] Fundamentally, though, the postwar Arctic was treated by these individuals and institutions as a potential and suitable stage for war, a characterization that ultimately transcended sovereignty anxieties. Combined with the apparent absence of "systematically collected" knowledge about the region, this representation justified the investments made by the U.S. and Canadian militaries in northern infrastructure, operations, and research projects across a wide range of physical, environmental, and human sciences.[5]

After juxtaposing the early histories of AINA and the DRB, in the conclusion I speculate on one result of those military investments: an individual with imperial ancestry, no doubt, but also someone just coming into focus in the late 1940s. He was inextricable from the men who in the same period moved assuredly between military and academic realms, and between northern field sites and southern boardrooms. But while largely their creation, he was not granted the status of an Arctic authority. Indeed, he was essentially anonymous relative to the broader, multifaceted militarization of the postwar north. Nor was he a permanent resident, although the bodies of Indigenous northerners, when examined and exploited, might hold clues to his survival, even as they were largely treated by his inventors as irrevocably different.[6]

"Man in the north," a common term across the spectrum of postwar military and scientific inquiry that concerns me here, was not an abstract phrase reflecting only the gendered grammar of the period. Instead, it consistently referred to a white, masculine soldier, posted to or traveling through the region. Although many opined on his fate, he was most directly a product and a subject of the human sciences—fields such as medicine, physiology, and psychology—as they turned to the conditions of northern military life, in the process creating or supporting new facilities and initiatives to better understand this life. Like the landscape that he was understood to occupy, his generalizability reflects the status and shape of northern expertise at the dawn of the Cold War.

The Musk-Ox alumni[7]

The first issue of *Arctic*, the periodical published by AINA, was released in the spring of 1948. Featuring a cover photograph, credited to Canada's National Film Board, of three Baffin Island Inuit women, the issue included no significant discussion of northern Indigenous communities.[8] Instead, it began with a message from the previous Chair of the Institute's Board of Governors, the University of Toronto geologist and geophysicist John Tuzo Wilson. The journal, he wrote, had been launched at a moment when "the exploration, mapping, and scientific study of the North American Arctic," the same "inhospitable regions" that had claimed John Franklin's expedition a century earlier, were "being pursued on a scale never before possible."[9]

Wilson's editorial adopted vaguely internationalist language, stressing the "goodwill" that would be generated by cooperation across political boundaries, and characterizing AINA as merely a "scientific society."[10] The Institute was certainly a product and a promoter of the expanded pursuits that so excited him. But AINA had been created near the end of the Second World War to draw two federal governments closer to academia, with respect to the "Arctic."[11] And in both Canada and the United States, the branch of government most concerned with that region, in the middle of the twentieth century, was the military.

In 1948, the year he turned forty, Tuzo Wilson was still in the early stages of a very distinguished career in the Earth Sciences.[12] Like others in the small world of North American Arctic "experts," his academic trajectory had been affected by service in the Second World War. For many in this community, their military ties continued after the War.

After receiving his PhD from Princeton in 1936, Wilson spent several years with the Geological Survey of Canada, including stints in the Northwest Territories, where he followed the eighteenth-century routes of Samuel Hearne. He also began to assemble "an arctic library that was the envy of many who shared his enthusiasm for the North."[13] Early in 1940, Wilson traveled to Europe as a member of the Royal Canadian Engineers. At the end of 1943, he was recruited to lead nascent operations research efforts at National Defence Headquarters in Ottawa.[14] It was in this capacity that Colonel Wilson and his Canadian Army Operations Research Group (CAORG) turned to the predicaments of winter and northern warfare. His photograph-rich February 1946 article for the *Canadian Geographical Journal*, "Winter Manoeuvres in Canada," surveyed a series of three exercises held during the final months of the War, from the coastal mountains of British Columbia to the lands west of Hudson Bay.[15] These led to a "culminating test" instigated in that same season, when Wilson directed Exercise Musk-Ox (February–May, 1946), a widely-publicized trial of "equipment and techniques" across a huge arc of central Canada, from Churchill, Manitoba to Edmonton.[16]

Musk-Ox prefigured Churchill's role, during the early Cold War, as a site for military training and DRB research, and it inspired similar endeavors across the Arctic and sub-Arctic.[17] Later, Wilson claimed that "it wasn't a military operation," stressing instead the more technical need to learn "how to operate in our own country."[18] But this was at most a matter of degree. As Wilson's summative employment report for CAORG and his public pronouncements in the wake of Musk-Ox both made clear, a *combination* of military and scientific activity was necessary to fill the northern "vacuum" that he and others diagnosed at the end of the War. As he put it, in language that cut across military, scientific, and colonial domains, and that anticipated bi-national desires to "civilianize" postwar northern militarization, "the settlement of more permanent stations and the execution of intensive studies around them are the next requirements."[19]

The leader of Musk-Ox's Moving Force—its central group—was the Scottish-born glaciologist Patrick Baird, who had roamed the north on various treks in the 1930s, including the British-Canadian Arctic Expedition of 1936–39. During the War, Baird was a member of the Royal Canadian Artillery, and conducted "Arctic and mountain warfare training" in Iceland, Scotland, and Canada.[20] He was then involved in the three tests that preceded Musk-Ox. According to Baird, the initial planning for Musk-Ox occurred on a train bound for Winnipeg from Churchill, after the last of these tests, Lemming, concluded in April 1945. He was joined in the train car by Wilson and the geographer and mountaineer Walter Wood, a U.S. military attaché in Ottawa. Three military men—and yet Baird recalled them as "scientists of a sort, fellow founders of the Arctic Institute of North America."[21]

The head of Musk-Ox's advance party was Graham Rowley, another participant in the British-Canadian Expedition.[22] Trained as an archaeologist,

Rowley also served in the Canadian Army during the War. He had met Wilson in the elevator at the National Museum of Canada in 1939. The two also shared a troopship across the Atlantic and became lifelong friends. In November 1945, still at Canadian Military Headquarters in London, Rowley received a cable from Wilson, inviting him to join Musk-Ox.[23] Not long after the exercise, both Baird and Rowley moved to the lively Arctic Section of the DRB, the organization created to direct scientific inquiry across Canada's Department of National Defence. Inspired in part by Wilson's CAORG, the DRB was officially launched in the spring of 1947, under the leadership of another wartime operations researcher, Omond Solandt.[24]

Baird soon shifted again, to direct the Arctic Institute's main office at McGill University in Montreal.[25] Meanwhile, less than a year after starting his academic position in Toronto, Wilson accepted a place on the DRB's Arctic Research Advisory Committee (ARAC), where he and a small number of academics and civil servants were joined by Rowley and other military representatives.[26] Early in 1948, in his ARAC role, Wilson drafted and distributed for discussion a 24-page classified memo, a document that struck a very different tone from his concurrent *Arctic* editorial. Titled "Defence Research in the Canadian Arctic (with especial reference to geophysical research)," it was significant for its nationalism, but also for the conflation of military and scientific pursuits. Beyond geophysics, Wilson stressed the importance of psychological studies, describing the poles as "inimical to life," and having "a depressing effect upon most men ... especially upon city dwellers who are not indoctrinated and upon troops who did not volunteer to go north."[27]

As this sketch suggests, the overlapping trajectories of Baird, Rowley, Wilson, and other Musk-Ox alumni are part of a larger conjuncture.[28] Its outlines are clearer when we consider the early years of two institutions that sustained the personal and professional relationships forged during the Exercise.

Headquarters, Montreal

Although the idea of an Arctic Institute was at least several years older, it was pieced together at a set of meetings in 1944, in Ottawa, New York, and Montreal, and legally incorporated the following year.[29] As one principal later put it, while "good work had been done in the Canadian North in the earlier years.... The North and its inhabitants scarcely existed in the Canadian consciousness."[30] Plainly using the language of colonial frontiers, the 1944 Proposal for AINA equated the "situation" in the Arctic to that "of the undeveloped west in the middle of the last century."[31] This premise was dramatically amplified by a global conflict that drew the "top of the world" into strategic schemes, logistical webs, and new cartographic perspectives, and the concomitant effort, particularly in the United States, to compile, archive, and synthesize "regional intelligence" to aid worldwide military operations.[32]

As Patrick Baird's train-car recollection implied, despite its useful status as a civilian institution, and claims to scientific exclusivity and "objective study," AINA was nonetheless steeped in war.[33] Its first two heads, the geologists Laurence Gould (acting director, 1944–45) and A. Lincoln Washburn (1945–50), were Americans tied to an organization with a self-explanatory name: the US Army Air Forces' (USAAF) Arctic, Desert and Tropic Information Center (ADTIC).[34] ADTIC's wartime head, William Carlson, was also a geologist with northern expeditionary experience; he led the US military delegation to the AINA planning meetings.[35] Gould led the Center's Arctic section, and Washburn was an ADTIC intelligence officer, known for his "comings and goings between New York and Ottawa."[36] Another Arctic section member, Yale's Richard Foster Flint, who was Washburn's PhD supervisor, joined AINA's first Board of Governors.[37]

In the midst of the War, these ADTIC employees were "in frequent contact with government and military circles in Ottawa," communication facilitated by Gould's friend Trevor Lloyd, a geographer who had been seconded from Dartmouth College to work in the Canadian capital. (Lloyd was also the first editor of *Arctic*.) Despite national (and no doubt other) differences, the ADTIC representatives believed that an organization like AINA would effectively preserve and circulate some of the "information" that the Center had assembled. A full set of ADTIC's Arctic and sub-Arctic maps was given to the Institute after the War.[38]

The first meeting of AINA's Board of Governors was held in Montreal in January 1945.[39] Later that year, as the wartime iteration of ADTIC was closing, Lincoln Washburn "climbed the stairs" of McGill's nineteenth-century Arts Building on Sherbrooke Street, to AINA's "first quarters in a couple of rooms in the administration wing."[40] In the subsequent days, a few months before Exercise Musk-Ox, he was in close contact with Tuzo Wilson, who invited Washburn to "call on me at any time." Washburn replied that he knew "how much [Musk-Ox] means to you and Pat [Baird] and I cannot help but feel that it will also mean much to the Arctic Institute."[41] Meanwhile, Richard Foster Flint was organizing an early AINA publication: *A Program of Desirable Scientific Investigations in Arctic North America*. Released in March of 1946, the compendium of comments from experts across numerous fields began and ended with remarks from Wilson.[42]

In addition to contributions from the Hudson's Bay Company, some of the most significant early support for AINA arrived from the Canadian War Technical and Scientific Development Committee—started in 1940 as a way of channeling private donations into wartime research—and the Carnegie Corporation of New York, known for its extensive support of area studies during and after the War.[43] But alongside the web of affiliations built, from 1947 onward, with the DRB, in its first two decades AINA also fostered an extensive relationship with the U.S. Office of Naval Research (ONR), which was launched in 1946. One account of the Institute's founding suggests AINA's "system of grants for research" was "largely made possible through initial contract support" from the ONR.[44] Along with Canada's DRB and the US Army, the ONR also funded some of AINA's initial flagship projects, including an *Arctic Bibliography* pitched to the Office by Washburn in 1947.[45] Simultaneously, a deep affiliation was established between the Institute and the ONR's Arctic Research Laboratory, set up in the same year at Point Barrow, Alaska. In the initial months of ONR support, the Office "suggested that the Services not be represented directly" on AINA committees; rather, "observers" from the Navy (the ONR's M.C. Shelesnyak) and the Army (the explorer and military geographer Paul Siple) "would attend the meetings."[46] But this may not have made much difference: the first issue of *Arctic* listed both of them, along with Omond Solandt, Tuzo Wilson, and Patrick Baird (who was Secretary), as members of the Institute's Research Committee.[47]

Unsurprisingly, AINA's physical presence at McGill was supplemented by other ties between the two institutions. McGill faculty were on the Institute's Board of Governors, its list of fellows, and its committees. Faculty and students received numerous AINA grants.[48] The Institute brought together disciplines across the university, including "meteorology, oceanography and geology." One of the most profound relationships was developed between the Institute and a new university department established in the same year.[49]

McGill's first professor of Geography was George Kimble, who was released from meteorological duties with Britain's Royal Navy to take up the position early in 1945. In January 1946, he was joined by Kenneth Hare, another British meteorologist (with the Royal Air Force), who became the most prominent proponent of northern research on campus, assuming the department chair after Kimble and then Dean of Arts and Science in 1962.[50] It was a time, one of Hare's students remembered, "when everything came together" in Arctic and sub-Arctic research at the McGill.[51] The relevant chapter in Hare's (unpublished) memoir was titled "The Northern Thrust, 1946-50."[52] Early on, Kimble invited Tuzo Wilson and the explorer Vilhjalmur Stefansson to give lectures on campus.[53] Hare quickly allied with AINA, "a group of vigorous enthusiasts whose interests I shared."[54] Meanwhile, the university's powerful Principal, F. Cyril James, encouraged Kimble to deepen his department's connections to Washburn and the Arctic Institute. In a November 1946 letter to James, describing Washburn's desire "to get geography students to do research on Arctic problems," Kimble noted that Washburn had "given us the freedom of the Institute," and had "represented our interests at Washington on more than one occasion."[55] Baird and Washburn both became honorary lecturers in Geography, and by the 1948–49 academic year, AINA's "Arctic Exploration" lecture series included talks by Hare and Trevor Lloyd, among others.[56]

With Hare's help, Kimble moved quickly to establish and direct McGill's annual Geography Summer School at Stanstead, near the Canada-U.S. border in Quebec's Eastern Townships. From its first instantiation, in 1947, the School "emphasized a programme of polar geography."[57] The *McGill Daily* reported in May that "one of the special features of the courses will be a series of lectures under the heading 'Man in the Arctic,'" held in conjunction with AINA. The advertised lecturers included "Lt.-Colonel Baird," Stefansson, Hubert Wilkins (like Stefansson, another military-allied adventurer), "and a physiologist still to be named." As this roster implied, the Summer School was not strictly an academic affair: "Among those already known to be registered for these lectures on adaptation of man to the Arctic and Subarctic conditions are twenty American Army officers."[58]

"Man in the Arctic" was actually a late addition to the list of courses, at the request of the U.S. military. As Washburn, a recent employee of the USAAF, characterized the situation in a February 1947 letter to a contact at Air University, James and Kimble had "been looking into the possibility of expanding the work of the Geography Summer School in co-operation with the Arctic Institute to include Arctic work along the lines suggested in your letter." Washburn promised that if "three or more students are sent by the Air University, arrangements can be made whereby it would be possible for the Geography Summer School to emphasize Arctic and sub-Arctic subjects in three of the courses already scheduled, and probably to include an additional six-week course devoted to the Far North."[59]

Baird, Stefansson, and Wilkins all returned to the Summer School staff in 1948 and 1949. (Tuzo Wilson was invited for 1948, but could not attend.[60]) In 1948, along with the English mountaineer and geologist Noel Odell, they led a popular graduate course on "Polar Problems."[61] Again, enrollees included "representatives from the Armed Forces of both Canada and the United States."[62] The U.S. Navy appealed directly to the McGill Principal to send five officers. Support arrived from the reliable Carnegie Corporation, whose funding was largely allocated to bursaries, distributed across a range of professional sectors. One 1948 grant went to the Laval University physiologist L.P. Dugal, who was also head of the DRB's Arctic Medical Research Panel (discussed below).[63] Along with Paul Siple, Dugal joined Baird, Stefansson and Wilkins to lead the Polar Problems seminar the following year, while Baird, still listed as "Colonel," oversaw the undergraduate Arctic course. At the end of the 1949 School, the Geography Department was "given to understand from several sources that the Services' interest in the Arctic side of our work will continue to grow."[64]

Stephen Bocking has suggested that for Kenneth Hare and his McGill colleagues, the complicated relationship between Canada and the United States, as it concerned the north, was "not so much a tension as an opportunity," particularly as the academics "benefited from"—and contributed to—the assertion of state power across the postwar Arctic. Bocking emphasizes Hare's use of aviation and aerial photography to "extend a biogeographical perspective over northern Quebec and Labrador." This effort, which Bocking (following Hare) dates to a 1949 invitation from the DRB, ultimately "encouraged a view of the northern environment as a landscape shorn of mystery," one that could instead by "surveyed and managed" by a combination of science and government.[65] As it happened, Hare's interest in and use of aerial photography for northern research was shared by Tuzo Wilson. While they went on to conduct distinct, substantial photo-based projects in the 1950s, in the early stages of this work, both "were able to benefit from the rapidly expanding interests" of the DRB and its military parent. It was apparently Graham Rowley who "obtained official approval for the Hare-Wilson venture to check the validity of an unfrozen Hudson Bay."[66]

Hare's contact with the DRB in 1949 was preceded by over a year of "research for the Federal Government on the dynamic and physical climatology of the eastern Canadian Arctic and Subarctic." The public centerpiece of this work was Operation Cariberg, described romantically by the *Geographical Journal* as an "air expedition."[67] Ostensibly focused on the dual study of caribou migration and Hudson Bay ice, Cariberg was much more physically and intellectually wide-ranging, sending a Royal Canadian Air Force (RCAF) North Star aircraft 7300 nautical miles from Montreal, across the Canadian north to Fort Nelson, British Columbia and back, over four and a half days in May 1948. The thirty-two people on board included representatives from the DRB and other branches of the Canadian military, "and scientists from various university and government laboratories," including Hare and Wilson. As the DRB's Kenneth Maclure wrote in a report for the *Arctic Circular*, while the "primary scientific" aims were realized, to many of the observers, the view of a "cross-section" of the country was similarly significant: Cariberg was an Air Force "contribution towards practical instruction in Canadian geography."[68] In Maclure's account, and in the extensive media coverage, Cariberg was a sort of airborne Exercise Musk-Ox, seemingly scientific and "practical" rather than military, even as it was also part of the military's intense effort to photograph much of the country from above.

Another Cariberg participant was Wilson's University of Toronto colleague Donald Solandt. A physiologist, his presence on an airborne reconnaissance mission might have seemed incongruous. However, according to a Canadian Press article on Cariberg, Solandt "went on the trip for the purpose of making life in the Arctic livable. He claims that present harsh conditions in Canada's backyard leave people living there time to do nothing but fight to survive."[69] This was also one major objective of his brother Omond's Defence Research Board, at least when it came to soldiers—and within the DRB, Wilson and Rowley's ARAC, which also included Donald Solandt as an early member.

Ottawa circles

In August 1945, as the Second World War concluded, the new Chief of the Canadian General Staff, Lieutenant-General Charles Foulkes, and C. J. Mackenzie, the President of the National Research Council, turned to the state of defense research. The existing situation—with programs and facilities scattered across three military services and the NRC—seemed haphazard. The efforts of the services occasionally overlapped, and they were subject to budgetary constraints. Meanwhile, even as the NRC, created during the First World War, had embraced defense work during the more recent War, the Council's leaders were reluctant to create a distinct and more permanent military division. Foulkes and his colleagues also recognized that the Canadian military was likely to continue operating in the shadow of the more substantial U.S. and British armed forces, a relationship that dictated the sort of research to pursue—including on winter warfare.[70]

As a new organization, ultimately called the Defence Research Board, was pieced together, in December Foulkes, seeking a suitable director, phoned the prominent medical scientist Charles Best at the University of Toronto, seeking his opinion on a Winnipeg-born graduate of three programs at the university. Best responded enthusiastically, and Omond Solandt was named Director General of Defence Research. Solandt moved to Ottawa early in 1946, and by the late fall, the DRB was in place (it received legal status in March 1947 and convened for the first time the following month).[71] For the first three decades of the Cold War, the vast majority of Canadian defense research was overseen by the DRB—whether directly, at the Establishments under its control, or indirectly, through the Board's program of extramural research in Canadian academia and industry. By 1951, the DRB employed over 1600 people.[72]

In a telling phrase, Solandt's enthusiastic biographer describes the Arctic as "the DRB's responsibility." As early as December 1946, the journalist Leslie Roberts called the region Solandt's "laboratory."[73] And Solandt was certainly active in the year before the DRB was formalized. In April, he had written to his superiors, proposing the creation of a Directorate of Arctic Research. Citing Exercise Musk-Ox and the establishment of AINA, Solandt stressed that the military need to "learn to live, move, and fight in the Arctic" was complimentary with "exploration and research."[74] By May, the Chiefs of Staff had created an Inter-Service Committee on Winter Warfare and a Sub-Committee on Winter Warfare Research. Solandt led the latter group, but with the formation of the DRB, it was transformed into an Arctic Research Advisory Committee.[75]

As Solandt was adjusting to Ottawa in 1946, U.S. officials "began to pepper" their Canadian counterparts with requests and proposals that would provide Americans with "greater knowledge of Arctic conditions." Many of the ensuing discussions, some contentious, occurred at the bi-national Permanent Joint Board on Defence (PJBD).[76] One January PJBD memo suggested that the two countries should pursue "Joint manoeuvres and joint tests of material of common interest." This led to plans, formalized later that year through the Inter-Service Committee on Winter Warfare, to establish an "arctic experimental and testing station" at Churchill, "under conditions representative of northern Canada."[77] Soon, this site attracted visitors from across the Canadian and U.S. militaries, including defence scientists passing through or posted to the DRB's Defence Research Northern Laboratory (DRNL), established in 1947 and built over the next two years. By the end of the decade, efforts at the Laboratory—which, while "symbolically important ... was never a major part" of the DRB's slate—were focused on the human sciences, to match the trials run through the adjacent Fort.[78] As Tuzo Wilson had remarked with respect to Exercise Musk-Ox, one civilian visitor reported in 1948 that Fort Churchill's "atmosphere was much more that of a scientific laboratory than of a military base." But if the priority was "experiment in the techniques of living and operating in the north," these were the "techniques" of military bodies, tied to the presumption that the "vast wilderness" abutting Fort Churchill was a violent landscape.[79]

Before the creation of ARAC, the Winter Warfare Research group identified a need to compile "a record of all the information on northern Canada, which is in possession of various agencies, departments, and individuals."[80] This was in keeping with the Advisory Committee's expanded status. In a March 1947 memo, Patrick Baird pondered the composition of a new "Advisory Panel", which would include a member from the National Research Council, one from each of the three services, Tuzo Wilson ("representing universities"), Donald Solandt ("representing Medical Sciences"), and Baird himself. The chair, Baird suggested, should be Dr. Hugh Keenleyside—not a military official or academic, but a prominent diplomat, civil servant, and AINA founder who had just been appointed to a significant twinned position in Ottawa: Commissioner of the Northwest Territories and Deputy Minister of Mines and Resources.[81] Although he modestly claimed in his memoirs that he was "forced to rely constantly and heavily on the knowledge and judgement of my more sophisticated colleagues," Keenleyside was undoubtedly a crucial conduit between the DRB's Arctic pursuits and the wider northern interests of the Canadian state. Conversations with people like his "admired friend" Trevor Lloyd guided Keenleyside's approach to "matters of policy and administration" in the late 1940s.[82]

Concerned about his many roles, Keenleyside nonetheless decided that chairing ARAC was "a responsibility which I should not turn down."[83] Within days, Baird sent him letters of invitation to approve and sign. These spelled out the hope "that close coordination will be assured between investigations of prime interest to National Defence and those which are mainly concerned with the scientific approach toward the development of the north." Recipients were invited to the first ARAC meeting on May 15, 1947.[84]

The Committee assembled sixteen times between 1947 and 1949 (and at least ten more times thereafter, until 1954).[85] The first, chaired by Keenleyside, included both Omond and Donald Solandt, along with the three Musk-Ox principals: Wilson, Rowley, and Baird (acting as Secretary). Aside from Baird, who left for AINA weeks later, and Omond Solandt, who did not regularly attend, this core group remained consistent through those sixteen meetings. At the initial meeting, the DRB's Chair introduced the Board, suggesting that the military would need "assistance from many sources and it was hoped that they would in turn be able to help general scientific progress in the Arctic by making available to scientists their transportation and other research facilities."[86]

Another feature of the first ARAC meeting was a description by Donald Solandt of an ad-hoc gathering, two days earlier, at the NRC "to discuss Arctic Medical Research." This group included U.S. military scientists and L. P. Dugal of Laval University: "Arrangements were made to co-ordinate research plans of these three at Churchill next winter." This was not only an opportunity for Donald Solandt's brother to follow with a description of the DRB's nascent Defence Research Northern Laboratory at Churchill, but it also foreshadowed the formation of the Board's Panel on Arctic Medical Research (PAMP), chaired by Dugal, in 1948.[87] Donald Solandt was also a member of this group, and—along with DRB employees like Rowley—acted as a liaison with ARAC.

ARAC and PAMP deserve their own study. Even in the bureaucratic form of agendas and minutes, the available records of their meetings, including Graham Rowley's regular "Recent Events in the Arctic" updates for each ARAC gathering, reflect the profound fusion of military and scientific "priorities" in the postwar Canadian north. While a third government body, the Advisory Committee on Northern Development (first started, rather ineffectively, in 1948, with Hugh Keenleyside also chairing) adopted a broader mandate of *responsibility*, it is nonetheless salient that the two DRB groups, replete with distinguished academics and other "experts", adopted such a restricted approach to northern societies.[88] Indigenous voices were absent from the spaces of their deliberations—which were occasionally held in northern locations, like Churchill and the overarching emphasis on soldiers, and on adaptation and survival, meant that northerners were reduced to (at most) comparative bodies

This narrow approach to the north is all the more striking when we consider the confluence of Arctic interest in Canada's capital after the Second World War, and the spillover from restricted institutions like the DRB into public venues—albeit spaces that were still dense with military presence. According to the anthropologist Nelson Graburn, concerns over the influence of U.S. funding at AINA led a group of Ottawa-based individuals to launch a "club" for discussion of northern matters.[89] The Arctic Circle was conceived in the fall of 1947 at the home of Tom and Jackie Manning (see note 20), in Ottawa's Glebe neighborhood, and organized by a small group that included the Mannings, Diana and Graham Rowley, Trevor Lloyd, and Ken Maclure.[90] The Circle's first meeting was held one month later, at the RCAF Mess on Gloucester Street in downtown Ottawa. Shifting thereafter to the Army Mess on Sparks Street, over the next three years, the Circle convened some two dozen times in this location.

The deep military affinities of the Arctic Circle, particularly in its early years, were signaled by the presentation that launched the first meeting: a "description," "illustrated … with an excellent kodachrome film," by another DRB representative, of the "establishment of weather stations at Eureka

Sound and Cornwallis Island." The terms of the club were then discussed, including the production of a newsletter, which Diana Rowley agreed to edit. Prepared in advance of the second Circle meeting in January, the *Arctic Circular* bore a resemblance to the soon-to-be-launched *Arctic*.[91] But like the meetings of the Circle, and the group's executive, the *Circular*'s initial blend of scientific and military activities seems even more taken-for-granted, echoing Graham Rowley's surveys of Arctic activity for ARAC and the DRB.

Conclusions

The first issue of *Arctic* also included "research reports" from a number of scholars who are now reputed for their postwar northern fieldwork. These reports are also notable, though, for the regular mention of military involvement—use of the RCAF aircraft for transportation, for example, and, in Point Barrow, Alaska, the beginnings of the ONR's Arctic Research Laboratory.

RCAF planes were used to facilitate a project that stood out, in that first set of *Arctic* reports, for its overt concern with the human sciences. In the summer of 1947, as McGill's first Summer School was launched, Dr. George Malcolm Brown led a team of Queen's University medical researchers to Southampton Island (*Shugliaq*), the large land mass in the northern expanse of Hudson Bay, to study some eighty percent of the island's Inuit residents, including children. As it was initially described for readers of *Arctic*, the ostensible intent of the Expedition was to assess "the morbidity of various diseases," "nutritional habits and status," and to conduct "certain dietary experiments." The last objective was accomplished by choosing a "small representative group ... to determine average daily food intake, tolerance for pemmican, and the rate at which they develop acidosis during starvation." It was in this final, grim effort that the objective and the sponsor of the Expedition were revealed, if only allusively in the initial *Arctic* account: Brown and his colleagues were interested in comparing the relationship between diet and acidosis among Inuit to that among "Canadian and American soldiers," as the scientists sought methods to medically screen personnel considered for northern duty.[92]

Born in the small Ontario town of Campbellford, Malcolm Brown was a brilliant student, completing medical school at Queen's in his early twenties and moving to Oxford on a Rhodes Scholarship. Following a stint at Oxford's Radcliffe Infirmary in the early years of the Second World War, Brown joined the Royal Canadian Army Medical Corps before returning to Queen's in 1946.[93] His teams traveled to Southampton Island five times between 1947 and 1950 (and returned in 1954, as well), to continue similar work beside the Coral Harbour airstrip.[94] The Queen's Expeditions were supported in part by the NRC, the Canadian Department of National Health and Welfare—and, in one instance, by an AINA grant. But the main sponsor was ultimately the Defence Research Board. In a report for the DRB, completed in December 1950, Brown acknowledged the Board's "Lt. Col.

G.W. Rowley" for "invaluable help in planning and making arrangements for several trips to the North." The "main purpose" of these trips, Brown acknowledged for his DRB readers, was actually "to study the effect of cold on man." This ambition justified five visits to the Eastern Arctic over four years—visits which were, Brown admitted plainly, "collecting parties."[95]

Brown was unquestionably familiar with similar medical and physiological research underway elsewhere in the North American Arctic, from Churchill to Alaska, by the late 1940s—and equivalent investigations at southern universities and military laboratories. In Canada, much of this work was funded or monitored by the DRB's Panel on Arctic Medical Research, where he was a stalwart attendee, rising to Chair in 1950.[96]

In an important article, Matthew Wiseman has shown how the Queen's Expedition was a particularly distressing piece of the DRB's early northern efforts.[97] It is also notable that despite Brown's limited time in the Arctic, his focus on one site, and on one modest "population" of northerners, these experiences nonetheless sealed his reputation as a southern authority on "man in the north"—a reputation expressed and expanded in a number of publications.[98] For Brown, this "north" truly appeared as a laboratory; it only seemed vital insofar as it maintained the "traditional" lives of authentic Inuit worthy of study as suitable contrasts to his white control group. As Wiseman argues, the conclusions of this research, which were largely murky, were less significant than the premise: a search for information that would aid military life in the Arctic.

This search was of course shared by Brown's sponsors at the DRB. And in the summer of 1949, Omond Solandt, Graham Rowley, Tuzo Wilson and several others joined Brown in an attempt to "extend" the Queen's Expedition's "investigations to an area where the Eskimo were less touched by civilization," flying from Winnipeg to Churchill, then to Coral Harbour, and on to the community of Igloolik, north of Southampton Island. Operation Lyon, as it was dubbed, was not particularly consequential, beyond a mixture of geographic encounters and a test for a RCAF Canso "flying boat." The Queen's team never returned to Igloolik. But Rowley summarized the initiative for the *Arctic Circular*, and he and Wilson reported on Lyon at an ARAC meeting, where the discussion turned to Inuit employment at military facilities.[99] Again, specific results were less important than the activity itself, and the resulting enhancement of a form of northern expertise.

If the Indigenous bodies scrutinized by the Queen's Expedition were exemplary in one specific sense, they were also seen as beset by disease and ill health, and their occupants were understood as less *substantial*, and certainly less modern, than the "normal" control group of university students (themselves standing in for soldiers who might eventually learn a thing or two from Indigenous habits). Moreover, the consequences of experimentation and loose comparison were essentially unaddressed in Brown's publications, and in the vast majority of the early discussions on the DRB's Arctic committees and panels (at least, in the available records). These consequences surely included the naturalization of military activity, or research supported by militaries, in the Arctic after the Second World War: an era of planetary militarization, led by the United States but aided, in this instance, by the Canadian Department of National Defence, that was intensely preoccupied with "hostile" environments. That academic science accompanied this military expansion is indisputable. Just as significant, however, are figures like "man in the north", who was impossible without the combination of the two.

Notes

1. In terms of Arctic projects, one major new institution mentioned below, but mostly outside the scope of this chapter, was the U.S Navy's Office of Naval Research (ONR). Although it was without "budgetary authority," the U.S. military's Research and Development Board (RDB), formalized in 1947, also merits more attention. Both are addressed in Fae L. Korsmo, "The Early Cold War and U.S. Arctic Research," in Keith R. Benson and Helen M. Rozwadowski, eds., *Extremes: Oceanography's Adventures at the Poles* (Sagamore Beach: Science History Publications/USA, 2007), 173–199 (the quote is from p. 180).
2. Stephen Bocking, "A Disciplined Geography: Aviation, Science, and the Cold War in Northern Canada, 1945–1960," *Technology and Culture*, 2009, 50(2): 265. At a circumpolar scale, the historical shift is discussed in Ronald E. Doel et al., "Strategic Arctic Science: National Interests in Building Natural Knowledge—Interwar Era Through the Cold War," *Journal of Historical Geography*, 2014, 44: 60–80.
3. Although my invocation of an "official landscape" is tied to shifting southern "ideas of north", I am drawing from Rob Nixon, *Slow Violence and the Environmentalism of the Poor* (Cambridge, MA: Harvard University Press, 2011), 17. I have also been inspired by *The Official Mind of Canadian Colonialism*, part of the *Qikiqtani Truth Commission: Thematic Reports and Special Studies 1950–1975* (Iqaluit: Qikiqtani Inuit Association, 2014).
4. See, for example, Matthew Farish and P. Whitney Lackenbauer, "High Modernism in the North: Planning Frobisher Bay and Inuvik," *Journal of Historical Geography*, 2009, 35(3): 517–544; Richard C. Powell, *Studying Arctic Fields: Cultures, Practices and Environmental Sciences* (Montreal and Kingston: McGill-Queen's University Press, 2017).
5. In language similar to J.T. Wilson's (see below), the ONR's M.C. Shelesnyak wrote that "the accumulation of systematically collected data" using new "highly specialized techniques ... requires long-term studies at fixed or semi-fixed bases." Shelesnyak, "The History of the Arctic Research Laboratory (Under Contract with Office of Naval Research), Point Barrow, Alaska," *Arctic*, 1948, 1(2): 100. See also Matthew Farish, *The Contours of America's Cold War* (Minneapolis: University of Minnesota Press, 2010), Chapter 4.
6. See Matthew Farish, "The Lab and the Land: Overcoming the Arctic in Cold War Alaska," *Isis*, 2013, 104(1): 1–29; Matthew S. Wiseman, "Unlocking the 'Eskimo Secret': Defence Science in the Cold War Canadian Arctic, 1947–1954," *Journal of the Canadian Historical Association*, 2015, 26(1): 1911–223.
7. Kevin Mendel Thrasher, "Exercise Musk-Ox: Lost Opportunities" (MA Thesis, Carleton University, 1998), 10, refers to the "three key alumni of Musk Ox": J. Tuzo Wilson, Patrick Baird, and Graham Rowley.
8. The most substantial mention of Indigenous Peoples in the first *Arctic* was the disconcerting initial report from the Queen's University Expedition to Southampton Island, discussed below. An alternate approach, of "Eskimo

assistants ... used extensively for field observation and collecting and for constructing traps and cages," was briefly noted in a discussion of what became the ONR's Arctic Research Laboratory: "Northern Research Reports: Scientific Research at Point Barrow, Alaska," *Arctic*, 1948, 1(1): 66. See also Karen Brewster, "Native Contributions to Arctic Science at Barrow, Alaska," *Arctic*, 1997, 50(3): 277–288.

9. J. Tuzo Wilson, "A Message from the Arctic Institute of North America," *Arctic*, 1948, 1(1): 3. A concurrent portrait of this moment, presented at an AINA dinner by the Canadian Ambassador to the U.S., is Hume Wrong, "The New Importance of the Canadian Arctic," 7 May 1948 (http://gac.canadiana.ca/; accessed 9 May 2018).

10. Wilson, "A Message." In this it echoed an earlier description of AINA as "exclusively scientific." Richard Foster Flint, "Arctic Institute of North America," *Science*, 29 September 1944, 100(2596): 292.

11. In this chapter, I do not trouble the distinctions between "north" and "Arctic".

12. Wilson's significance is summed up by the title of Ali Polat, "John Tuzo Wilson: a Canadian who revolutionized Earth Sciences," *Canadian Journal of Earth Sciences*, 2014, 51: v–viii.

13. Graham Rowley and Diana Rowley, "John Tuzo Wilson, (1908–1993)," *Arctic*, 1994, 47(1): 104.

14. Rowley and Rowley, "Wilson"; see also G.D. Garland, "The Life of John Tuzo Wilson" (https://www.physics.utoronto.ca/physics-at-uoft/history/the-life-of-john-tuzo-wilson; accessed 10 April 2018); J. Tuzo Wilson, Oral History interview with Ronald Doel, American Institute of Physics, 16 February 1993 (https://www.aip.org/history-programs/niels-bohr-library/oral-histories/4371; accessed 8 May 2018). In the latter interview, Wilson indicated that before leaving England for Ottawa, he "went to see" some of the British leaders of operational research, including Omond Solandt.

15. J. Tuzo Wilson, "Winter Manoeuvres in Canada," *Canadian Geographical Journal*, 1946, 32(2): 88–100. These were Exercises Eskimo, Polar Bear, and Lemming, all held in the winter of 1944–45. Wilson's CAORG led the last two, and the subsequent Musk-Ox. Thrasher, "Exercise Musk-Ox," 2–9.

16. The first quote is from Wilson's public summary of Musk-Ox: J.T. Wilson, "Exercise Musk-Ox, 1946," *The Polar Record*, 1947, 5: 14. (This was a revised version of a speech given to the Empire Club of Toronto in April 1946.) The second is from Rowley and Rowley, "John Tuzo Wilson," 104. Musk-Ox has received some attention from military historians. One useful discussion is P. Whitney Lackenbauer, Peter Kikkert, and Kenneth C. Eyre, "Lessons in Arctic Warfare: The Army Experience, 1945–55," in Lackenbauer and Adam Lajeunesse, eds., *Canadian Arctic Operations, 1941–2015: Lessons Learned, Lost, and Relearned* (Fredericton: Gregg Centre for the Study of War and Society, University of New Brunswick, 2017), 47–104.

17. The post and airport of Fort Churchill, not far from the colonial fort of the same name, and a few kilometers east of the town, was established at the behest of the United States in 1942, as part of the Crimson Route that would transport equipment to Europe in the midst of the Second World War. This scheme was never fully executed, and in 1944 the Crimson facilities at Churchill and elsewhere were purchased by the Canadian government; Churchill then fell under the control of Canada's Department of Transport. After Exercises Lemming and Musk-Ox, in October 1946, the Canadian Army assumed control. Starting in 1956, Churchill was also the site of a rocketry facility, the Churchill Research Range. C.J. Taylor, "Exploring Northern Skies: The Churchill Research Range," *Manitoba History* 44 (2002-03) (http://www.mhs.mb.ca/docs/mb_history/44/exploringnorthernskies.shtml; accessed 21 June 2018).

18. Wilson, Oral History interview with Doel. Wilson made the same claim in a Canadian Broadcasting Corporation radio address before Musk-Ox: "Talk on Canadian Army and Royal Canadian Air Force Exercise 'Musk-Ox'," 15 December 1945, in MG28-I79 (Arctic Institute of North America fonds), Volume 105, Folder "Op. Muskox 1944–48", Library and Archives Canada (hereafter LAC). U.S. Army observers also favored the description of "a scientific expedition" over "a military operation": U.S. War Department, "Canadian Winter Exercise MuskOx," 3 June 1946, RG 24, Volume 8151, Folder 1660-11 V.1, LAC, 3.

19. Wilson, "Exercise Musk-Ox, 1946," 25. See also "Report of retiring Director of Operational Research," 31 May 1946, J. Tuzo Wilson Personal Records (B1993-0050), University of Toronto Archives, Box 47, Folder 12 (hereafter JTW). On 'civilianizing' military activity, see, for example, "Civilian Operations in Support of Defence Projects," prepared for discussions in Ottawa between prominent U.S. and Canadian bureaucrats in December 1946 (RG 25, Volume 5749, Folder 52-C(s) pt. 3, LAC).

20. Svenn Orvig, "Patrick Douglas Baird, 1912–1984," *Arctic*, 1984, 37(2): 193. See also Baird's curriculum vitae, in the James Patrick Croal fonds (MG 31-G34), Volume 12, Folder 5, LAC. The leader of the British-Canadian Expeditions was Thomas Manning, a biologist and geographer, who during the war spent time in the Canadian Navy and consulted with the U.S. Army Corps of Engineers on the construction of an airfield at Southampton Island, a location he knew from his pre-war travels. Manning also participated in Exercise Musk-Ox. After working for the Canadian Geodetic Survey, from 1948 to 1953 he was a consultant with the DRB and helped to found the Arctic Circle (see below); a few years later, he became the Executive Director of AINA. Andrew H. Macpherson, "Thomas Henry Manning (1911–1998)," *Arctic*, 1999, 52(1): 104; Jason Delaney, "'He was Writing the Book'—Lieutenant Commander James P. Croal: The Royal Canadian Navy's Cold War Arctic Specialist," *The Northern Mariner*, 2015, 25(4): 401.

21. Quoted in Thrasher, "Exercise Musk-Ox," 16. Baird's comment slightly blurred the roles each played in AINA's founding. See "Founders and Governors of the Arctic Institute," *Arctic*, 1966, 19(1): 102–7. On Wood, see Peter H. Wood, "Walter Abbott Wood (1907–1993)," *Arctic*, 1994, 47(2): 203–204, which also notes that Walter Wood worked for the U.S. Army Air Forces' Arctic, Desert, Tropic Information Center (see below) during the Second World War.

22. John MacDonald, "Graham Westbrook Rowley (1912–2003)," *Arctic*, 2004, 57(2): 223. Graham Rowley, *Cold Comfort: My Love Affair with the Arctic*, 2nd Ed. (Montreal and Kingston: McGill-Queens University Press, 2007), includes a photo featuring Baird, Rowley, and Wilson at Churchill (p. 257).

23. Rowley, *Cold Comfort*, 249, 255. Rowley and Wilson crossed paths in London "from time to time" during the War (*Cold Comfort*, 253).

24. Thrasher, "Exercise Musk-Ox," 5. In 1947–48, Rowley was also appointed acting director of the new Joint Intelligence Bureau, which fell under the DRB. See Rowley, *Cold Comfort*, 269–270; Kurt F. Jensen, *Cautious Beginnings: Canadian Foreign Intelligence, 1939–51* (Vancouver: UBC Press, 2008), 144–146.

25. Orvig, "Patrick Douglas Baird," 194. In 1948, Walter Wood established an AINA office in New York—the first of a few AINA outposts.

26. For Wilson's academic position, see Wilson to Sidney Smith [President of the University of Toronto], 15 May 1946, Box 50, Folder 2, JTW. For Wilson's ARAC invitation, see H. L. Keenleyside to J.T. Wilson, 28 April 1947, RG 24, Volume 4233, Folder DRBS 3-750–43, LAC.

27. J. T. Wilson, "Defence Research in the Canadian Arctic (with especial reference to geophysical research)," 27 February 1948, RG 24, Volume 35558, Folder 1200-A1 (Pt.1), 13. The memo, which has been misattributed to Graham Rowley, is mentioned (and correctly credited) in Edward Jones-Imhotep, *The Unreliable Nation: Hostile Nature and Technological Failure in the Cold War* (Cambridge, MA: The MIT Press, 2017), especially pp. 51–52. In March 1948, Wilson crafted another, less scientific confidential assessment: "The Geopolitical Significance of the Arctic," Box 47, Folder 5, JTW.
28. Another Musk-Ox participant, James P. Croal, was the Royal Canadian Navy's (very active) observer. After demobilization in 1946, he joined the DRB, and was the Board's first employee dispatched to the new Defence Research Northern Laboratory. Croal later traveled to Greenland and joint US-Canadian Arctic weather stations alongside the US Navy, trained military survival instructors, and contributed to survey and sea-lift work in the construction of the Distant Early Warning (DEW) Line. See "Biographical Sketch," Folder 1, Volume 1, Croal fonds, 1–2; Delaney, "'He Was Writing the Book'".
29. These meetings are documented in MG 28-I79, Volume 1, LAC. The best secondary discussion of AINA's formation remains Shelagh D. Grant, *Sovereignty or Security? Government Policy in the Canadian North, 1936–1950* (Vancouver: UBC Press, 1988).
30. Raleigh Parkin, "The Origin of the Institute," *Arctic*, 1966, 19(1): 6.
31. "Proposal for an Arctic Institute of North America," 8 September 1944, MG-28-I79, Volume 1, LAC, 1. This language was repeated in Flint, "Arctic Institute of North America," 292.
32. The reference is to M.C. Shelesnyak, *Across the Top of the World: A Discussion of the Arctic* (Washington, DC: Office of Naval Research, 1947). See also Farish, *Contours*, Chapter 2.
33. An Act to Incorporate the Arctic Institute of North America (1945), quoted in Robert MacDonald, "Challenges and Accomplishments: A Celebration of the Arctic Institute of North America," *Arctic*, 2005, 58(4): 440.
34. According to McGill's Max Dunbar, Washburn left Montreal to establish AINA's Washington office in March 1951: "A. L. Washburn," *Arctic*, 1952, 5(1): 3. Washburn later headed the US Army's Snow, Ice, and Permafrost Research Establishment (SIPRE), which eventually became the still-extant Cold Regions Research and Engineering Laboratory (CRREL). Carl S. Benson, "Albert Lincoln Washburn (1911–2007)," *Arctic*, 2007, 60(2): 212–214. ADTIC was started at Eglin Field, Florida, in 1942; moved to New York City in October 1943; transferred back to Florida (Orlando Field), under the name Arctic, Desert and Tropic Branch (Army Air Forces Tactical Center) in April 1944; and deactivated in October 1945. It was reactivated in 1947 at Maxwell Air Force Base in Alabama. Paul H. Nesbitt, "A Brief History of the Arctic, Desert, and Tropic Information Center and its Arctic Research Activities," in Herman R. Friis and Shelby G. Bale, Jr., eds., *United States Polar Exploration* (Athens, OH: Ohio University Press, 1970), 134–145.
35. Carlson went on to lead four universities after the War. His northern interests and work were documented in his book *Lifelines Through the Arctic* (New York: Duell, Sloan and Pearce, 1962). See also "About William S. Carlson," http://www1.udel.edu/research/polar/carlson.html; accessed 20 December 2017.
36. The quote is from Parkin, "The Origin of the Institute," 13. See also Benson, "Albert Lincoln Washburn"; Walter Sullivan, "Laurence McKinley Gould, a Polar Explorer and Innovative College President, Dies at 98," *New York Times* 22 June 1995 (www.nytimes.com; accessed 30 October 2017).

37. Parkin, "The Origin of the Institute," 11, recalls the conclusion at the September 1944 "that the majority of the Institute's Governors need *not* be scientists" (original emphasis). For a full list, see "Founders and Governors of the Arctic Institute." Flint's supervisory role is mentioned in Benson, "Albert Lincoln Washburn," 212. Another significant polar researcher employed at ADTIC during the War (and later by the US Army) was the biologist Carl Eklund. See Paul Siple, "Carl R. Eklund (1909–1962)," *Arctic*, 1963, 16(2): 147–148.

38. Parkin, "The Origins of the Institute," 7–8, 13. See also J. Brian Bird, "Trevor Lloyd (1906–1995)," *Arctic*, 1995, 48(3): 308; Grant, *Sovereignty or Security?* 122–123. Dates differ on the handover of the maps, but references to the delivery are in LAC MG28-I79, Volume 2, LAC, and "AINA Progress Report, June 1948," Series III, Subseries A, Box 42, Folder "Arctic Institute of North America," Carnegie Corporation of New York Archives, New York City.

39. See the photograph in Parkin, "The Origin of the Institute," 12.

40. John C. Reed, "Yesterday and Today," *Arctic*, 1966, 19(1): 19. By 1947, AINA had moved – interestingly – to the Ethnological Museum in McGill's Medical Building. *Ibid.*, 20. "Also on display in the museum are scientific instruments commonly used in Arctic work, and many articles of Eskimo and Indian clothing." Fred Chafe, "Arctic Institute Probes Life in Canadian North," McGill Daily 18 November 1947, 1.

41. J. T. Wilson to A. L. Washburn, 10 October 1945, and Washburn to Wilson, 11 October 1945, both in Box 50, Folder 2, JTW.

42. *A Program of Desirable Scientific Investigations in Arctic North America* (Montreal: Arctic Institute of North America Bulletin No. 1., March 1946). In the first section, Wilson was quoted on the "mapping of the whole Arctic region," which he characterized as "in an almost unbelievably primitive state" (p. 3).

43. On the Committee (later called the Banting Fund), see Donald H. Avery, *The Science of War: Canadian Scientists and Allied Military Technology During the Second World War* (Toronto: University of Toronto Press, 1998), 47–48; Terrie M. Romano, "The Associate Committees on Medical Research of the National Research Council and the Second World War," *Scientia Canadiensis*, 1991, 15(2): 71–87. On the Carnegie support, see Carnegie Corporation of New York: *Annual Report: 1947* (New York, Carnegie Corporation, 1947).

44. Parkin, "The Origin of the Institute," 16.

45. Henry B. Collins, "Arctic Bibliography," *Science*, 26 March 1954, 119 (3091): 3A. The Bibliography's Project Director was Marie Tremaine, who moved from the Toronto Public Library to the Library of Congress to oversee the work. See Nora T. Corley, "Marie Tremaine, 1902–1984," *Arctic*, 1985, 38(2): 165–166; Marie Tremaine, "Bibliography of Arctic Research," Arctic, 1948, 1(2): 84–87. Tremaine, p. 85, refers to the DRB as "the Canadian Government." This matched Omond Solandt's request, in 1948, for the language to be used "in connection with" DRB-supported AINA projects. Solandt suggested that the ONR, meanwhile, "would probably prefer that it be referred to by name," because it was "well known as a research organization as well as a military agency." "Minutes of the Meeting of the Executive Committee, January 3, 1948," MG28-I79, Volume 114, Folder "Lloyd, T. – Part 5 – 1947–1948," 4, LAC.

46. "Report on the Bibliography and Roster Projects, October 9, 1947," in MG28-I79, Volume 2, Folder "Board of Governors 1946–1947—Miscellaneous," 1–2, LAC. More discussion of the AINA-ONR relationship is warranted – including the work of Siple. Washburn, Siple, Gould, and other familiar figures were among those consulted by the Navy in 1946 on the establishment of the Laboratory. Shelesnyak, "The History of the Arctic Research Laboratory," 97–98. See

also John C. Reed and Andreas G. Ronhovde, *Arctic Laboratory: A History (1947–1966) of the Naval Arctic Research Laboratory at Point Barrow, Alaska* (Washington, DC: Arctic Institute of North America, 1971), an extensive volume prepared under an ONR contract.

47. *Arctic*, 1948, 1(1), front matter.
48. Reed, "Yesterday and Today," 24; MacDonald, "Challenges and Accomplishments," 441.
49. Matthew L. Wallace, "Reimagining the Arctic Atmosphere: McGill University and Cold War Politics, 1945–1970," *The Polar Journal*, 2016, 6(2): 367.
50. J. Brian Bird, "Geography at McGill University: A 50 Year Perspective: 1945–1955" (http://geog.mcgill.ca/ggs/others/GEOGRAPHY%20AT%20 McGILL%5B1%5D.pdf; accessed 20 June 2018), n.p.; Wallace, "Reimagining the Arctic Atmosphere," 363. See also Bocking, "A Disciplined Geography."
51. Jack Ives, *The Land Beyond: A Memoir* (Fairbanks: University of Alaska Press, 2010), 10. Ives refers to the period from 1945–54.
52. F. Kenneth Hare, "I Always Take a Window Seat! A Chronicle of My Life," Box 183, F. Kenneth Hare Fonds (F2016), Trinity College Archives, University of Toronto.
53. Bird, "Geography at McGill," n.p. Among other military-oriented activities, from late 1946 the Office of Naval Research was supporting Stefansson's *Encyclopedia Arctica* initiative. See Vilhjalmur Stefansson, "Encyclopedia Arctica," *Arctic*, 1948, 1(1): 44–46.
54. Hare, "I Always Take a Window Seat!" 60.
55. George H. T. Kimble to F. Cyril James, RG 2, Box 131, Folder 03752, 26 November 1946, McGill University Archives (hereafter MUA). Washburn offered a graduate seminar on the Arctic in 1948; see "The Record," *Geographical Journal*, 1948, 112(4/6): 252.
56. "Arctic Institute of North America, Arctic Exploration Series 1948–1949," Box, 8, Folder 1, David C. Nutt papers (MSS-71), Rauner Special Collections Library, Dartmouth College.
57. Bird, "Geography at McGill," n.p.
58. "Geography, Music to Hold Summer Schools; French, Regular Courses also Set," *McGill Daily* 23 May 1947, 1. Given that the Department of the Air Force was not established until later in 1947, it is likely that some of the "officers" were from the U.S. Army Air Forces.
59. Washburn to Colonel Charles G. Kirk, 28 February 1947, RG 2, Box 131, Folder 03752, MUA.
60. J. Tuzo Wilson to George Kimble, 27 November 1947, Box 47, Folder 4, JTW.
61. McGill University Geography Summer School (1948) brochure, RG 2, Box 131, Folder 03752, MUA.
62. "The Record," 251.
63. E. K. Walker to F. Cyril James, 22 June 1948; "A Report on the McGill University Summer School in Geography, Stanstead College, July 4 – August 15, 1948," both in RG 2, Box 131, Folder 03752, MUA.
64. McGill University Geography Summer School (1948) brochure; "A Report on the McGill University Summer School in Geography, Stanstead College, July 4 – August 13, 1949," both in RG 2, Box 131, Folder 03752, MUA.
65. Bocking, A Disciplined Geography," 266. See also Wallace, "Reimagining the Arctic Atmosphere."
66. Ives, *The Land Beyond*, 10. Much of this 'checking' was done by Margaret Montgomery, who came from the DRB to earn McGill's first graduate degree in Geography in 1949, under the supervision of Hare, after participating,

104 *Farish*

with the DRB's Moira Dunbar, in Operation Cariberg (see below) and other ice-reconnaissance flights. Montgomery, who returned to the Board, and Dunbar were two important exceptions within the overwhelmingly masculine spaces of military scientific research in and on (and over) the north. See Turner, "The Defence Research Board," 257.

67. "The Record," 251. In "I Always Take a Window Seat!" 61, Hare recalled a phone conversation with the DRB's Rowley in the winter of 1947–48: "Graham proposed a spring flight across the centre of [Hudson] Bay in May, 1948, and invited me to go along. I accepted at once." Hare's memoir (p. 62) also mentions other work with the DRB.

68. K.C. Maclure, "Operation Cariberg," *The Arctic Circular*, 1948, 1(6): 60–62. On Maclure, see Keith R. Greenaway, "Kenneth C. Maclure (1914–1988)," *Arctic*, 1988, 41(3): 258–259. Like Maclure, Greenaway was a prominent military aviator who worked for the DRB. See Buzz Bourdon, "Keith Greenaway was one of the world's leading authorities on Arctic navigation," *The Globe and Mail* 18 May 2010 (https://www.theglobeandmail.com; accessed 21 June 2018).

69. "5-Day Air Jaunt to Study Problems of Arctic Life," undated and unattributed Canadian Press article, Box 47, Folder 5, JTW.

70. Captain D. J. Goodspeed, *A History of the Defence Research Board of Canada* (Ottawa: Queen's Printer, 1958), viii, 8, 20–24. See also Jason Sean Ridler, *Maestro of Science: Omond McKillop Solandt and Government Science in War and Hostile Peace, 1939–1958* (Toronto: University of Toronto Press, 2015), 114.

71. Goodspeed, *A History*, 42, 46; Ridler, *Maestro of Science*, 120, 122.

72. O.M. Solandt, *Defence Research in Canada* (Montreal: The Engineering Institute, 1951), 1.

73. Ridler, *Maestro of Science*, 142; Leslie Roberts, "Canada Fears Being Ham in U.S.-Soviet Sandwich," *PM* 22 December 1946, 7. See also Bocking, "A Disciplined Geography," 275.

74. O.M. Solandt, Division of Defence Research, to Chiefs of Staff Committee, "Proposed Establishment of a Directorate of Arctic Research," 16 April 1946, RG 2, Volume 101, Folder R-100-D (4) 1946–47, LAC.

75. Goodspeed, *A History*, 177–78. See also Wiseman, "Unlocking," 203; and Wiseman, "The Development of Cold War Soldiery: Acclimatisation Research and Military Indoctrination in the Canadian Arctic, 1947–1953," *Canadian Military History*, 2015, 24(2): 138.

76. Peter Kikkert, "1946: The Year Canada chose its Path in the Arctic," in P. Whitney Lackenbauer, ed., *Canadian Arctic Sovereignty and Security: Historical Perspectives* (Calgary: Centre for Military and Strategic Studies, University of Calgary, 2011), 76.

77. The first two quotes are from *Canada's Post-War Defence Policy, 1945–50* (Report No. 90, Historical Section, Army Headquarters, 1 April 1960), 60–61. The third is from Winter Warfare Committee, "Churchill Report," 16 August 1946, RG 24, Volume 8104, Folder 1280-124, LAC, 1. See also Andrew Iarocci, "Opening the North: Technology and Training at the Fort Churchill Joint Services Experimental Testing Station, 1946–64," *Canadian Army Journal*, 2008, 10(4): 76.

78. Jonathan Turner, "The Defence Research Board of Canada, 1947 to 1977" (PhD Thesis, University of Toronto, 2012), 93. See also A.M. Pennie, *Defence Research Northern Laboratory, 1947–1965* (Ottawa: Department of National Defence, 1966).

79. R.G. Riddell, Memorandum for the Under-Secretary of State for External Affairs, 23 August 1948, in RG 25, Volume 3346, Folder 9061-40, LAC, 2.

80. Arctic Research Group, "Defence Research: Progress Report No. 1" [October 1946–March 1947], RG 24, Volume 84, Folder 1280-123, LAC, 2.
81. Lt. Col. P.D. Baird to Major MacNeil, 28 March 1947, RG 24, Volume 4233, Folder DRBS 3-750-43, LAC, 1. On Keenleyside, see Shelagh D. Grant, "Hugh Llewellyn Keenleyside: Commissioner of the Northwest Territories, 1947–1950," *Arctic*, 1990, 43(1): 80–82.
82. Hugh L. Keenleyside, *Memoirs of Hugh L. Keenleyside, Volume 2: On the Bridge of Time* (Toronto: McClelland and Stewart, 1982), 316, 308. Concurrent with the first meeting of ARAC in 1947, and against some objections from the military (particularly Rowley's Joint Intelligence Bureau), Keenleyside led the creation of a Geographical Bureau in his department, and Lloyd became the first head. See Trevor Lloyd, "The Geographical Bureau," *Canadian Geographical Journal*, 1948, 36(1): 39; Grant, *Sovereignty or Security?* 208, 220–221.
83. H. L. Keenleyside to O.M. Solandt, 15 April 1947, RG 24, Volume 4233, Folder DRBS 3-750-43, LAC.
84. Baird to Keenleyside, 28 April 1947, *Ibid*; Keenleyside to J.T. Wilson, 28 April 1947, *Ibid*.
85. This is my own estimation, compiled from files across multiple record groups at LAC, and from JTW.
86. Minutes of the 1/47 Meeting of the Arctic Research Advisory Committee, 15 May 1947, RG 22, Volume 853, Folder S84-11-4A pt. 1, LAC, 1.
87. *Ibid.*, 2. Omond Solandt, Paul Siple, and Graham Rowley attended the Panel's first meeting, where Donald Solandt proposed "a study on the special senses" of the "'racially pure'" Inuit. Minutes of the First Meeting of the Panel on Arctic Medical Research, 16 December 1948, RG 24, Volume 4129, Folder DRBS 4-78-53 pt.1, LAC, 3.
88. See P. Whitney Lackenbauer and Daniel Heidt, *The Advisory Committee on Northern Development: Context and Meeting Minutes, 1948–66* (Calgary: Centre for Military and Strategic Studies, University of Calgary, 2015). For all of the paternalism of groups like the Advisory Committees, sources like Rowley's *Cold Comfort* also indicate that the individuals introduced in this chapter held varying views of northern communities and relationships between Indigenous Peoples and the state, views often shaped by the nature and extent of their experiences in those communities.
89. Nelson H. H. Graburn, "Canadian Anthropology and the Cold War," in Julia Harrison and Regna Darnell, *Historicizing Canadian Anthropology* (Vancouver: UBC Press, 2006), 251.
90. *Arctic Circular*, 1948, 1(1): 2; Diana Rowley, "A Short History of the Arctic Circle" *The Arctic Circular*, 1984, 32: 1. On another Circle founder, Frank Davies, and his role at the DRB's Radio Physics Laboratory, see Jones-Imhotep, *The Unreliable Nation*.
91. *Arctic Circular*, 1948, 1(1): 1.
92. "Northern Research Reports: Queen's University Expedition to Southampton Island," *Arctic*, 1948, 1(1): 65. See also Wiseman, "Unlocking," 194.
93. Brown went on to become one of the most prominent medical professionals in postwar Canada. "Obituary: G. Malcolm Brown," *British Medical Journal*, 9 July 1977, 2(6079): 131; Royal College of Physicians, "George Malcolm Brown" (http://munksroll.rcplondon.ac.uk/Biography/Details/592; accessed 21 June 2018).
94. Wiseman, "Unlocking," 192–193. Like that at Churchill, the Southampton airstrip was built during the Second World War as part of the Crimson Route.
95. G. Malcolm Brown, *Progress Report on Clinical and Biochemical Studies of the Eskimo* (Ottawa: Defence Research Board, 1951), 8.

96. See the material in RG 24, Volume 4129, Folder DRBS 4-78-53 pt. 1, LAC, including discussion of the AMRP's visit to Churchill in the winter of 1949.
97. Following Wiseman, "Unlocking," more work is required to link military research in Indigenous communities with that discussed in (for example) Ian Mosby, "Administering Colonial Science: Nutrition Research and Human Biomedical Experimentation in Aboriginal Communities and Residential Schools, 1942–1952," *Histoire sociale/Social History*, 2013, 46(91): 145–172. See also Farish, "The Lab and the Land."
98. The specific reference here is to G. Malcolm Brown, "Man in the North," in V.W. Bladen, ed., *Canadian Population and Northern Colonization* (Toronto: University of Toronto Press, 1962), 136–147. The chapters in this book derived from a 1961 symposium at the Royal Society of Canada.
99. Minutes of the 4/49 Meeting of the Arctic Research Advisory Committee, 26 September 1949, RG 24, Volume 35558, Folder 1200-A1 pt.2, LAC, 4; Graham Rowley, "Operation Lyon," *Arctic Circular*, November 1949, 2(7): 99–101. As Rowley, *Cold Comfort*, indicates, he had an extensive relationship and a deep affinity for Igloolik.

Part 3
Cold War economies

7 Arctic pipelines and permafrost science: North American rivalries in the shadow of the Cold War, 1968–1982

Robert Page

In North America the history of the Cold War is commonly defined in terms of tensions across the Arctic with the Soviet Union. While these tensions were real, they varied in time and in relation to other policy concerns. Economic rivalries between Canada and the United States at times overshadowed tensions with the Soviet Union; economic interests in pipelines allowed prudent cooperation; and in spite of political tensions there was collaboration on permafrost science for the common benefit of Siberia, Alaska, and Arctic Canada. This paper considers the nature of the Cold War, the role of pipelines in energy supply, and scientific collaboration. The author was both a historian of these events and a participant, and brings with him to this analysis the strengths and weaknesses of his role in these events forty years ago.[1]

The 1970s marked a significant interlude in the Cold War years, different from the Korean War era of the 1950s or the Cuban Missile Crisis of the early 1960s. The closing years of the Vietnam War, the vigorous street protests of the Peace Movement, and public distaste for Richard Nixon's foreign policy initiatives meant that these were not normal Cold War years. In Canada, a strong nationalist movement denounced American ownership of Canadian industry (all oil and gas "majors" were in American or European hands). The Arctic was a particular area of nationalist focus with the voyages in 1969 and 1970 through the Northwest Passage (claimed by Canada) of the S.S. *Manhattan*, an American oil tanker that had been converted into an icebreaker, and the Exxon-led Mackenzie Valley Pipeline debate. Nationalists argued that this pipeline would benefit American producers and consumers while leaving serious environmental and social impacts in Canada. It was felt that the *Manhattan* voyages threatened Canada's marine sovereignty and the pipeline, land sovereignty. While these conclusions were exaggerated, the debate reflected the times and the intensity of Canadian feelings. Prime Minister Pierre Trudeau, for example, feared that any public discussion of nationalism would impede his efforts to defeat Quebec separatism. Yet the nationalist umbrella brought together a variety of causes in opposition to the pipeline. John Livingston (an academic and environmental activist) writing in the *Ontario Naturalist* magazine in 1970 caught some of the nuances of the campaign: "I propose for Canada a thoroughly respectable form of neo-nationalism which might be described as ecological independence."[2] Here he was talking about the freedom to make our own policy decisions.

This paper therefore contributes to Arctic historiography by linking nationalist aspirations in the 1970's with developments in permafrost science and engineering technology. Together, these three forces defeated one of Canada's largest private sector consortiums and one of the largest projects in North American business history. In Alaska, there was a different outcome for reasons to be discussed later. With this paper I am trying to fill the gap in our Arctic literature; scientific accounts of this episode do not explain the historical context and historians have largely avoided the complexities of the relevant science and engineering.[3] The background archival records are huge. The National Energy Board hearing transcripts include over 200 volumes, 40,000 pages and thousands of separate documents, and the Mackenzie Valley Pipeline Inquiry records are equally extensive. My own involvement in the process helped me to navigate these complex records. I also have my own notes from discussions with other participants and regulatory officials. I got to know Thomas Berger (who led the Mackenzie Inquiry) and his staff as well as key players for both corporate applicants who on occasion used me as a sounding board.[4] This huge process documented well the weaknesses and complexities of megaproject development in Arctic conditions.

Soviet/North American scientific links

The pipeline debate provides a window into the scientific links between scientists on both sides of the Iron Curtain. Soviet scientists working on permafrost issues had working linkages to European and North American scientists that were expanded during the 1970s because of the common interest in permafrost issues for pipeline design. These connections were forged through science journals, conferences, public lectures and personal contacts. In 1965, the Soviets were the world leaders in permafrost science, but they still lacked a definitive understanding of many aspects of the forces at work when applied to pipeline engineering. There were institutional bridges like the Scott Polar Research Institute at Cambridge and individuals like Dr. Terence Armstrong who traveled frequently to Siberian research stations. I got to know Armstrong during his visits to Trent University (where I was teaching at the time), and the Soviet scientists I met were enthusiastic about sharing their research results and getting our reactions. There appeared to be a common bond of scientific inquiry, which had not been eroded by Cold War constraints and tensions. I also had the privilege of meeting several Russian historians who had read some of my pipeline and oil articles. They explained to me that energy had played a central role in the success of Soviet Communism and therefore was a priority for Soviet historiography.

Hydroelectricity had brought modern living to Soviet citizens and, during the Second World War the Battle of Stalingrad had saved the essential oil supply from Baku for the victorious Red Army. In the post-war period, the Siberian oil reserves had become increasingly important and oil exports were critical in financing Soviet weapons and space programs. However, this capital also needed to be invested back into oil and gas production as well as pipeline technology. The Soviets needed to redesign and reconstruct the Soviet pipelines, which suffered from extensive leakages. They desperately needed access to North American pipeline technology and were prepared to share their own pure science of the nature of permafrost for access to the applied science and engineering of North America. It was a collaborative business deal that largely ignored the tensions of the Cold War.

Prudhoe Bay and Arctic delivery systems

1968 saw one of the great events in global energy history when, on the barren northern slopes of Alaska at Prudhoe Bay, the largest oil reserves in American history were discovered. The immediate challenge for the American government was to design and execute a safe delivery system for this region to supply American consumers and lessen their dependence on the turbulent Middle East. The political and economic excitement was huge in Washington and on Wall Street in New York. The excitement was also great in Ottawa because the geological formations in which the oil had been found stretched eastward into Canadian territory. The North American Arctic, many hoped, would become the new Middle East!

Regulatory processes were announced in both countries to handle corporate applications for pipelines. In the US, the Federal Power Commission and the Department of the Interior were involved and in Canada, the National Energy Board (NEB) was legally responsible. But the Trudeau government decided that was not enough given the unique Arctic First Nations and environmental issues. So, in 1974, the federal government appointed Mr. Justice Berger to lead a special inquiry with a wider mandate to complement the work of the NEB. There is a wide literature on the First Nations and wildlife issues, which will not be dealt with in this paper. The NEB hearings got off to a rocky start when the Supreme Court of Canada forced its Chair and panel to step down on the basis of "reasonable apprehension of bias." A new panel was formed, and the NEB started over again after nearly eight months. This restart increased public controversies over the pipeline and fuelled concerns about the effective execution of Canadian control and sovereignty over the archipelago.

The NEB and Berger Inquiry hearings attracted wide popular and media attention. This was exceptional for regulatory hearings, given their popular image and technical focus. Participants included a wide range of business, environmental and social organizations including some church groups challenging the ethics of capitalism.[5] At the NEB there were over 100 legal counsel representing business and other parties. All three major TV networks provided live coverage of the opening of the NEB hearings, which I have never seen before or since for such a process. Outside of the formalities were many players in the background including representatives of the Soviet Embassy. The Soviet Cultural Attaché invited me to lunch and to visit the North American Institute in Moscow. Given that his activities were being monitored, I received a phone call from Canadian officials warning me of security risks in this Cold War setting. I ignored the warning.

The US State Department also took notice of these Arctic pipeline events. Later I was invited to Washington to lecture on these issues to the US War College, a security think tank of senior US officials with "strategic" responsibilities. They were interested in the role of Canadian nationalism in limiting US energy plans in the Arctic. They gave away their bias with several jokes afterwards about Trudeau as the Castro of the North for his launching of Petro-Canada (a state-owned petroleum company). My reflection on my contacts with Canadian and American security officials was that they had too much time on their hands if they were concerned about activities like mine dealing with permafrost science and pipeline engineering. Also, like many other academics it gave me a rather cynical view of the Cold War rhetoric—which was less than realistic but quite natural given our attitude to President Nixon and his foreign policy.

Permafrost science and the Alaskan pipeline

The strategic geography of US energy supply drastically changed in 1968 with the Prudhoe Bay oil discoveries. Now the challenge was to design and implement an effective, secure, and low cost delivery system from this isolated location to southern refineries and consumer markets. With the huge size of the reserves they were looking at two million barrels per day in delivery capacity either by tanker or pipeline. In the politics of the Cold War, these were strategic new resources for the US industry and a significant contribution to energy self-sufficiency and balance of payments stability. But the creation of this delivery system turned out to be much more difficult than anyone initially expected. The pipeline experts were confident that they could solve any technical problem. The Arctic, however, would inspire a new scientific humility. There were four options to be considered: two tanker and two pipeline.

The first option triggered immediate Canadian concern over Ottawa's claims to marine sovereignty. Exxon proposed a tanker route eastward from Prudhoe Bay, through the Northwest Passage where Canada claimed jurisdiction (which the US rejected under the principle of freedom of the seas beyond the three-mile limit). Humble Oil for Exxon refitted the super tanker *Manhattan* for ice prone conditions and the US government informed Canada of their intentions. The Canadian government, in turn, cooperated by providing an icebreaker escort for the 1969 and 1970 voyages. However,

Canadian public opinion reacted angrily to the US action in violation of the Canadian claims and the perceived environmental risk of tanker oil spills.

Ottawa rushed into action by offering a Mackenzie Valley Oil Pipeline route as an alternative transportation route without any serious environmental or other research. This reflected opposition in British Columbia to West Coast tankers from Alaska and some US Midwest and Chicago interests who wanted the oil to be transported there instead of to the West Coast. Annoyed with the US action, the Trudeau cabinet announced talks with the Soviet Union on Arctic cooperation, which seemed to ignore Cold War tensions. But given the lack of substance to the Canadian pipeline proposal, the Americans quite rightly never took the Mackenzie valley oil pipeline option seriously. This option, however, was not without some support in official circles in Washington. Both the Defense and State Department were concerned that in the event of war, Soviet submarines—which regularly patrolled Canadian water during the 1970s—could intercept and sink US ships in the Pacific or in the Arctic.

The *Manhattan* voyages in 1969 and 1970 received global media attention as an alleged challenge to Canadian marine sovereignty claims. The voyages also drew attention to the boundary dispute in the Beaufort Sea with its potential oil. The Trudeau government, however, was saved further embarrassment when the *Manhattan* failed to achieve its navigational objectives. Multi-year ice ridges in McClure Strait at the west end of the Northwest Passage stopped any forward movement and according to the Captain created "real danger" for the tanker and escort. They were then forced to retreat and proceed south through the narrow Prince of Wales Strait, which raised issues about territorial waters.[6] For reasons of both ice and politics, Exxon concluded the tanker option was not a viable and secure supply route for Alaskan oil to US east coast ports.

The second tanker route suffered from similar ice, but dissimilar Cold War political complications. This route proposed passage westward from Prudhoe Bay, swinging north around Barrow, west and then south through the narrow Bering Strait between Siberia and Alaska, then south into the open Pacific to west coast US ports. The first part of the route had difficult and unpredictable wind-driven ice conditions, which made the passage unreliable. But even more serious for Exxon was the narrow Bering Strait with the Soviets' Diomede Island perched in the center. This presented a Cold War threat to the secure passage of this strategic US oil supply. This option was quickly dropped.

Exxon now turned to consider land-based pipeline routes. According to US law, regulators had to consider alternative routes. In this case it meant the Mackenzie Valley route down across Canada. For the Federal Power Commission in Washington, there were three powerful arguments for rejecting this option. First, US officials wished to avoid wherever possible allowing foreign regulators to have any control or influence over US domestic oil production. Secondly, there had been a dispute with Canada over BC gas exports. When gas production had dropped, BC regulators had made up the shortfall by cutting exports to the United States. Thirdly, there did not appear to be any Canadian or American companies prepared to build the line—only ambiguous statements of support from the Ottawa government, which, quite rightly, were not taken seriously by US officials.

The fourth and final option had been Exxon's favorite from the start, but faced technical and financial challenges. The Alyeska pipeline involved an overland route south from Prudhoe Bay, up over the Brooks mountain range, across 800 miles of continuous and discontinuous permafrost, buried or elevated over major Arctic rivers like the Yukon, with ice scour then through the earthquake zone recently active in southern Alaska, to the ice-free port of Valdez, then out by tanker through Prince William Sound to the Pacific for transport to west coast ports. There were huge questions about a geotechnically unprecedented "hot" oil pipeline crossing 800 miles of shallow and deep permafrost. However, with their usual high confidence, the design engineers did not predict any serious problems.

The initial concept was to lay the pipeline on a soil or gravel berm above the permafrost, similar to some Siberian systems. But field tests quickly showed the serious risks of permafrost thaw settlement. The basic problem was the temperature of the oil. It came out of the well at 80 degrees Celsius (176 degrees Fahrenheit) and had to be kept up to 65 degrees Celsius (149 degrees Fahrenheit) to ensure easy flow and avoid viscosity. Tests clearly showed insulation was not adequate and the pipeline had to be elevated sufficiently above the permafrost to avoid heat transfer. A further problem was the fragility of the "Active Layer" on the surface, which melted down a foot or two each summer and froze again in the fall. As a result, construction was limited in the summer to avoid deep ruts in the active layer. The main construction activities were focused in the winter months where −40 degrees Celsius (−40 degrees Fahrenheit) temperatures reduced the productivity of workers and equipment.

The design process was a lengthy one, as scientists had to map out the permafrost conditions and model the heat exchange barriers. This proved to be much more difficult and complicated than expected in terms of both the basic science and the innovative technology needed to counter the permafrost conditions. The geophysical forces had to be comprehended before technical designs could be produced. Discussions with some of the participants made me aware of the unprecedented nature of the engineering challenges as well as the extent to which we were breaking new ground for the profession and for the industry. I also noted that academic scientists were being drawn into adversarial cross-examination in regulatory processes. In these contexts many felt quite uncomfortable, with in some cases their science being used for commercial objectives and therefore not eligible for academic credit, irrespective of its scientific merit for permafrost science.

Research and testing required several years and the costs began to mount. The process would never have succeeded without Exxon's deep pockets.

The initial cost estimates of $900 million exploded to a final capital cost of $8.5 billion or a nearly 1000 percent cost overrun. The final design was unique because of the nature of the Arctic physical conditions. Much of the pipeline route consisted of an elaborate above ground H-structure with the hot oil pipeline resting on a trolley on the cross piece between the two poles. The pipeline was sufficiently elevated that I could walk underneath it, and caribou could migrate past it. The trolley allowed for expansion and contraction of the pipeline in the extreme temperatures. It was also designed as a safety feature for the earthquake zone. Each of the poles was buried deep into the permafrost and contained a self-regulating refrigeration unit to keep the pole permanently bonded to the surrounding permafrost in spite of friction vibrations from Arctic winds. When I visited the construction, I was amazed at the intricate complexity of the design required to meet Arctic conditions. You almost needed an engineering degree to follow some of the explanations.

There were a number of scientific questions to be answered as a prerequisite for the design team for the pipeline. They had to be able to estimate the rate and factors controlling heat transfer into the permafrost, the volume of ice in permafrost soils and the seasonal and yearly migration of permafrost into unfrozen soils, and the impact of extreme winter temperatures on the metallic properties of equipment and performance of workers. In the construction of the pipeline there were particular problems with welding and a major scandal with the doctoring of the inspection x-rays. Overall, from a scientific and engineering point of view, there was a huge increase in our understanding of permafrost science and the necessary engineering solutions. Without the huge capital resources of Exxon, however, the pipeline with its capacity of two million barrels per day would never have been completed. The pipeline remained a controversial aspect of Alaskan life with continuing battles between industry and environmental groups. Today, 40 years later, the pipeline is still operating well but with the decline in oil production the daily throughput is below one-third of its original capacity.

The Mackenzie Valley pipeline debate in Canada

While the oil pipeline debate proceeded in Alaska, the equally intense natural gas pipeline battle proceeded in Canada but with a very different context for permafrost science. The Prudhoe Bay oil reserves were accompanied by equally large gas reserves seeking a market as well as much smaller gas reserves in the Mackenzie Delta areas. While the most cost-effective means of transporting oil was by marine tanker, gas was best transported by high-pressure pipelines. Given these economies, Exxon quickly indicated that the Mackenzie Valley was its preferred route.

The actual route crossed virtually the whole of North America. It ran eastward from Prudhoe Bay through the Arctic National Wildlife Refuge, entering Canada along the north slope of the Yukon, then accessing the

Canadian gas at the Delta before heading south up the Mackenzie Valley and then overland into southern Alberta with one lateral from there to California and the other to Chicago. It would have been the longest and most costly pipeline system in the world. Wall Street and Bay Street (the center of the Canadian financial community) eagerly anticipated the financing premiums.

The Arctic Gas consortium was a huge group of 28 large Canadian and American companies led by Exxon and its Canadian subsidiary Imperial Oil. They sought to bring together all interested parties to avoid competing applications and speed regulatory approvals. However, the project quickly triggered environmental, First Nations and national concerns who argued that the proposed new energy corridor would unduly influence northern affairs in Canada. It would, they argued, erode land sovereignty in the same way the *Manhattan* voyages had allegedly eroded marine sovereignty. The pipeline corridor would link the western Canadian Arctic into Alaska. Unknown to most at this time, it sparked a strong reaction from Peter Lougheed, the premier of Alberta, who wished to capture the energy corridor to the North for Alberta interests.

The Arctic Gas pipeline also created some interesting engineering issues, which helped to expand our scientific knowledge of discontinuous permafrost. The project called for a chilled natural gas pipeline to operate at the ambient temperature of the surrounding permafrost to avoid heat transfer and permafrost melting. But by solving one problem they created another. From the central area of the Mackenzie Valley down into northern Alberta, the permafrost was not continuous but included large patches of unfrozen ground. These unfrozen zones were unmapped. In these unfrozen areas the chilled pipeline would create ice lensing in moist soils, the ice would cause frost heave, upward pressure on the underside of the pipeline, and even fracture the pipeline with a potential explosion. The ice lensing potential was greatest in very moist water-laden soils such as under major rivers where access would be very difficult if repairs were required. Given these scientific challenges the call went out for scientific input from universities, government bodies and consulting companies. Arctic Gas had to understand the science before they could create a design innovative enough to work in conditions unprecedented in pipeline history.

If Arctic Gas made one mistake, it was to underestimate the strategic determination of the Alberta government to protect its own pipeline interests. The government of Peter Lougheed owned Alberta Gas Trunk Line (AGTL), the provincial pipeline system for gas, which was also an initial member of the Arctic Gas Consortium. The President of AGTL, Bob Blair, was a close advisor to Lougheed. Their strategic goal was to have their own company and Alberta be the funnel for northern gas to reach southern markets. This would make Alberta and AGTL a major player in North American energy affairs.

However, Exxon had no interest in anything except a stand-alone system from Prudhoe Bay to Chicago, which they would control. When Blair could not achieve the Alberta strategy he withdrew and formed a separate project, Foothills Pipeline, with West Coast Transmission of Vancouver. Initially it was only for a Mackenzie Valley Pipeline to the Delta but they added later the Alaska Highway system to tap US Prudhoe Bay gas.

Blair was quite an unconventional oil and gas executive. He was a Canadian nationalist who called Foothills the "Maple Leaf Line" to catch the nationalist Canada First mood of the 1970s. He was a prominent Liberal who would later run for the party. He spent considerable time with environmental groups and personally visited many First Nations communities. He frequently appeared in his buckskin jacket. However, he was not given much hope of success by those involved. One of my colleagues summed up his choices: "It was David vs. Goliath. With no sling shot." These qualities made him very controversial in the eyes of his industry colleagues.

Imperial Oil, which played a key role in the Canadian section of the project, was familiar with the Mackenzie Valley. They had been at Norman Wells (the site of a small production facility in the valley) since 1920. Arctic Gas launched a major permafrost research and field testing program in both the Northwest Territories and Alaska. This effort was expensive and failed to produce a conclusive solution for the frost heave problem. After extensive effort, they came forward with their "Shut Off" theory. They proposed to construct a three-metre (ten feet) berm above the potential ice lensing and argued from their field testing that the weight of the berm would create sufficient downward pressure to stop the ice lensing process and protect the pipeline. Some Arctic scientists immediately challenged their pressure calculations but there was no empirical evidence from comparable testing to allow for any definitive challenges.

The frost heave pipeline debate was taken up by the Berger Inquiry in a way that proved significant. He had created a special structure for his Counsel as an independent party free to seek out evidence in the "public interest." Ian Scott, his Counsel, accordingly sought independent research results on frost heave. Carleton University Professor Peter Williams and a graduate student had tested the "Shut Off" theory but their experiments showed ice lensing many times the magnitude reported from the Arctic Gas experiments. According to their results, generating the necessary downward pressures for "Shut Off" would require a berm 10 or more times the size proposed. Given the fragility of the active layer at the sites, constructing such enormous structures would be impossible.[7] When confronted with the new evidence, Arctic Gas defended the accuracy of their experiments. Then, in the last days of the Berger hearings, they made a dramatic announcement: they admitted that there had been a "malfunction" in the test chambers of their equipment and they withdrew their evidence.[8] The National Research Council had discovered the problem when it was using the test facilities for its own experiments. Their three-metre berm was no longer adequate to control the projected frost heave pushing up on the pipeline and their design was technically invalid.

The debate now moved to the National Energy Board (NEB) hearings still underway. Arctic Gas rushed to amend their technical design to manage the increased ice lensing pressures on the integrity of the pipeline. They proposed new insulation as well as heat probes and heat tracing to melt ice forming in the moist soils around the pipeline. These additions now required an electricity source along the pipeline right of way as well as sensitive equipment to turn the heating on or off with the migration of the permafrost. The evidence supporting the validity of this design was theoretical. At the hearings, we called this approach "the electric blanket around the refrigerator" model, but everyone recognized the difficult dilemma Arctic Gas was in with the challenges of discontinuous permafrost and the loss of their technical design. Berger's independent research had found the flaw in the pipeline proposal. When the NEB report came down it concluded, as had Berger, that the Arctic Gas application should be rejected.

Lurking behind the technical considerations was another issue seldom mentioned in the commentaries: Canada's national interest and commercial aspirations. The failure of the Arctic Gas application opened the way for approval of the smaller Canadian alternative, the Foothills Alaska Highway project. It was to be built along the existing highway right-of-way, it had fewer technical, environmental and land claims issues and offered clear economic benefits for the Foothills partners and Peter Lougheed, the premier of Alberta.

Bob Blair was one of the shrewdest political strategists I ever met. In his communications to the Trudeau cabinet he stressed the linkage between the policies behind his pipeline and other Trudeau initiatives including Petro-Canada, the Foreign Investment Review Initiatives, and the Canada Development Corporation. He also liked to point out the Ontario, Alberta and federal equity investments in Syncrude (the oil sands company that was then foreign owned), as well as his public endorsement of the Committee for an Independent Canada. In an era of nationalist enthusiasm, he had built his own credentials.

In the public campaign for nationalist objectives, two large country-wide organizations led the debate. On one side was the Waffle, a radical public ownership wing of the New Democratic Party, and on the other was the Committee for an Independent Canada (CIC), which advocated a more liberal approach through government regulation to reverse the impacts of foreign ownership. While their messaging was different they cooperated privately in both Ottawa and in the intellectual context around Berger in Yellowknife. The CIC had members in the Trudeau cabinet and in the parliamentary caucuses of all three federal parties as well as branches across the country. The Waffle had a powerful following in universities across the country. For both organizations, the oil and gas sector was the worst example of foreign ownership with no major companies in Canadian hands and the northern pipelines the key battle for control. In the media, the *Toronto Star* openly advocated the Walter Gordon nationalist wing of the Liberal Party and the Canadian Broadcasting Corporation was clearly in the nationalist camp.

In the early months of 1977, the pipeline battle reached a zenith of intensity. Arctic Gas allies like Wood Gundy, the stock brokerage firm, complained bitterly to the federal cabinet about media bias. Earle Gray, the head of public affairs for Arctic Gas, later denounced the nationalist coalition against his company, describing them as radical environmentalists, Marxist church groups, separatist First Nations, anti-business and anti-American nationalists.[9] There was certainly some media bias against Arctic Gas. There were strong emotions in the campaign to ensure Canada kept control of its northern resource riches. Linking the disparate groups opposing the Arctic Gas pipeline was the belief that Canada had to have democratically controlled domestic public policy decision-making and that US influence within Canada was too high. While some of this seems exaggerated today, there were times when *Lament for a Nation* (a book by philosopher George Grant that warned of the Americanization of Canada) was discussed as if it forecast the end of Canada.[10] When I was interviewed for a job at a Canadian university, the first question from the university president was: "where do you stand on Grant's *Lament?*" Today this seems so distant. I remember well my good friend Flora MacDonald who in the 1970s was Executive Director of the Committee for an Independent Canada, yet just over a decade later was Canada's Minister of External Affairs under Prime Minister Brian Mulroney (who negotiated a Free Trade Deal with the United States). In understanding the nationalist context of the 1970s, one must accept the exceptional nature of some public attitudes during those years for pipelines and other areas of public policy.

At the Trudeau cabinet table, there were representatives of both the continentalist and the Walter Gordon nationalist wings of the party. When the pipeline debate began in 1969/70, the continentalist wing had the upper hand and so Canada offered Washington the Mackenzie Valley route for the early oil pipeline. But in the years following, the power-center shifted in the cabinet to nationalism, which stopped the US-backed Arctic Gas option. The cabinet, like the country, had come to believe: "The true North Strong and Free" was a battle slogan not just a cliché.

The new regulations

The Mackenzie Valley pipeline issue had a huge impact on regulatory processes in Canada, reflecting a public appetite for accountability and transparency. The Berger Inquiry broke new ground in terms of the precision and comprehensiveness of environmental and social evidence as well as the considerable weighting they received in decision making. The NEB processes were forced to change in the 1980's, by reaching out to stakeholders.

While the NEB had been focused on economic interest, Berger forced much greater consideration of the broader "public interest." Berger also set new standards for public participation with his community hearings, his definition of eligible parties to the process, and his relaxed rules of evidence.

The Berger Inquiry also presented serious challenges to the role of academic science, including from traditional First Nations knowledge. The community hearings brought forth a wealth of observational data that were integrated with conventional science, although not without some tension in issues such as caribou herd numbers. Berger also accorded respect and standing to Non-Government Organizations (NGO's), then known as Public Interest Groups. Berger ended debate over their role with his ruling: "They do not represent the public interest, but it is in the public interest that they be heard." The Berger Inquiry was a watershed in the evolution of regulatory thinking in Canada to a greater extent than has been publicly recognized even in the legal literature on regulation.

Conclusions

This chapter documents the close interaction between Arctic permafrost science, new pipeline technology, and international scientific collaboration, even amidst Cold War tensions. Corporate needs triggered huge investments in basic permafrost research, which greatly increased our knowledge of frost heave and thaw settlement. This has serious implications beyond pipelines given that half of Canada is covered by continuous or discontinuous permafrost. This interaction was achieved by an unlikely partnership of academic, corporate, and government scientists working together for a perceived common good. This combined effort was intense but brief; the funding evaporated with the construction or rejection of pipelines. It showed, as never before, the complex forces which were involved with permafrost and Arctic economic development. The Arctic was a different world for pipelines.

While Alyeska was successfully designed, its cost discouraged further Arctic oil and gas development. The revolutionary design for construction and operation of the pipeline sent shockwaves through the engineering profession. No similar pipeline was built in Russia although there are huge Arctic oil reserves in which Exxon has been involved in developing (it is currently withdrawing from much of its Russian activities in response to American and European Union sanctions). Today with oil at about $50.00 per barrel, it is financially out of the question: revenues could not amortize the capital or operating costs. The technology issues for the Arctic Gas concept are even more negative. The challenges of frost heave have still not been solved. In spite of several efforts, no Mackenzie Valley gas pipeline has been built, particularly because the North American price of gas today is a fraction of what it was in 1980, and huge low cost reserves have been found further south. The failure of projects like Arctic Gas provides a lesson about the constraints that Arctic conditions impose on corporate experiments.

This pipeline case study is an intriguing example of energy collaboration in the Cold War context, in which both sides saw economic benefit. Companies like Shell and Exxon already had links to the Kremlin so this cooperation was not unique, but part of a wider long-term strategy. The Soviets had huge oil and gas reserves in western Serbia but without the capital and technology to develop them. Western oil companies, already experienced in Alaska, and with ample capital and technology, wanted to get access to those reserves. Increased oil production would greatly strengthen Russian oil exports (their principal source of foreign exchange) while responding to Exxon shareholders' demand for greater reserves. Near the end of this era of Cold War cooperation on pipelines, President Ronald Reagan relaxed the US ban on oil and gas technology exports to the Soviet Union, on the advice of the industry. I am not sure if this was Karl Marx's version of socialism and the class struggle.

Notes

1. Robert Page, *Northern Development: The Canadian Dilemma*, (Toronto: McLelland and Stewart, 1986).
2. John Livingston, *The Ontario Naturalist*, 8, December 4, 1970.
3. Peter Williams, *Pipelines and Permafrost: Science in a Cold Climate*, (Ottawa: Carleton University Press, 1986); Shelagh Grant, *Polar Imperative: A History of Arctic Sovereignty in North America*, (Vancouver: Douglas & McIntyre, 2010).
4. Thomas Berger. *Northern Frontier, Northern Homeland: The Report of the Mackenzie Valley Pipeline Inquiry*, vols. 1 and 2, (Ottawa: Minister of Supply and Services Canada, 1977).
5. Roger Hutchinson, *Prophets, Pastors and Public Choices: Canadian Churches and the Mackenzie Valley Pipeline Debate*, (Waterloo: Wilfrid Laurier University Press: 1992).
6. William Smith, *Northwest Passage, the Historic Voyage of the S. S. Manhattan*, (New York: American Heritage Press, 1970).
7. Williams. *Pipelines and Permafrost*.
8. Berger, *Northern Frontier, Northern Homeland*.
9. Earle Gray, *Super Pipe: The Arctic Pipeline—World's Greatest Fiasco?* (Toronto: Griffin House, 1979); Hutchinson. *Prophets, Pastors and Public Choices*.
10. George Grant, *Lament for a Nation: The Defeat of Canadian Nationalism*, (Toronto: McClelland and Stewart, 1965).

8   Cold oil: linking strategic and resource science in the Canadian Arctic

Stephen Bocking

Strategic priorities exerted a decisive influence on Arctic science during the Cold War. But other factors were also important. Among these was interest in natural resources, as the region became increasingly integrated within industrial economies. Oil and gas discoveries in Alaska and northern Canada in the late 1960s sharpened this interest, encouraging (as Robert Page discusses in his chapter) research aimed at solving the challenges involved in extracting and transporting resources while managing impacts on the environment. These new priorities transformed the contexts of Arctic science.

Historians have understandably tended to consider these episodes separately: focusing either on science that linked to Cold War strategic priorities or to resource development and environmental protection. But the relations between these also merit inquiry. How should our understanding of Cold War Arctic science encompass the emergence of resource and environmental priorities? How did scientific activities in the Arctic evolve as priorities shifted from the strategic to the economic? These questions imply a need to link the historiographies of the Cold War and of northern resource economies. By doing so we can enhance our understanding of both strategic science and resource and environmental science, particularly by demonstrating the significance of the Arctic environment and of social, political, and economic issues in shaping these research activities.

That Cold War strategic science exhibited relations with global environmental perspectives has become well-established: study of the transboundary movement of contaminants (through tracking of fallout and other materials, and analysis of atmospheric dynamics), of climate change (through modeling and ice core analysis), and the formation of the view itself of the Earth from space, all testify to these relations. However, this chapter argues for a broader view of the links between strategic and environmental priorities, to encompass those between strategic science and environmental administration (including regulation of industry)—a realm of activity that formed within its own social and institutional space, largely distinct from that of global science. I will begin my discussion by noting how strategic research in the Arctic often incorporated economic priorities; thus, the notion of a distinction between these priorities needs to be reconsidered. I then survey aspects of the research provoked by oil and gas discoveries after the late 1960s: the relative roles of industry, government and other parties, the factors that shaped their involvement, and the research that resulted. This provides the basis for identifying several features shared by strategic and resource and environmental science, including scientific practices, views of the northern environment, attitudes regarding the capabilities of science and technology, and the occupational categories of scientific workers. These features also, I suggest, reflected shared perspectives on how decisions should be made, and by whom, particularly in the context of uncertainty. In sum, while the immediate objectives of strategic and of resource science were distinct, scientific activities in both eras were underpinned by shared assumptions regarding the practice and politics of science.

Strategic-industrial collaboration during the Cold War

During the early Cold War much of the science taking place in the Arctic exhibited both strategic and economic motivations that were linked to similar operational challenges. For example, the properties of permafrost and the engineering challenges it posed were relevant to building both Distant Early Warning (DEW) Line stations and other strategic facilities and to the extraction of oil and gas from beneath the tundra environment. Knowledge of sea ice contributed to both the movement of naval forces and the shipping of natural resources. Accordingly, an array of military and civilian agencies, including the Arctic Research Laboratory in Alaska and the National Research Council of Canada (in collaboration with Imperial Oil) pursued studies of permafrost.[1] Similarly, resource mapping by the Geological Survey of Canada drew on aerial surveys, mapping, and transportation infrastructure that had been justified in terms of strategic priorities.[2] The Polar Continental Shelf Project, which after 1959 served an essential role in supporting research in the Canadian Arctic, was originally motivated by both strategic knowledge needs and by the resource potential of the Arctic seabed.[3] At McGill University in Montreal, geographical research by Kenneth Hare in northern Quebec combined strategic objectives (including planning of the Mid-Canada radar line), and the perfecting of aerial survey techniques that could support resource activities. In return, industry helped Hare and his colleagues establish field research facilities, including a station at Schefferville, the site of a new iron mine.[4]

These activities exemplified the Cold War consensus shared by government and business: of defense and industrial development as dual priorities, both to be achieved through the application of technocratic expertise. Cold War science was applied not just to assuring strategic security, but to pushing (as Rafico Ruiz notes in his chapter) the geophysical boundaries of resource extraction. Such a view was also compatible with the imperative of development through modernization that was part of the global Cold War—the ambition to incorporate "Third World" countries within the Western industrial system.[5] This encompassed an image of the earth as both a knowable space and a resource storehouse; in the Arctic, water, ice and land was no longer a "hostile" wilderness, but a space to be controlled and exploited.[6] This consensus came together within the restricted community of Arctic research, in which scientists like Hare, moving easily between government, academia and industry, demonstrated how a well-connected individual could link strategic and economic considerations. These links also implied a new kind of scientific worker, able to bridge strategic and industrial contexts: specialists that drew from but were distinct from academic science, working in firms that served both the state and corporate capitalism, embedded within the midcentury research complex that combined military, industrial, and academic knowledge production.[7]

Strategic and corporate research in the Arctic also had features specific to this region. One was of exploitation as the primary frame for human relations with the environment. This reflected the long-standing image of the region as a resource frontier—in effect, an economic colony of southern industrial centers. Another was the link drawn (particularly by Canada) between development and sovereignty—that is, the role of industrial activities as a means of asserting a national presence in the north.[8] The distinctive challenges that this environment posed to strategic and industrial activities, and the kinds of information that responses to these conditions were thought to require, constituted a third feature specific to the Arctic. In Canada, these challenges encouraged a specific view of the role of the federal government: to facilitate development by providing services such as resource surveys, aerial photography and mapping, as well as infrastructure and other services that the private sector could not provide on its own—a perspective exemplified in the late 1950s by Prime Minister John Diefenbaker's "Roads to Resources" program.[9] The center-periphery structure of Arctic research was also significant: strategic and industrial initiatives formed similar relations between the Arctic and the rest of North America, in which sites of interest in the North became tied to corporate, military or government centers of decision-making and their associated scientific and technical communities in the south. Through strategic, industrial and scientific activities the Arctic became effectively part of North America. Thus, just as, for example, the DEW Line represented the assimilation of the Arctic into continental surveillance, so did resource development represent the Arctic's assimilation into a continental market, and scientific activities its assimilation within the North American knowledge economy.

A new Arctic oil and gas research economy

As Page describes in his chapter, the discovery in 1968 of oil in Alaska immediately heightened the prospects for oil and gas elsewhere in the Arctic, including Canada. Plans for accelerated exploration, followed by extraction and transport to southern markets by pipeline or tankers, led conservationists to demand that environmental impacts receive careful consideration. The wider emergence during this era of environmentalism, and the perception of the Arctic as a distinctively "fragile" environment, amplified their concerns.[10]

Scientific research formed a major part of debates about northern development, as both proponents and opponents agreed (albeit for different reasons) that more knowledge was needed about the Arctic environment. As a result, discovery of oil and gas provoked not just proposals for resource development, but expanded environmental research efforts. A new research landscape quickly emerged in the Canadian Arctic. Industry attempted to develop the technical basis for resource extraction, while meeting the requirements of an emerging regulatory regime. The Canadian government funded its own research, both to support industry efforts, and to provide the basis for regulatory decisions. On a smaller scale, non-profit organizations based in the south, as well as northern Indigenous agencies, initiated their own research activities. Throughout the 1970s and 1980s research activity rose and fell with interest in resource development, peaking between 1971 and 1974, and again in the early 1980s. Industry and government research efforts were massive: costing millions of dollars, and producing not only individual papers, but multiple volumes of reports. For example, study by the Arctic Gas consortium of the challenges and impacts of its proposed gas pipeline along the Mackenzie Valley was said to have cost between $20 and $30 million over four years, and generated a 41 volume report series. Industry also experimented with novel ways of organizing research: companies collaborated on shared research challenges; and Arctic Gas established the Environmental Protection Board, a short-lived experiment in the corporate support of independent science.[11]

The federal government had initially preferred, after 1968, to let industry take the lead in environmental research relating to oil and gas development. Soon, however, it initiated its own Arctic Land Use Research Program (ALUR), aimed at providing a scientific basis for land use regulations. This was followed in 1971 by a larger "crash" research program, organized by the Task Force on Northern Oil Development (the agency assigned to facilitate federal support for northern development) through its newly-formed Environmental-Social Committee. The effort was intended to operate independently of industry, and focused on practical issues relating to assessing pipeline applications in the Mackenzie Valley corridor. Through this program numerous federal agencies became involved in research, including some, such as the Canadian Wildlife Service and the Fisheries Research Board of Canada, that had traditionally led federal scientific efforts in the Arctic. The Polar Continental Shelf Project was similarly redirected towards supporting studies relating to development. In subsequent years the government developed a series of additional research programs, often in collaboration with industry, that examined the environmental dimensions of oil and gas development, focusing on the knowledge needed to ensure efficient regulatory approvals.[12]

Research by industry was ostensibly guided by regulatory requirements. In 1972 the federal government imposed its Expanded Guidelines for Northern Pipelines, specifying requirements for applicants to produce extensive biological and socioeconomic data regarding their proposals. In practice, however, much of the research agenda relating to oil and gas development was set by industry. Research responded to industry intentions: in the early 1970s plans for a Mackenzie Valley gas pipeline encouraged a focus on terrestrial issues, including the engineering feasibility of a pipeline built atop tundra and its potential impacts on landscapes and wildlife. However, plans formed later that decade for exploration, production and transportation in the Beaufort Sea led to a shift towards studies of the challenges and risks of offshore development. Industry also often served as the dominant partner in joint research programs, providing much of the funding and insisting on setting the agenda.[13]

Much of this research examined issues that had been of concern throughout the Cold War, such as the engineering challenges posed by the Arctic environment. Industry generally used their own studies for operational purposes: to reduce their impacts on the environment, or to overcome the environmental hazards that threatened their projects. But more novel questions also received attention. These included the potential impacts of development on "fragile" northern ecosystems. Scientific activities were framed in relation to specific sites, particularly those considered of special value or vulnerability. By the early 1970s the susceptibility of tundra and permafrost to disturbance had become a special focus of concern to environmentalists.[14] The proposed Mackenzie Valley pipeline raised other issues as well: impacts on streams and on birds and wildlife (including the Porcupine caribou herd and their habitat), as well as effects on beluga whales and other marine mammals. Resources important to northern communities, such as furbearers (the basis for the local trapping industry), also received attention. So did aspects of the environment that were of continental significance, including waterfowl, and endangered bird species such as peregrines, newly perceived as vulnerable to pesticides.[15] Research was also shaped by scientists' disciplinary perspectives, including views of nature in terms of species populations or ecosystems, or in terms of scales only visible through technology—from the very large (through aerial surveys and satellite remote sensing) to the very small (such as trace contaminants).

These research activities reflected the construction of new northern industrial environments. These environments became especially evident, for example, in the novel orientation of ocean science: not only would it focus on strategic priorities such as assuring mobility through or under sea ice, it would also assess the impacts of operations on the marine environment.[16] Thus, the Arctic Ocean became not only a difficult and hazardous environment, but one that was itself vulnerable to hazards. These new environments also encouraged new combinations of expertise. For example, assessments of the impacts of ice-breaking drew on the work of marine mammal biologists, hydrocarbon chemists, underwater acousticians, physical oceanographers, and Inuit hunters.[17] Much of this expertise became organized in relation to objects or aspects of the environment of special concern, such as permafrost, caribou, or sea ice. These environments also became places of tension between different forms of expertise, such as engineering and ecological expertise. These tensions reflected, in part, challenges specific to the Arctic, as projects planned on the basis of knowledge gained in temperate environments encountered unexpected consequences. Engineering expertise accordingly became more readily implicated in controversies than it had in the context of Cold War strategic challenges. A noteworthy instance of this (as Page discusses in his chapter) took place during the Mackenzie Valley Pipeline Inquiry (1974–1977), when frost heave emerged as an unsolved technical challenge to building a pipeline on permafrost, undermining the technical credibility of the scientists involved.[18]

Cold War science in the resource era

Arctic oil and gas developments and the environmental issues they provoked constituted a novel chapter in the history of Arctic science and technology. Research objectives were transformed: instead of enabling strategic operations the aims now were to develop resources and meet environmental regulations. Yet if we examine not just scientific agendas but the practice of science and its justification in relation to political and economic contexts, several features emerge that together exhibit continuities between the eras of strategic science and of resource and environmental science. As oil and gas became a focus of interest, government and industry sought to maintain important features of the consensus that had shaped Arctic science during the early Cold War. These included the definition of the challenges of resource exploitation in the Arctic as technical issues, the formation of a "government-consultant-industrial" complex founded on shared objectives and close cooperation, and a consensus on the appropriate relations between science and politics. The shared assumption that overcoming the challenges of the northern environment demanded collaboration reinforced this consensus.

These continuities were immediately evident in the practices of resource and environmental science. Research techniques that were applied in the 1970s: aerial surveys and remote sensing of terrain, ocean, and ice; radio-tagging of wildlife; and tracing the movement of contaminants in the environment—had originated in response to Cold War demands for surveillance and operational capabilities. Much of this research was focused on the physical hazards and challenges encountered by industry, such as permafrost (particularly in relation to pipelines), and ocean ice, currents, and waves. Research in the Beaufort Sea focused almost entirely on the conduct of drilling operations, including marine engineering and studies of environmental hazards (especially ice) and of the ocean floor. "Environmental forecasting"—predicting the impacts of the environment on industry—became a focus of interest. This demanded the capacity to evaluate the technical feasibility of projects in difficult environments, illustrating how development plans, such as pipelines and offshore development platforms, presented unsolved engineering challenges—continuing the Cold War concern with the collision between technology and the Arctic environment.[19]

In contrast, industry was slower to demonstrate interest in the ecological aspects of northern development. Even in November 1970, as protection of the northern environment was emerging as a significant political issue, no biologists participated in the industry-organized Fifth National Northern Development Conference. And in the early 1970s, as research plans for the Beaufort Sea were being negotiated, only when Environment Canada noted how little was known about its biology and ecology did industry decide to invest not only in studies of the hazards that this environment posed to industry, but how to protect the environment itself from industry.[20] Further, much of the research on environmental protection conducted by industry was in engineering and the physical sciences, as it was argued that knowledge of the physical environment was required to ensure the integrity of drilling and transport systems. In fact, engineers and other technical experts saw environmental hazards and environmental protection as two aspects of the same issue: hazards threatened to damage facilities, which could then cause environmental harm—and conversely, if everything worked properly, then the environment should also be protected.[21] Even as biological aspects of the environment became of greater concern, environmental impact studies often discussed geology and other physical features in more detail, simply because those were the features about which most was known.[22] Overall, therefore, this research extended the Cold War distinction between the physical and biological environmental sciences.[23]

This orientation towards engineering and physical sciences supported a view of the north as a laboratory or proving-ground, in which southern technology could be tested and adapted to northern conditions. Research on environmental hazards was often conducted using what were, in effect, experimental industrial facilities. An early instance of this appeared in 1969, when the Mackenzie Valley Pipeline Research Company set up in Inuvik a 600-meter test pipe to evaluate the feasibility of transporting Alaskan oil down the Mackenzie Valley. Similarly, the voyages in 1969 and 1970 of the *Manhattan* tanker were conceived as a research operation: gathering ice and other navigational data while testing the feasibility of transporting oil by tanker.[24] A few years later, discoveries of natural gas in the Arctic Islands led to plans for a pilot project to move liquefied gas from Melville Island; this was intended to serve as both an industrial project and a study of its technical feasibility. These projects demonstrated how industry viewed research as part of the process of development. The idea was to learn while doing, while defining environmental concerns not as obstacles demanding resolution before development could proceed, but as problems to be solved during the process of development. This expressed what quickly became a dominant discourse of science, development, and the environment: "problem solving" in which industry and governments collaborated in devising efficient solutions. Another term applied to this approach was "phased development": a cautious incrementalism, in which pilot projects functioned, in effect, as controlled experiments.[25] This also reflected how the role of science in development decisions was defined: it would not determine whether projects should go ahead—it was normally assumed that, subject to economic considerations, they would—but how to implement that decision as efficiently as possible, while minimizing adverse effects. This view that science would define the means, but not the ends, had also described its role in relation to Cold War strategic priorities.

A confidence, verging on hubris, in the problem-solving capabilities of science and technology formed another common element in strategic and resource science. Just as science could enable soldiers to fight or the DEW Line to be built, so could it provide the means with which to overcome the challenges facing industrial development. Industry officials often expressed a "can do" attitude: confidence that whatever problems were encountered in the Arctic environment, technology could be developed to solve them. Even the expression of uncertainty was considered "unscientific".[26] Allied to this view was a fascination with engineering virtuosity—a view akin to the "technological sublime" evident in accounts of Cold War technologies. Challenges were not to be considered unresolvable, but merely symptoms of a lack of knowledge, readily corrected through more research. As a result, all parties—industry, government, even some environmentalists—framed these challenges in terms of science. Regulations usually included requirements that industry conduct research—implicitly defining environmental protection as a technical, not a political issue. Advisory processes normally drew on experts from industry, universities, and government, but, particularly in the early years, much less often on northerners or Indigenous Peoples. A similarly technocratic perspective was evident in calls for northern environmental governance to rely not on "political" processes, but on the collective professional judgement of experts.[27]

As previously noted, another feature of strategic science and technology had been the formation of a new occupational category: technicians that could mediate between strategic imperatives, technological systems, and Cold War environments. This occupation also became, in the form of consultants, characteristic of regulatory science in the context of the Arctic petroleum industry. After 1975 the federal government reduced its funding for research, and many of its scientists migrated to the private sector. University scientists also saw their role in research relating to petroleum development decline (although some, such as engineering experts at the University of Alberta, continued to make important contributions).[28] Instead, the new occupation of consultants emerged, and quickly dominated research relating to the Arctic petroleum industry. Northern environmental research became itself an industry, with dozens of consulting firms supported by contracts from government and resource developers. Consultants were considered better able than academic scientists to organize large-scale, continuous, multidisciplinary research efforts. Their task was not just to conduct research, but to manage a close, collaborative relationship between scientists, government, and industry. Consistent with this role, their success was measured not primarily in terms of generating knowledge, but ensuring the efficient and timely approval of development proposals. Large consulting firms developed novel strategies for managing research, forming along the way a new professional community, dedicated to cultivating collaborations between industry and government, through informal working relationships built up over time. Consultants therefore represented another new category of scientific worker—situated at the nexus of science and economic imperatives—that was analogous to the technicians that had linked science and strategic imperatives. Thus, by the 1970s there were at least two broad categories of scientists active in the north, exhibiting divergent priorities and ways of conducting and communicating science: those doing basic research who communicated primarily with other scientists, and consultants who communicated mostly with government and industry.

The formation of this community of consultants reflected the emerging importance in northern research of what Cyrus Mody has referred to as the "excluded middle": scientists who were concerned about social issues (in this case, protection of the northern environment), but who chose to express this not by opposing industrial development, but by working with industry to minimize its impacts.[29] Our understanding of consultants and their role in mediating relations between industry and government can also be enhanced by considering aspects of the historiography of science in the private sector. The organization and direction of research in private industrial labs, as well as business views of society and the environment are particularly relevant. After 1970 a new regime of business and science formed: alongside the privatization of research, as industry gradually displaced government as the larger funder of science relevant to technological innovations, corporations outsourced their research activities, contracting them to other firms.[30] This trend also became evident in Arctic resource science, as firms—rather than building up their own northern environmental expertise—instead contracted these tasks out to consultants. In addition, the concept of "industrial-hazard regimes," as proposed by Christopher Sellers and Joseph Melling to describe the political economy of risk, can provide a framework for understanding the perspective on the relations between the environment and economic activities shared by public and private institutions, that shaped how consultants would create and contest knowledge.[31]

Some aspects of the industrial-hazards regime formed through the working relationships linking consultants, industry and government were specific to the Arctic. One was the imperative for collaboration that the distinctive environmental challenges facing industry seemed to imply—just as the challenges of this "hostile" environment had elicited a particular organizational response earlier in the Cold War. Another was the predominance of scientists and technicians who came from elsewhere. Northern scientific activity stimulated by oil and gas prospects was largely the work of transients from the south, with few staying in the north long enough to learn, in the words of one experienced scientist, "northern manners." At least initially, consultants often had little experience in the North, or in specific northern environments, and more experienced scientists often dismissed their work, considering it to be of low quality, based on an inadequate understanding of northern conditions, too focused on the client's priorities rather than the public interest, and not subjected to peer review.[32]

Finally, scientific studies that supported petroleum development or assessed their environmental impacts implied a particular perspective on Arctic landscapes, akin to that imposed during the most active period of Cold War strategic science. Aerial surveys and inventories of resources or environments that might be affected by development exhibited a decontextualized, impersonal understanding of landscapes that privileged the visual over other senses. This was a space devoid of human presence—in effect, a natural rather than the cultural and social space familiar to northerners (as Andrew Stuhl notes in his chapter). The northern environment was also redefined as an homogenous space, characterized by its physical factors, such as ice, permafrost, and cold, that presented similar engineering challenges everywhere in the region. This perspective was reflected, for example, in the establishment of engineering test facilities intended to perfect practices and technologies that would be appropriate across the region. This view also obscured those perspectives that were from the region itself, including those of Indigenous people—in effect, marginalizing the diverse and complex ways in which people experience and attach meaning to the landscape. They therefore expressed the cognitive and ideological work involved in preparing these territories to be incorporated within larger systems of power—whether these were defined in terms of strategic surveillance or resources.

Forming industrial impact research

Cold War strategy and industrial development also converged on a strong interest in avoiding public controversy. The development of Cold War surveillance and operational capabilities was, of course, never considered a matter for public debate. Similarly, as they made plans for oil and gas development, industry and the Canadian government collaborated closely, restricting decision-making to a small circle of senior officials, and imposing secrecy whenever possible. This discretion was considered necessary not because of security concerns, but because northern development was intrinsically uncertain, and it was felt controversy would only add more uncertainty. And in an industrial context, uncertainties were expensive.

Many uncertainties were unavoidable. Some stemmed from nature: the difficulties involved in assessing the size and accessibility of oil and gas deposits, and the challenges involved in working in an environment of permafrost, ice, and cold. There were also economic uncertainties: the pricing decisions of the Organization of Petroleum Exporting Countries, the vagaries of the North American energy market, the unpredictable costs of northern facilities. The regulatory process introduced additional uncertainties. Many within industry and government saw the Mackenzie Valley Pipeline Inquiry as a leading example of the production of unwelcome uncertainty. As I have noted, the Arctic Gas consortium had spent millions on studies while preparing its application to build a Mackenzie Valley pipeline; however, the Inquiry then recommended against approval of this project—making this an expensive experience that industry did not want to repeat.[33] Public criticism of research conducted in support of development constituted another source of uncertainty. According to some academic scientists and environmentalists, these studies often failed to provide the information necessary for credible regulatory decisions—one prominent scientist described environmental impact assessments as a "boondoggle."[34] By undermining confidence in project decisions, these criticisms contributed to controversy and uncertainty. Even the prospect of public discussion of development projects would, it was feared, introduce uncertainties. As one scholar of environmental administration noted, if development projects became subject to a debate between northerners and the federal government, that could "upset" a decision-making process that, ideally, it was felt, should be outside of politics.[35]

A chief aim of engineering and environmental research, accordingly, was to reduce uncertainties by understanding environmental hazards and developing technologies and practices to overcome them. However, reducing uncertainties that stemmed from the regulatory process and enhancing predictability for development called for other strategies. One was close collaboration between industry and government in the study of northern environmental challenges. This occurred at venues such as the Canadian Northern Pipeline Research Conference of 1972, as well as Arctic Environmental Workshops sponsored by the Arctic Petroleum Operators Association and the Canadian Petroleum Association.

This collaboration became especially evident in the formation by scientists from industry, government, and consulting firms of a formal process for planning and conducting environmental impact research. This effort began in the early 1970s, and continued over several years and numerous workshops.[36] Among their objectives was to focus research on the issues that were of most concern, and so most likely to pose obstacles to decisions to approve a project. The aim was to ensure that regulatory decisions would be less readily challenged, or even subjected to debate—with the experience of the public examination of science that took place during the Mackenzie Valley Pipeline Inquiry serving as an example of what should be avoided.

The ultimate goal was to build a sense of trust and confidence, to enable resolution of difficult scientific and regulatory issues, thereby achieving a predictable, stable working environment. Industry-government collaboration became formalized in processes of environmental administration that tended to exclude other parties and their concerns.[37] With knowledge considered an essential political resource in northern environmental controversies, secrecy became a significant feature of these arrangements. Industry and government restricted access to information about their development intentions, such as plans for offshore drilling.[38]

By the early 1980s an environmental assessment process had been developed that could be described as a "fairly sophisticated planning review mechanism".[39] It had two key components: a procedure for setting research priorities in relation to the specific characteristics and requirements of proposed development projects, and formal criteria for evaluating the quality and relevance of the results of research. These would be implemented through several steps: identification of those aspects of the local or regional environment that were of particular interest or concern (to be designated as VECs—Valued Ecosystem Components); construction of hypotheses regarding how these VECs might be affected by the development or activity (these hypotheses functioned, in effect, as summaries of current knowledge relevant to the VECs); and design of research to test these hypotheses. By following these steps, the research process would not only generate new knowledge, but would set out the reasoning through which regulatory decisions could be justified. Further, by defining these decisions, including project approvals, as the outcome of a scientific process (as attested to by the procedures of constructing hypotheses and then testing them), the regulatory process would be able to draw on the aura of objectivity and authority customarily attached to science.[40] Strict definition of VECs and hypotheses would help ensure control over this process—in effect, bringing consultants' field studies closer to the ideal (and the authority) of the controlled laboratory.[41]

This formalized approach to impact science also constructed a boundary between the impact assessment process and the political issues raised by development. As was realized from the outset, regulatory decisions and project approvals would be less readily challenged if the science of environmental impact assessment was kept separate from political and ethical issues—a process of boundary formation observed throughout the development of environmental and health regulation over the last fifty years.[42] Accordingly, "difficult" issues, such as the intangible values of wilderness and of Indigenous landscapes, and the panoply of northern political and social issues that had been aired through the Mackenzie Valley Pipeline Inquiry, land claims negotiations and other processes, were ruled out of bounds, as they could not be incorporated into the framework of ecosystem components and hypotheses. Instead, knowledge would be constructed that would represent not the diverse perspectives evident in northern politics, but a single point of view, with conflicts resolved not through public hearings, but through more informal and less public workshops. These forums were consistent with viewing development not as a matter of public debate, but as a series of problems to be solved. In this way, the environment in which development occurred was defined in terms of features relevant to imperatives flowing from the south, rendered legible, and brought within the realm of rational administration. This perspective exemplified an orientation akin to that seen throughout Cold War strategic science, in which decisions were kept within a restricted circle of agencies that shared a consensus on objectives and how they should be achieved.

This approach also discouraged involvement by the public, including Indigenous communities. Processes of public engagement were often ill-suited to the distinctive social and cultural realities of northern communities, and unable to encourage appropriate discussion of peoples' concerns about development, or draw from their knowledge of local environments.[43] Indigenous groups criticized these scientific activities accordingly: for example, the Inuit Tapirisat perceived a closed partnership between government and industry, that facilitated mechanistic decision-making, without sufficient involvement by other parties. Other observers suggested that the environmental impact assessment process had turned science into a tool of the petroleum industry.[44]

While the formation of environmental impact research enacted similarities between Cold War strategic science and the science of petroleum exploitation, there were also significant differences. One was that, unlike Cold War strategic objectives, the view of resource exploitation as a technical issue—a series of problems to be solved—was not uncontested, but was itself the product of debate and controversy. Northern development posed several novel tensions. To the challenge posed by the Arctic environment was added the emergence of environmental concerns as a political issue (including concerns specific to impacts on a "fragile" Arctic environment), as well as the particular social and political contexts of northern development, especially land claims negotiations and movement towards Indigenous self-government. These tensions had become particularly evident in the course of the Mackenzie Valley Pipeline Inquiry. They also became evident in views of science. Critics of the industry, whether on environmental or other grounds (and including some scientists) were less confident about the capabilities of science and technology. They doubted that it could solve the many conflicts—not just environmental, but social and cultural—that resource exploitation could provoke and that had received so much attention during the Mackenzie Inquiry and other public processes.[45] They also dismissed as unscientific what they called the "engineering approach," including optimism regarding unsolved problems. Instead, these critics saw no contradiction between asserting one's expert status and acknowledging the scientific and technical uncertainties of northern development.[46] In the early 1970s some ecologists and activists even urged a moratorium on resource development, arguing that science needed to "catch up" to development plans, by building a full understanding of arctic ecosystems before projects could be approved. They also resisted other aspects of resource development and impact science: the assumption that a boundary could be constructed between politics and expertise; the close relations between government and industry, including a tendency towards secrecy, that in their view impeded a balanced approach to development and environmental protection; and the view of the landscape that obscured the presence of communities and long-standing relations between humans and other species.[47] These critiques, it must be noted, were voiced largely outside the community of environmental assessment professionals.

Conclusions

Cold War Arctic research was not only dedicated to strategic imperatives. By the late 1960s much of the urgency attached to these imperatives had dissipated. Instead, oil and gas exploration and production became a focus of research activity. This shift generated several novel features in Arctic science. Most obviously, its objectives changed, to now include attention to the potential environmental impacts of resource exploitation.

Yet there were several continuities in research practices and organization between these periods. Environmental assessment science perpetuated the emphasis in Arctic research on the physical sciences (while only with difficulty accommodating biological phenomena), as well as the view of science as essentially a problem-solving activity, seeking consensus while avoiding the clash of opposing views, particularly by asserting a boundary between science and politics. Similar occupational categories—technicians or consultants—were also maintained. Overall, therefore, ideas about research that had been formed in response to strategic imperatives continued to be influential. Certain kinds of research questions also remained important, including those relating to engineering and mobility in this environment, as did particular ways of answering these questions, and of relating them to decisions and to the political contexts of northern resource development.

But the extension into an era of environmental and Indigenous politics of forms of research nurtured during the early Cold War had its own consequences. One lesson of resource controversies during the 1970s was that science alone could not resolve development issues: contrasting values and attitudes regarding the environment and the future of the North were also at stake. But as this chapter has discussed, the subsequent history of environmental science attempted to deny that lesson. The outcome was a continuing tension between attempts by government and industry, aided by consulting scientists, to impose narrow definitions of knowledge and the public interest, and assertions by northerners, especially Indigenous People, of a role in northern knowledge generation and decision-making. On the one hand, scientific research priorities became more specific, reflecting the emergence of the VEC concept and impact hypotheses, and the overall demand for quantitative, defensible results. On the other hand, public debate regarding resource developments focused on their broader political or social implications. This tension was also expressed through different ways of organizing science and its application to resource development: through a corporatist collaboration between government and industry, mediated by consulting scientists, or through processes that blended expert evidence and the views of citizens (a hallmark of the Mackenzie Valley Pipeline Inquiry). The former represented an approach inherited from Cold War strategic science, but which now collided with more recent modes of relating knowledge to politics, as advanced by environmentalists (including some scientists), and, especially, by Indigenous communities and agencies. Hence, while the science of environmental assessment was pushed in one direction, the politics of northern development moved in another, rendering the science increasingly unable to resolve controversies, even as environmental and Indigenous concerns became more important.

Notes

1. Andrew Stuhl, *Unfreezing the Arctic: Science, Colonialism, and the Transformation of Inuit Lands*, (Chicago: University of Chicago Press, 2016).
2. Morris Zaslow, *Reading the Rocks: The Story of the Geological Survey of Canada, 1842–1972*, (Toronto: Macmillan, 1975).
3. Richard C. Powell, "Science, Sovereignty and Nation: Canada and the Legacy of the International Geophysical Year, 1957–1958," *Journal of Historical Geography*, 2008, 34: 618–638.
4. Stephen Bocking, "A Disciplined Geography: Aviation, Science, and the Cold War in Northern Canada, 1945–1960," *Technology and Culture*, 2009, 50: 320–345.
5. Gilbert Rist, *The History of Development: From Western Origins to Global Faith*, 4th Ed., (New York: Zed Books, 2014).
6. Simone Turchetti and Peder Roberts, "Introduction: Knowing the Enemy, Knowing the Earth," in: Simone Turchetti and Peder Roberts, eds., *The Surveillance Imperative: Geosciences during the Cold War and Beyond*, (New York: Palgrave, 2014), 1–19.
7. Matthew Farish and P. Whitney Lackenbauer, "Western Electric Turns North: Technicians and the Transformation of the Cold War Arctic," in: Stephen Bocking and Brad Martin, eds., *Ice Blink: Navigating Northern Environmental History*, (University of Calgary Press, 2017), 261–292.
8. Shelagh Grant, *Polar Imperative: A History of Arctic Sovereignty in North America*, (Vancouver: Douglas & McIntyre, 2010).
9. Powell, "Science, Sovereignty and Nation"; K. J. Rea, *The Political Economy of the Canadian North: An Interpretation of the Course of Development in the Northern Territories of Canada to the Early 1960s*, (Toronto: University of Toronto Press, 1968).
10. Stephen Bocking, "Science and Spaces in the Northern Environment," *Environmental History*, 2007, 12: 867–894.
11. Robert Page, *Northern Development: The Canadian Dilemma*, (Toronto: McClelland and Stewart, 1986).

12. Edgar J. Dosman, *The National Interest: The Politics of Northern Development 1968–75,* (Toronto: McClelland and Stewart, 1975); Page, *Northern Development*; Advisory Committee on Northern Development, *Government Activities in the North, 1972* (Ottawa, 1973).
13. Douglas Pimlott, Dougald Brown and Kenneth Sam, eds., *Oil Under the Ice,* (Ottawa: Canadian Arctic Resources Committee, 1976).
14. Stuhl, *Unfreezing the Arctic.*
15. Slaney & Co: "Interim Report: 1972 Environmental Field Program, Taglu - Richards Island, Mackenzie Delta" (1973); Williams Brothers Canada Ltd.: "Outline of Wildlife Studies, Northwest Project, June 1971" (1971); L. C. Bliss, "Oil and the Ecology of the Arctic," *Transactions of the Royal Society of Canada,* 1970, ser. 4, 8: 361–372.
16. J. W. Langford, "Marine Science, Technology, and the Arctic: Some Questions and Guidelines for the Federal Government," in Edgar J. Dosman, ed., *The Arctic in Question,* (Toronto: Oxford University Press, 1976), 163–192.
17. David L. VanderZwaag and Cynthia Lamson, eds., *The Challenge of Arctic Shipping: Science, Environmental Assessment, and Human Values,* (Montreal & Kingston: McGill-Queen's University Press, 1990).
18. Page, *Northern Development,* 155–164.
19. Allen R. Milne, "Arctic Marine Environmental Studies," in: S. J. Jones, *Proceedings of the Eighth Arctic Environmental Workshop* (Calgary: Arctic Institute of North America, 1979), 28–43; K. Sam, J. MacPherson and G. Thompson: "Polar Gas and the International Biological Programme," *Nature Canada,* 1977, 6(3): 2–3, 33–37; Richard M. Hill, "Petroleum Pipelines and the Arctic Environment," *North,* 1971, 18(3): 1–5.
20. Douglas Pimlott, "Offshore Drilling in the Canadian Arctic: Elements of a Case History," in: Canadian Arctic Resources Committee, *Mackenzie Delta: Priorities & Alternatives* (Ottawa; CARC, 1976); Douglas Pimlott, "The Hazardous Search for Oil and Gas in Arctic Waters," *Nature Canada,* 1974, 3(4): 20–28.
21. Eastern Arctic Marine Environmental Studies, "Eastern Arctic Marine Environmental Studies (EAMES): Outline of a Program," (Ottawa: Department of Indian Affairs and Northern Development, 1977); Milne, "Arctic Marine Environmental Studies," 33–35.
22. Richard G. B. Brown, "Seabirds and Environmental Assessment" in: VanderZwaag and Lamson, *The Challenge of Arctic Shipping,* 97.
23. Ronald E. Doel, "Constituting the Postwar Earth Sciences: The Military's Influence on the Environmental Sciences in the USA after 1945," *Social Studies of Science,* 2003, 33(5): 635–666.
24. Langford, "Marine Science, Technology, and the Arctic," 168.
25. R. A. Hemstock, "Industry and the Arctic Environment," *Transactions of the Royal Society of Canada,* 1970, Ser. 4, vol. 8: 387–392; W. Speller, "Approaches to Environmental Problem-Solving" (1980); A. R. Lucas and E. B. Peterson, "Northern Land Use Law and Policy Development: 1972–78 and the Future," in: Robert F. Keith and Janet B. Wright, eds., *Northern Transitions: Second National Workshop on People, Resources and the Environment North of 60°,* Volume II (Ottawa: Canadian Arctic Resources Committee, 1978), 63–95: Brian D. Smiley: "Marine Mammals and Ice-Breakers", in: VanderZwaag and Lamson, eds., *The Challenge of Arctic Shipping,* 70–71.
26. Brian Lewis Campbell, *Disputes Among Experts: A Sociological Case Study of the Debate over Biology in the Mackenzie Valley Pipeline Inquiry,* (Ph.D. thesis, McMaster University, 1983), 288.
27. J. C. Ritchie et al.: "Report of Second Task Force to Mackenzie Delta" (1970); Lucas and Peterson, "Northern Land Use Law," 81.

28. Norbert R. Morgenstern, "The University View" (1981), 44–51.
29. Cyrus C. M. Mody, "Square Scientists and the Excluded Middle," *Centaurus*, 2017, 59: 58–71.
30. Christine Meisner Rosen and Christopher C. Sellers, "The Nature of the Firm: Towards an Ecocultural History of Business," *Business History Review*, 1999, 73(4): 577–600; Philip Mirowski and Esther-Mirjam Sent, "The Commercialization of Science and the Response of STS," in: Edward J. Hackett et al., eds., *The Handbook of Science and Technology Studies*, (Cambridge: The MIT Press, 2008), 658–659; Rebecca Lave, Philip Mirowski and Samuel Randalls, "Introduction: STS and Neoliberal Science," *Social Studies of Science*, 2010, 40(5): 659–675. The relations between business and science have received surprisingly little attention from Cold War historians; for example, no chapters focus on this relation in either J. R. McNeill and Corinna R. Unger, eds., *Environmental Histories of the Cold War*, (New York: Cambridge University Press, 2010), or Naomi Oreskes and John Krige, eds., *Science and Technology in the Global Cold War*, (Cambridge: MIT Press, 2014).
31. Christopher Sellers and Joseph Melling, "Towards a Transnational Industrial-Hazard History: Charting the Circulation of Workplace Dangers, Debates and Expertise," *British Journal of the History of Science*, 2012, 45(3): 401–424.
32. "Northern manners" quote from Peter Larkin, "Science and the North: An Essay on Aspirations," in: Robert F. Keith and Janet B. Wright, eds., *Northern Transitions: Second National Workshop on People, Resources and the Environment North of 60°*, Volume II (Ottawa: Canadian Arctic Resources Committee, 1978), 122; Douglas Pimlott, Dougald Brown and Kenneth Sam, *Oil Under the Ice*, (Ottawa: Canadian Arctic Resources Committee, 1976), 65; Brown: "Seabirds and Environmental Assessment," 96.
33. Page, *Northern Development*, 261; Edgar J. Dosman, "Arctic Seas: Environmental Policy and Natural Resource Development," in: O. P. Dwivedi, ed., *Resources and the Environment: Policy Perspectives for Canada* (Toronto: McClelland and Stewart, 1980), 198–215; J. C. Ritchie, "Northern Fiction—Northern Homage," *Arctic*, 1978, 31(2): 69–74.
34. D. W. Schindler, "The Impact Statement Boondoggle," *Science*, 1976, 192 (4239): 509; see also G. Rempel and N. B. Snow, "Offshore Environmental Studies Panel," in: *Proceedings of the Seventh Arctic Environmental Workshop* (1978), 74–85.
35. Gordon Beanlands, "Introduction," in VanderZwaag and Lamson, eds., *The Challenge of Arctic Shipping*, x.
36. Gordon E. Beanlands and Peter N. Duinker, *An Ecological Framework for Environmental Impact Assessment in Canada*, (Halifax: Dalhousie University, 1983).
37. Robert R. Everitt and Nicholas C. Sonntag, "Environmental Issue Resolution in Canadian Frontier Oil and Gas Exploration," in: Environment Canada, *Audit and Evaluation in Environmental Assessment and Management: Canadian and International Experience. Volume I: Commissioned Research* (Ottawa, 1987), 46–64.
38. In one instance in the late 1970s, Inuit were given only three days to review 40 volumes of technical reports relating to petroleum development in Lancaster Sound: Milton M. R. Freeman, "Traditional Land Users as a Legitimate Source of Environmental Expertise," in: James Nelson, et al., eds., *The Canadian National Parks: Today and Tomorrow Conference II: Ten Years Later*, (Waterloo: University of Waterloo, 1979), 366.
39. Beanlands, "Introduction," ix.

40. Robert R. Everitt, D. A. Birdsall and David Stone, "The Beaufort Environmental Monitoring Program," in: Reg Lang, ed., *Integrated Approaches to Resource Planning and Management*, (Calgary: University of Calgary Press, 1986), 251–266.
41. Robert E. Kohler, *Landscapes and Labscapes: Exploring the Lab-Field Border in Biology*, (Chicago: University of Chicago Press, 2002).
42. Stephen Bocking, *Nature's Experts: Science, Politics, and the Environment*, (New Brunswick, NJ: Rutgers University Press, 2004), 19–21.
43. Karen Roberts, ed., *Circumpolar Aboriginal People and Co-management Practice: Current Issues in Co-management and Environmental Assessment*, (Calgary: Arctic Institute of North America / Joint Secretariat—Inuvialuit Renewable Resource Committees, 1996); Carly A. Dokis, *Where the Rivers Meet: Pipelines, Participatory Resource Management, and Aboriginal-State Relations in the Northwest Territories*, (Vancouver: UBC Press, 2015).
44. Terry Fenge and Brad Wylynko, "Hydrocarbon-related Research in Northern Canada: The Case of the Environmental Studies Revolving Funds," in: Milton M. R. Freeman and Charles W. Slaughter, eds., *Arctic Science Policy and Development; Proceedings: A UNESCO-MAB International Conference*, (Washington: United States Man and the Biosphere Program, 1986); Canadian Arctic Resources Committee, *The Environmental Studies Revolving Funds and Offshore Oil and Gas: Workshop Report*, (CARC, 1982). Workshop, May/June 1982" (1982).
45. Thomas Berger, Commissioner of the Mackenzie Valley Pipeline Inquiry, described the "feeling that, with enough studies and reports, and once enough evidence is accumulated, somehow all will be well" as an "assumption that does not hold in the North"; Thomas Berger, *Northern Frontier, Northern Homeland: The Report of the Mackenzie Valley Pipeline Inquiry*, Volume 1 (Ottawa: Minister of Supply and Services Canada, 1977), xi.
46. Campbell, *Disputes Among Experts*, 289.
47. James Woodford, *The Violated Vision: The Rape of Canada's North*, (Toronto: McClelland and Stewart, 1972); John B. Sprague, "Aquatic Resources in the Canadian North: Knowledge, Dangers and Research Needs," in: Douglas H. Pimlott, Kitson M. Vincent and Christine E. McKnight, eds., *Arctic Alternatives: A National Workshop on People, Resources and the Environment North of 60°*, (Ottawa: Canadian Arctic Resources Committee, 1973), 168–189; Pimlott, "The Hazardous Search".

9 Icebergs in Iowa: Saudi dreams, Antarctic hydrologics and the production of Cold War environmental knowledge

Rafico Ruiz

On October 1st, 1977, the *Des Moines Register* reported on the moving of a large piece of ice from Portage Lake in Alaska to the somewhat unlikely destination of Ames, Iowa. Deemed "the most expensive ice cube in the world," the 2,500-pound block of ice, roughly 6 by 5 by 4 feet in size, would travel wrapped in insulation and dry ice via helicopter, plane, and truck to Ames at a cost of \$8,500.[1] This elaborate logistical feat, requiring divers from the U.S. Arctic Naval Research Laboratory to select the berg, an aerial harness to secure it, a refrigerated truck to move it, and, finally, a walk-in refrigerator at the Iowa State Memorial Union to store it, was meant to convey the feasibility of approaching glacial ices of various kinds as sources of potable water. The ultimate, landlocked, destination of the berg, Ames and Iowa State University, is explained by the fact it was to be the site of the First International Conference on Iceberg Utilization, to be held from October 2nd to 6th. This was to be a meeting of some two hundred scientists (glaciologists, marine biologists, amongst other natural scientists), engineers, as well as corporate and political representatives, coming from eighteen nations, all of whom were committed to the idea of treating Antarctic icebergs as a source of fresh water for the world. Largely sponsored by Prince Mohammed Al-Faisal, the official in charge of Saudi Arabia's desalination program at the time, and the National Science Foundation, the Iceberg Utilization conference displayed how getting icebergs on the move was an international concern that motivated Saudi Arabian royalty, American military engineers, and Rand Corporation physicists, amongst a number of other actors on a spectrum of interests running from the capitalist to the techno-utopian.[2]

My aim in this chapter is to demonstrate how this event contributed to the formation of water-based resource imaginaries in the late 1970s that were used to justify incursions into unconventional sites of resource extraction such as Antarctica by "dry" nations such as Saudi Arabia. By extending recent work between environmental and media history, specifically through their articulation of the varied materialities of water, I attempt to characterize the archival fonds of the Iceberg Utilization meeting, from keynote addresses to scientific papers, and photographs of ice sculptures to Alaskan glacial ice, as a form of evidentiary claim in the emergence of schemes on the part of water "poor" regions to achieve viable forms of water provision.[3] The conference's active production of a body of documentary evidence underpinning this need for resource expansion gave rise to a hydrologic that could prove, justify, and ultimately allow for the harvesting of Antarctic icebergs as a global fresh water supply. In this optic, Antarctica became a remote if central node in Cold War scientific networks aiming to push the geophysical boundaries of legal resource extraction. These networks were shaped by geopolitical interests pursuing diverse forms of what I call environmental mediation—the mobilization of such documentary evidence in order to legitimize an emergent natural resource through epistemological if speculative means. It is valuable to examine the ways in which these individual and institutional actors attempted to expand the global resource base by justifying incursions into emerging resource frontiers through the construction of such resource logics—what I call here an Antarctic "hydrologic"—as they shed light on the specific, situated, and material epistemologies of extraction that underlie their production.

Saudi waters

For participants of the Iceberg Utilization conference, it would have seemed as though ice was omnipresent—a motivational substance that would decorate their plenary hall, cool their drinks, and provide instructional, hands-on evidence of the possibilities inherent in its very materiality. As Richard Cameron, the then head of the NSF's Glacial program, claimed, the objective of the iceberg demonstration project, as made manifest in the block of ice that landed in Ames in early October, was to "provide the conference participants with the opportunity to examine the nature, properties, and viability of suggested utilization techniques of icebergs through visual perception."[4] In other words, in order to test their evidence-seeking theories on how to go about towing, cutting, and melting Antarctic ice, they could benefit from having the real thing on hand, and, in the process, consolidate the conference's claim to being a privileged site of new forms of water-related knowledge production.

While the moving of the Alaskan berg might have been a press-savvy way to accomplish this aim, it also fit within the conference's original logic of harnessing icebergs as a potential resource from a systems viewpoint, notably with reference to Ludwig von Bertalanffy's general system theory and its categorization of natural phenomena into complex if coherently interacting and functional systems.[5] In the research proposal submitted by Dr. Abdo Husseiny, professor of nuclear engineering at Iowa State, Principal Investigator of the proposal, and a close friend of Prince Al-Faisal's, the conference was meant to be a strong commitment to iceberg harvesting as a global resource practice, and one that could present itself as a viable and cost-effective alternative to desalination.[6] While the conference proceedings would ultimately be published by Pergamon Press, they were originally promised to *Desalination, the international journal on the science and technology of desalting and water purification*, along with the simultaneous publication of Arabic-language proceedings in the *Al-Ahram* newspaper of Cairo.[7] While Prince Al-Faisal was the official in charge of Saudi Arabia's desalination program, he saw it as costly and inefficient, requiring vast amounts of capital to build desalination plants, and equally vast amounts of energy to run them. By initiating and funding the conference, he was looking towards a future wherein Saudi Arabia would no longer have its non-renewable oil wealth, and would be facing recurring and increasing demands for fresh water. As a result, one of the guiding, water-derived logics of the conference was found in its establishment of a reliable and institutionally-sanctioned form of knowledge surrounding iceberg utilization, one that came about through its differentiation from established forms of water provision such as desalination. While this was accomplished through conventional means such as conference papers on topics as diverse as the tracking and selection of icebergs and their impact on weather modification, keynote addresses by the likes of glaciologist Henri Bader, the former Director of the US Army Snow, Ice, and Permafrost Research Institute, and the collecting and circulating of their findings to interested groups of scientists, investors, governments, and the press at large, this consolidation of the conference as the site of a nascent hydrologic also lay in its active production of what could be thought of as a "second" water: a host of media, from photographs to films to algorithmic models, destined to document and present the emergent, elemental medium of iceberg water as a legitimate form of resource provision.[8] It was indeed a form of environmental mediation provided by scientific networks that were attempting to establish and consolidate icebergs as part of the uncertain and very much contested Antarctic resource frontier. At the time, Antarctica, as it is today, was considered a common, non-commercial scientific region to be shared by the signatories of the 1959 Antarctic Treaty. While the US Geological Survey claimed in the mid-1970s that the continent might hold roughly 45 billion barrels of oil, icebergs as potential water resources could more easily circumvent the perceived strictures of the Antarctic Treaty and also possibly lay the foundations for future transnational forms of Antarctic resource exploitation.[9] As Mia Bennet notes, "the enclosure and operationalization of a resource frontier depends on two processes: the initial and ongoing representation of a region as frontier and its actual enclosure and exploitation."[10] This process of environmental mediation, as I argue in this section, was predicated on fostering polar scientific networks of knowledge exchange that could legitimize Antarctic icebergs as extractable natural resources through such future-oriented media, those photographs, films and algorithmic models noted above, of the continent's "second" water. That, in essence, could produce an extraction-based "hydrologic" specific to the continent itself.

While the principal conference organizers, Iowa State and Abdo Husseiny, Prince Al-Faisal, and the NSF's Glacial program, wanted this knowledge to circulate, with the conference garnering an astounding degree of press coverage running the gamut from the sensationalistic to the pun-seeking, it was also a form of knowledge marked by substantial proprietary concerns: the backdrop to the conference was the open question of to whom the Saudi Arabian iceberg contract would go. To whoever could figure out how to move an iceberg from near an ice sheet in Antarctica to the coast of Saudi Arabia, with minimal melting and at an efficient speed and cost, there would likely be a considerable financial windfall—with speculation at the time placing the figure at a hundred million US dollars.[11] Roughly one year prior to the Iceberg Utilization conference, a group of engineers at the *Centre d'Informatique Commerciale et Economique et de Recherche Operationelle* (CICERO), based in Paris, produced a study, commissioned by Al-Faisal, on the feasibility of towing an eighty five million ton berg from Antarctica to Saudi Arabia. CICERO concluded that they could deliver a cubic meter of potable water at half the cost of water produced via desalination.[12] They presented their findings at a symposium held in Paris in June, 1976, with Dr. Wilford Weeks, a glaciologist with the U.S. Cold Regions Research Laboratory, and the co-author of an influential 1973 study on iceberg towing and harvesting, claiming that "[t]hose guys are hard at it."[13] With Prince Al-Faisal and Saudi Arabia essentially the sole "customers" for the delivery of an iceberg, it was claimed by faculty representatives at Iowa State that CICERO colluded to force Al-Faisal into signing a contract prior to the commencement of the conference.[14] While the allegations were never proven, they signal the degree to which commercial concerns were monitoring and actively influencing the outcomes of their deliberations. On the American side, the Washington-based firm International Six, represented by former White House advisor General Brent Scowcroft, and largely specializing in oil production in the Middle East, attended and partly financed the Iceberg Utilization conference.[15] More broadly, certain factions of the American press viewed the Saudi incursion into Antarctica as a threatening precedent, with an editorialist in the *Wall Street Journal* calling for "the UN [to] hold a 'law of the iceberg' conference forthwith, before it is too late," and the *Washington Star* claiming that "the Saudi Arabians owe the world an environmental impact statement:" "We are almost, but not quite, speechless at the sheer exuberance of the transplanted-iceberg scheme. It does, however, make us somewhat uneasy wondering what other fantastic futuristic notions are being contemplated in the thinktanks of Jidda."[16] While such editorializing no doubt also echoed Saudi Arabia's role in the 1973 oil crisis, these three competing discursive claims also demonstrate to what extent the active production of proprietary, and even popular, iceberg water-related knowledge in the mid-1970s was a fractious meeting of commercialism and resource nationalism, and their attendant influence on a "scientific" meeting intending to address the prospect of iceberg utilization should not be underestimated.

If French and American commercial concerns were the most prominent players in trying to vie for the Saudi contract, overlaying this commercializing dynamic were also the possibly competing interests of the NSF. As Dwayne Anderson, a Chief Scientist at the NSF, wrote in a letter to Abdo Husseiny a few months prior to the start of the conference: "We are confident that the completion of this project will be a substantial contribution to the U.S. effort in Antarctica."[17] Anderson's claiming of the conference's iceberg-related knowledge production for the Americans' consolidating influence in Antarctica displays to what extent the conference was indeed a prime generator of documentary, scientifically-informed claims to an emergent aquatic resource that, and this is the crucial point, required a form of scientific mediation prior to being economically sanctioned, deemed technically feasible, and even conceptually grounded as a legitimate practice of resource extraction. Relatedly, in a preparatory meeting prior to the start of the conference, Glenn Murphy, a professor of engineering at Iowa State, describes convening with NSF representatives in Washington and relates how they requested that the conference "must present both [sic] pros and cons of iceberg utilization so [sic] NSF doesn't appear in the position of an advocate of either position," with the NSF going so far as to request a letter from Iowa State summarizing the manner in which the conference would also include an adversarial approach to the possibility of iceberg harvesting.[18] As such, this process of scientific and, ultimately, environmental mediation was actively shaped by agencies such as the NSF and the underlying commercial concerns of Al-Faisal and Saudi Arabia. The result was the emergence of a form of consensus between scientific processes of legitimation and the commercial forecasting of imminent feasibility, both of which informed the programming and ultimate shape the conference would take, and placed it firmly within the overlapping institutional parameters of the United States' well-established Cold War military-industrial-academic complex.[19]

The argument I would like to emphasize in this first section is that when it comes to apprehending the human and non-human co-production of environmental phenomena, here water and ice, particularly in this context of the making of a speculative claim on an emergent "natural" resource, it is of fundamental importance to note how the active production and shaping of the documentary media that these speculative claims rely on is a relationship building process that is ultimately tied to critical understandings of environmental mediation. This form of mediation could equally comprise the competing discursive claims being made at the Iceberg Utilization conference itself, but also the ways in which, to circle back to the materialities of ice that were present at Iowa State in early October, 1977, tangible, testable, and present ice could newly become, through this process of environmental mediation, a sort of foundation material in the conference's production of a spectacle of the feasible (and consumable) water-commodity form.

Icebergs in Iowa

On Monday, October 3, the conference's "dinner speaker" was Dr. H. Guyford Stever, a former science advisor to Presidents Nixon and Ford, as well as a former director of the NSF. Addressing "New Patterns of Cooperation in International Science and Technology," Stever projected the iceberg utilization scheme at the intersection of what he saw as the complimentary sciences of oceanography, meteorology, glaciology, solid mechanics, and fluid mechanics, with the added influence of new technologies of ship design and at-sea-operations in both hot and cold weather climates. He deemed it a promising example of "Big Science," but one that could retain the sense of entrepreneurial risk and venture that characterized past technological developments related to space exploration, global air transportation, or international shipping, and that, in his assessment, had been lost to more cautious approaches to ambitious technical systems.[20] It was precisely in the substance of freshwater, "an essential element for all of humankind," that Stever read the possibilities of international cooperation that the iceberg utilization project might hold: "The program has another feature, perhaps its best. More fresh water is needed by many nations, rich and poor, developed and developing, all over the world. So [sic] goal of the undertaking is clear and good. The project should bring out the best of our science and technology, and reflect our respect for human needs everywhere."[21] In his address, Stever succinctly placed icebergs within the domain of a conception of the polar sciences that would operate on the basis of shared need, and the new awareness of "global problems" that had emerged, in his estimation, following the Club of Rome's influential study *The Limits to Growth*, published in 1972. It was to a new "internationalism" that science and technology now had to turn, and iceberg water had an important contribution to make, particularly to exchanges surrounding the development of arid agriculture. It is noteworthy that Stever's address served as the opening contribution to the published conference proceedings, as it signaled to what degree his framing of the iceberg utilization scheme echoed the quasi post-geopolitical approach the conference's organizing committee was attempting to espouse. Freshwater was precisely the substance within which a "global issue" could reside, and in its lack, one that could align with the other "shocking events" that Stever saw as characterizing the world's novel sensitization to "global problems": "the oil embargo of 1973; the blackouts of electric power in New York City; and the grain shortage brought on by one or two poor harvests in the Soviet Union and the People's Republic of China."[22] In his estimation, science and technology would be the bedrock for an emergent form of international cooperation whose "exchange medium" would be "knowledge and know-how, rather than product."[23] It was precisely to this bolstering of a truly "international" form of scientific knowledge production that the Iceberg Utilization conference was seeking to contribute, and Stever's somewhat editorializing address served to highlight how a novel sense of common human needs could supplant entrenched geopolitical interests. While it remained in tension with the commercializing and nationalistic concerns noted above, it added another dimension to the process of environmental mediation the conference was constructing, notably, as I will outline below, in its contributions to conceptions of the polar sciences as media of exchange that could, in the element of water, look beyond the entrenched positions of the Cold War. "The Iceberg Utilization Project represents a field," as Stever claimed, "where in doing 'good to others' we help many countries, rich and poor, including ours, which are badly in need of fresh water."[24]

In his "Preface" to the published proceedings, A.A. Husseiny, one of the conference's principal organizers, indirectly situated the geopolitical locales across which the iceberg harvesting scheme had emerged: "In a land of plenty, people have started to realize that a time of scarcity is approaching."[25] While Stever's internationalism implied an outward-looking western lens through which to apprehend this emergent polar resource, Husseiny wrote from the position of a western nation, here the United States, and also a western technocratic and scientific establishment, his "people," that had tasked itself with the responsibility of monitoring the earth's resources (from energy to fresh water), its demographic growth, rises in standards of living, rates of industrialization, and their attendant effects on ecological systems (noting in particular the spread of drought and deserts in some regions of the world). Icebergs could figure here as long-standing objects of scientific knowledge production, having come under examination since the early twentieth century as threats to navigation, and later in the century, to oil drilling platforms. Husseiny saw this as a propitious time to utilize the iceberg-related knowledge that had accrued, and recognized how it had indeed served as a form of "available information" that spurred and sustained international interest in the scheme from the outset.[26] While Husseiny and the other conference organizers sought to project the conference as an inclusive "research forum" that could bring together industry, consultant-based research, and scholarly knowledge production, it nonetheless built upon an existing narrative surrounding iceberg harvesting that had largely claimed it as an emergent, transnational resource practice that to some degree countered the Saudis' proprietary claims around Antarctic ice.[27]

The two most prominent studies that provided the original legitimacy and impetus for the 1977 conference were those produced by Wilford F. Weeks of the U.S. Army Cold Regions Research Laboratory and William J. Campbell of the U.S. Geological Survey, as well as another conducted by established Rand Corporation physicists John L. Hult and Neill Ostrander, with both reports published in 1973.[28] Each study undertook detailed calculations that made iceberg-derived irrigation water a seemingly convincing cost-competitive source compared to existing forms of water provision, notably through desalination. They also emphasized how the technical feasibility of the scheme was entering a phase in which humans could begin to account for the energy requirements needed to move such massive weights— in other words, they made the case that an iceberg could be towed if there was a country, institution, corporation, or individual willing to project such a scheme into a scientifically actionable present. The role of the two studies was addressed at the Iceberg Utilization conference by the well-known glaciologist Henri Bader, who saw in them a certain evidence of engineering hubris, noting their glaring omissions when it came to considering the glaci-ological and oceanographic dimensions of the scheme. Bader, in his keynote address, mapped out a competing genealogy, situating what he saw as the "modest" plan to tow an iceberg from the western flank of Antarctica to the Chilean coast that had originated from the Snow, Ice, and Permafrost Research Establishment of U.S. Army Corps of Engineers:

> The original modest concept was to sail south from Valparaiso, Chile, with a couple of tug boats, locate a good looking iceberg, attach cables, tow it into the Humbolt current and keep it on course while it floated up along the Chilean coast. There it would be nudged into a convenient cove. A fabric curtain would then be strung across the mouth of the cove. The promoter would install pumps and pipelines and establish a hopefully prosperous farming community.[29]

Bader, in his indictment of the 1973 studies, noted the prevailing tendency to take "nice, simple innovations, and by a process of complication and expansion, [convert] them into monstrous constructs"—with Weeks and Campbell serving as the "expansionists," and Hult and Ostrander as the "complicators."[30] Thus, for Bader, a proposal whose scale was that of a "regional affair" became a "global proposition."[31] In this competing narra-tive of environmental mediation, Bader notes how none of the governments of the territories covered by the two studies (largely the United States for Hult and Ostrander, while Weeks and Campbell focused on the potential benefits that could accrue to Chile and Australia) reacted to the proposed form of water provisioning, and so "the iceberg project seemed destined for early filing in the archives" until the government of Saudi Arabia decided to commission CICERO, the aforementioned French engineering consultancy, to pursue the idea of delivering icebergs to their coast.[32]

Bader's keynote address consistently hit a skeptical tone, particularly with regards to the current state of scientific knowledge surrounding iceberg behav-ior across different currents and water temperatures, and would have served as a sobering counterpoint to Stever's calls for cooperative efforts surround-ing icebergs as sources of a common resource provisioning. For Bader, the polar sciences were indeed "media of exchange" in the sense that they either gave objective scientific authority to the scheme or withheld it. Much like the emergence of oceanography in the 1950s and its approach to the ocean-as-frontier, when it came to projecting Antarctica as a resource horizon, Cold War polar scientific networks, though diverse and divergent even within the western sphere of influence, sought to establish Antarctic icebergs as legiti-mate sources of freshwater for those nations (and corporations) with the eco-nomic and technological wherewithal to undertake the task. As Al-Faisal put it in an episode of "Dimension 5," a state-affairs television program which aired on *WOI-TV* in central Iowa around the time of the conference: "The iceberg is really a derelict—whoever captures it owns it."[33] It was this process of "capture" that the Iceberg Utilization conference was attempting to define and project. Thus, by distinguishing between iceberg-related knowledge production pertinent to the domain of engineering as opposed to that of relevance to glaciology and oceanography, Bader established a hierarchy of knowledge that divided the scheme into its immediate concern of securing and moving a berg, as opposed to coming to know how icebergs would behave in the ocean.

In one of Bader's final comments of his address, he noted how in going from the "regional" to the "global" too quickly, iceberg harvesting schemes ran the risk of taking on a geo-political scale that, in his estimation, was detrimental to what was needed: a form of knowledge specific to iceberg behavior. While Bader's narrative of environmental mediation elides the proprietary and capital-intensive stakes behind the Iceberg Utilization conference, it nonetheless serves to highlight how the scheme's process of internationalization was a contentious projection of multiple and overlapping spheres of interest. For those invested in the iceberg harvesting scheme in the late 1970s, the production of an "appropriate" trans-national scientific knowledge was predicated on the ability to balance the possibly incommensurate domains of "snow and ice physics and mechanics" and the global need for fresh water.[34] In the closing moments of that same episode of "Dimension 5," Al-Faisal and Richard Cameron, the aforementioned head of the NSF's Glacial program, disagreed over the exclusive nature of the 1959 Antarctic Treaty. While Al-Faisal saw in icebergs "the future development of a world resource" that would respond to a pressing need for freshwater, he also told Cameron that they were not of course only talking about icebergs in Antarctica, but oil as well, and noted that international regulation had to keep pace with the continent's forecasted resource horizon. Cameron responded by rhetorically asking what might lie beyond 1989 and the conclusion of the thirty-year Treaty: "If nothing were found of value, the Treaty would go along. If metals are found, there would be some difficulties." Al-Faisal had the last word in pointing out that thirteen nations could not make claims on a world resource that could benefit the globe's one hundred and forty-four other nations.[35] While this is a far cry from the post-geopolitical approach the conference attempted to espouse, it signals how Antarctica as a real and projected resource frontier could be a malleable tool for renegotiating Cold War conceptions of the polar sciences as both, in Stever's estimation, cooperative and instrumental media of exchange.

Conclusions: Saudi dreams

In the lead up to the 1977 conference, Al-Faisal, who was, as much of the contemporaneous American press coverage emphasized, to some degree a cultural and economic Americanist (for instance, he was educated in the United States, attending the Menlo Park Business School), resigned from his position as head of the Saudi Saline Water Conversion Corp.[36] In the mid-seventies, prior to the Iceberg Utilization conference, he founded Iceberg

Transportation Co. International Ltd., an entity that sought to consolidate his efforts to pursue iceberg harvesting and towing as not only viable forms of water provision, but also as a means of transforming the climate and vegetation in certain coastal regions of Saudi Arabia.[37] This ambitious scheme was one that the broader Saudi government apparatus was to some degree skeptical of. In an interview with the *Des Moines Register* that appeared on October 3, 1977, when asked what the Saudi Arabian government made of the scheme, Al-Faisal responded: "They're saying 'Show us,' he said with a smile."[38] This progressive marginalization of Al-Faisal from his duties as Saudi Arabia's official in charge of desalination hints at how iceberg water was really being mediated into being as an emergent resource—it had to be decoupled from desalination as a competing water technology, and in a sense made into a discursive reality and a potential materiality that would be made to address Saudi Arabia's resource future.

In an interview with the *Christian Science Monitor* from November 1975, Al-Faisal claimed that "What we are trying to do in our most advanced projects, is to skip the current era of technology, which in most countries means nuclear development. But we are not ordering nuclear reactors. We feel we must go beyond that in order to be equal with the rest of the world. But we will share all our accomplishments with the world."[39] While the comment was made in relation to Saudi plans surrounding alternative energy sources, notably those related to fusion and solar technologies, it echoed Al-Faisal's commitment to treating icebergs as a resource of potential that required suspension of belief in current forms of resource provision. While this may seem like the natural, goal-oriented mode of operation of the natural resources sector, that is, to continually expand the base of resources upon which it can draw, Faisal's fashioning of iceberg water was rather exceptional in its turning towards an uncertain future of untested resource provision, with the relatively lawless oceans surrounding Antarctica serving as a testing ground for the continent's emergence as a legitimate resource frontier in its own right. What Faisal was after was a credible solution to Saudi Arabia's chronic and forecasted water shortages. As such the 1977 Iceberg Utilization conference figured as a strategic event that could actively generate versions of "icebergs" that were simultaneously embodying a time-consciousness of a future without oil, as well as the very potential of a nascent commercial hydrologic—that is, a water resource system that could be controlled from its resource base to its delivery site.

There were competing schemes, notably on the part of two aforementioned Rand Corporation physicists, John Hult and Neill Ostrander, who advocated for the scaling up of iceberg harvesting through the use of then present-day energy technologies, as well as the easing of legal forms of ownership around water resource rights.[40] But what is of interest in Al-Faisal's role in treating icebergs as a discursively and materially emergent natural resource is how we can track that treatment as a form of progressive production of media-derived evidence for its own potential as a future water resource. In other words, what I am suggesting we can perform is an inter-rogation of how historical epistemologies of extraction get made, and how they are intimately tied to contextual sites of knowledge production that are themselves forms of speculation—in this case, the forms of Cold War polar scientific knowledge exchange that were created by the Iceberg Utilization meeting, and that led to practices of environmental mediation that would foster the consolidation of Antarctica as a legitimate resource frontier. As such, the scale of the epistemology of extraction I am pursuing holds the pos-sibility of being intensely individual and local, with the Iceberg Utilization conference itself serving as an unusual if symbolic marker of how real and imagined resource frontiers can be created and sustained, and what sorts of relationships they establish with the array of documentary media that legit-imate them in the first place. And yet this limited and local scale of an epis-temology of extraction tied to the Iceberg Utilization meeting is also very much in tension with the ways in which iceberg harvesting schemes relied on broader networks of communication between nation states and major commercial concerns. As we know, Antarctica remains, amidst seemingly constant contest, a continent devoted to scientific research with the aim of serving the world as a whole. In the late 1970s, the Cold War polar scientific networks of western nations were mobilized to forecast where and how a new source of freshwater might emerge. It was left to the somewhat unlikely and marginal figure of a Saudi prince, looking to a future without oil, to dream up the horizon of an Antarctic resource frontier through enticing if speculative hydrologics tailor-made for arid regions, from Australia to California. That such hydrologics have turned away from the polar regions is equally the result of the geopolitical destinies and technological hubris of the past half century. Yet the environmental knowledge that Antarctic icebergs yielded across these Cold War polar scientific networks made the case for broadening what could count as a resource, whom it might benefit, and how it could be extracted. By contrast, icebergs today, as the recent calving of the Larsen C ice shelf made clear, are no longer real or imagined resources full of potentially lucrative freshwater—they represent global sea level rise. While the Cold War's planet was infused with the cumulating effects of global warming, it has taken the real and imagined fate of ice-bergs to make manifest how one resource dream can become the world's uncertain hydrological future.

Notes

1. Otto Knauth, "Iceberg Bound for Ames – At Last," *Des Moines Register*, October 1 1977; IC Des Moines Register "World's Most Expensive Ice Cube?" 1977 1/77.
2. For a genealogy of these competing attempts at global icebergs harvesting, see Rafico Ruiz "Media Environments: Icebergs/Screens/History," in *Journal of Northern Studies* (Special Section: Northern Environmental History), 2015, 9(1): 33–50.

3. See, by way of example, Mark Carey, "The History of Ice: How Glaciers Became an Endangered Species," in *Environmental History*, 2007, 12(3): 497–527; Julie Cruikshank, *Do Glaciers Listen? Local Knowledge, Colonial Encounters, and Social Imagination* (Vancouver: University of British Columbia Press, 2005); Paul Edwards, *A Vast Machine: Computer Models, Climate Data, and the Politics of Global Warming* (Cambridge, Mass.: MIT Press, 2010); Stefan Helmreich, "Nature/Culture/Seawater," in *American Anthropologist*, 2010, 113(1): 132–144; Melody Jue, "Proteus and the Digital: Scalar Transformations of Seawater's Materiality in Ocean Animations," in *Animation: An Interdisciplinary Journal*, 2014, 9(2): 245–260; Max Ritts and John Shiga, "Military Cetology," in *Environmental Humanities*, 2016, 8(2): 196–214; and Astrida Neimanis, *Bodies of Water: Posthuman Feminist Phenomenology* (London: Bloomsbury, 2017).
4. IC Iowa State Media Releases 1977, 1/91.
5. See Ludwig von Bertalanffy, *General System Theory: Foundations, Development, Applications*. New York: George Brazilier, 1968.
6. The idiosyncratic location of the Iceberg Utilization conference is only partly explained by the personal connection between Husseiny and Al-Faisal. In various conference-related materials, Husseiny and the other organizers also emphasized the relatively convenient location of Ames, Iowa, particularly for participants travelling from the East and West coasts of the United States.
7. IC Research Proposal to NSF 1976 2/20.
8. See A. A. Husseiny, ed., *Iceberg Utilization: Proceedings of the First International Conference and Workshops on Iceberg Utilization for Fresh Water Production, Weather Modification and Other Applications held at Iowa State University, Ames, Iowa, USA, October 2–6 1977* (New York: Pergamon Press, 1977).
9. Young Kim, "The Real Cold War," *Harvard International Review*, Summer 1994, 16(3): 56–57.
10. Mia M. Bennett, "Discursive, Material, Vertical and Extensive Dimensions of Post-Cold War Arctic Resource Extraction," *Polar Geography*, 2016, 39(4): 261. See also Adrian Howkins, "The Significance of the Frontier in Antarctic History: How the US West has Shaped the Geopolitics of the Far South," in *The Polar Journal*, 2013, 3(1): 9–30.
11. Robert Hullihan, "Of Icebergs at Iowa State," *Des Moines Register*, n.d.; IC News Clipping 1969-1977 1/81.
12. "Saudi Arabia Commission Iceberg Study," Special to *The New York Times*, 2 November 1976; IC News Clipping 1969–1977, 1/81.
13. News clipping AP article Thurs., 16 June 1977; IC 1/76; Robert Garrett, "- have water" [title cut off], *New York Daily News*, 7 May 1977; IC News Clipping, 1969–1977 1/81.
14. Robert Hullihan, "Of Icebergs at Iowa State," *Des Moines Register*, n.d.
15. Norman Sandler, UPI, "Former White House Aide Wants to Manage Iceberg Affair," *Ames Daily Tribune*, Fri, 7 October 1977.
16. "The $80 Million Ice Cube," *Wall Street Journal*, 22 October 1976; IC Iceberg News Clippings 1975-76 1/80; "Thinking Big – Very Big," *The Washington Star*, Editorial, 20 October 1976.
17. Letter from Dwayne Anderson, Chief Scientist, NSF, to Abdo Husseiny, 19 August 1977; IC Division of Polar Programs – NSF 1977 2/24.
18. IC Iceberg Committee Meeting Minutes 1977 2/18.
19. See Stuart A. Leslie, *The Cold War and American Science: The Military-Industrial-Academic Complex at MIT and Stanford* (New York: Columbia University Press, 1993).

20. H. Guyford Stever, "New Patterns of Cooperation in International Science and Technology," in *Iceberg Utilization: Proceedings of the First International Conference and Workshops on Iceberg Utilization for Fresh Water Production, Weather Modification and Other Applications held at Iowa State University, Ames, Iowa, USA, October 2-6, 1977*, ed. Abdo Husseiny (New York: Pergamon Press, 1977), 2.
21. Stever, "New Patterns," 2.
22. Stever, "New Patterns," 3-4.
23. Stever, "New Patterns," 7.
24. Stever, "New Patterns," 7.
25. Abdo Husseiny, "Preface," in *Iceberg Utilization*, XI.
26. Husseiny, "Preface," XI.
27. Husseiny, "Preface," XI.
28. For a more detailed assessment of these reports see Ruiz, "Icebergs on Screens."
29. Henri Bader, "A Critical Look at the Iceberg Project," in *Iceberg Utilization*, 35.
30. Bader, "Critical Look," 35, 36.
31. Bader, "Critical Look," 36.
32. Bader, "Critical Look," 36.
33. Dimension 5. "Iceberg Utilization." October 1977, *WOI-TV*.
34. Bader, "Critical Look," 43.
35. Dimension 5. "Iceberg Utilization." October 1977, *WOI-TV*.
36. For Al-Faisal's educational background: IC Prince Al-Faisal's schedule 1977 2/3; Associated Press, "A Saudi Who Sought Iceberg Quits Water," *The New York Times*, Sunday, 10 July 1977; IC News Clippings 1969–1977 1/81.
37. "Can you Lead an Iceberg to the Desert?" *Christian Science Monitor*, 21 November 1975; IC Iceberg News Clippings 1975-76 1/80.
38. Eileen Ogintz, "Ames Gives Its Iceberg Warm Greeting," *Des Moines Register*, 3 October 1977; IC Press Clippings Book 1 TD353 I53 1977b book 1 c.1.
39. "Can you Lead an Iceberg to the Desert?" *Christian Science Monitor*, November 21 1975; IC Iceberg News Clippings 1975-76 1/80.
40. See John Hult and Neil Ostrander, "Antarctic Icebergs as a Global Fresh Water Resource," R-1225-National Science Foundation, 3 October 1975 (Santa Monica: Rand Corporation, 1975); IC Water Rights and Assessments (Hult) 1974 2/25; In the report, Hult and Ostrander claimed that the "adoption of the proposed water rights and assessment principles providing unrestricted, nondiscriminatory access to water anywhere would greatly facilitate the initiation and evolution of the use of water for greatest world-wide economic benefit" (13). With regards to the logistics of Antarctic iceberg harvesting as a resource practice, they proposed a global auction-based approach with the appropriate bergs identified and rendered claimable by the Earth Resources Technology Satellite system.

10 Science and Indigenous knowledge in land claims settlements: negotiating the Inuvialuit Final Agreement, 1977–1978

Andrew Stuhl

Land claim settlements in Arctic North America constitute a transformative episode in circumpolar history. Over the last third of the twentieth century, Indigenous Peoples and national governments sought to resolve title to lands that had never been ceded.[1] By the end of the 1980s, negotiations among these parties produced the Alaska Native Claims Settlement Act in 1971, the James Bay and Northern Quebec Agreements in 1975, and the Inuvialuit Final Agreement in 1984, while several other treaties regarding Arctic territories in Canada reached advanced stages of deliberation. In Arctic Canada, the focus of this chapter, the Crown and Indigenous Peoples reterritorialized over 3.7 million square kilometers (1.4 million square miles) of land into seven different settlement areas and Nunavut between 1975 and 2008 (Table 10.1). However, land claim settlements were more than real estate transactions. In many cases, Indigenous signatories received bundles of specified rights, defined in the language of provisions and protected by enabling legislation. Consequently, a new legal basis emerged for structures of governance that included Indigenous representation alongside territorial, provincial, and national officials.[2] Because of their implications for social policy, these land claim settlements in Canada are called "comprehensive land claim agreements" or "modern treaties" to distinguish them from the fifty-six treaties signed between 1701 and 1923.[3] Negotiations in Alaska and Greenland, while playing out differently than in Canada, similarly reshaped domains such as health care, education, social development, economic development, hunting, fishing, and trapping, environmental protection, and mineral extraction.[4]

The production of knowledge has long been recognized as crucial to the implementation of land claim settlements and, thus, to their enduring social significance. In Arctic Canada, many land claim settlements created cooperative management boards with responsibilities for renewable resources, non-renewable resources, and land use planning (Table 10.1). Anthropologists, political scientists, conservation biologists, and Indigenous scholars have treated attempts to integrate science and Indigenous knowledge (or Traditional Ecological Knowledge) by Indigenous and non-Indigenous representatives on these boards as indices of progress on societal goals, like effective environmental management, respect for cultural traditions, and decolonization.[5]

Table 10.1 Land claims settlements in Arctic Canada, 1975–2008*

Settlement	Year	Indigenous Lands/ Settlement Region (Km² Of Land & Water)	Cash Transfer (Cad)**	Co-Management/ Self Government
James Bay and Northern Quebec Agreement[a]	1975	14,000/1,058,929	$225 m	Yes/No
Inuvialuit Final Agreement	1984	90,600/435,000	$152 m	Yes/Negotiated Later
Gwich'in Comprehensive Land Claim Agreement	1992	22,422/90,379[b]	$132 m	Yes/Negotiated Later
Nunavut Land Claims Agreement	1993	350,000/1,900,000	$1,900 m	Yes/Creation of Nunavut Territory
Vuntut Gwich' in First Nation Final Agreement	1993	7,744/129,499[c]	$19.1 m	Yes/Negotiated Later
Labrador Inuit Land Claims Agreement	2005	15,800/75,520[d]	$250 m	Yes/Yes
Nunavik Inuit Land Claim Agreement	2008	5,100/16,375[e]	$54.8 m[f]	Yes/No
TOTAL		**505,666/3,705,702**	**$2.72 bil**	

* compiled through reference to settlement documents (legislation and treaties)
** dollar values are consistent with the year the agreement was signed
a includes the 1978 Northeastern Quebec Agreement
b includes Gwich'in Settlement Area and the Primary and Secondary Use Areas in the Yukon Territory
c size of traditional Vuntut lands as identified by Yukon Bureau of Statistics
d does not include 44,030 km² of sea rights held by Labrador Inuit
e includes designation of Tongat Mountains National Park (10,000 km²)
f   does not include implementation funding of $40.1 mil over 10 years

In this chapter, I demonstrate how land claims settlements can serve as a site for critical historical inquiry. Behind each land claim settlement sits a complex process of proposal, negotiation, agreement, and ratification that unfolded over time. Actors from a range of backgrounds worked out final terms and conditions by invoking diverse knowledge systems, crafting provisions, and weighing the importance of those provisions against desired outcomes, like Indigenous self-determination or economic growth.[6] In the context of scholarship on the Cold War Arctic, including that presented in this volume, histories of land claims settlements are significant because they present another set of relationships among knowledge, Indigenous Peoples, government, and the scientific community. These relationships both complement and contrast those found in other sites for Cold War Arctic research.

My case study will be the negotiations carried out between 1977 and 1978 for what came to be called the Inuvialuit Final Agreement of 1984.[7] This case demonstrates the potential for broadening historical understanding of the Arctic during the Cold War, while exposing the diverse roots of contemporary circumpolar affairs. In pursuit of land claim settlements in Arctic Canada, scientists and Indigenous leaders documented knowledge about Indigenous land use and occupancy across Inuit homelands. The Inuit Land Use and Occupancy Project, carried out between 1973 and 1976, described hunting, fishing, trapping, and traveling activities of more than 170 Inuit across 34 towns and villages in Arctic Canada. This empirical data became the foundation of the Inuvialuit Final Agreement and the Nunavut Land Claim Agreement, precisely because the information was not readily available in existing governmental records. Moreover, the collaboration among Indigenous political organizations, the non-Indigenous scientists they hired to synthesize data, and the Indigenous fieldworkers and residents who collected and provided it became a model for other land claims settlements in Arctic Canada and Indigenous political organizations charged with land claim implementation.[8] These interactions between scientists and Indigenous political leaders, as well as the research areas they advanced, point to the Cold War as a defining period in the institutionalization of Indigenous perspectives on land and livelihood in Arctic Canada.

In 1973, the Department of Indian Affairs and Northern Development recognized Aboriginal title for the first time. However, it would not halt non-renewable resource development in the territorial north while negotiating land settlements. Indigenous associations felt pressured to reach agreements before industrial activity altered their homelands, especially given their unease over the Department's ability to protect the environment and renewable resources. As a result, Indigenous political leaders and scientists concerned about environmental degradation began to view land claims negotiations as a mechanism for achieving forms of land management. The Inuvialuit hired scientists to help delineate arguments about the biological productivity of the tundra and the economic value of surface and subsurface resources in order to support proposals for provisions such as wildlife management boards, impact assessment boards, and wilderness designations. Unlike information on land use and occupancy, this data was available in scholarly and governmental communities, namely conservation organizations and federal agencies with regulatory authority over natural resources.[9] Inuvialuit and consultants convinced federal negotiators of their arguments by articulating the value of hydrocarbon deposits and wildlife habitat through the lenses of Indigenous cultural values and Canada's national interests. In these ways, land claims negotiations constituted a space for the collection of Arctic knowledge during the Cold War distinct but not separate from the military, industry, universities, and non-governmental associations.

The case of the Inuvialuit Final Agreement

Land claim settlements in Arctic Canada reflect historical forces, including changes in federal treaty policies as well as their distinctive geographies and cultures.[10] What follows is an examination of some of the negotiations relating to one modern treaty in northern Canada, the Inuvialuit Final Agreement (IFA). Proposed in May 1977, the IFA was the first comprehensive land claim agreement north of 60°N to reach an Agreement-in-Principle (October 1978) in Canada and also the first to be finalized (June 1984). Accordingly, it defined what kinds of provisions could be included in land claims settlements. While federal policy today largely prescribes the areas open to negotiation and provides Crown representatives their bargaining positions, this was not the case in the 1970s. When the federal government received the Inuvialuit proposal in May 1977, it had settled only one other comprehensive land claim: the James Bay Agreement in 1975. But as federal land claims negotiators pointed out at the time, negotiations for the territorial north were free from the federal-provincial relations that had constrained their position with Cree and Inuit in northern Quebec. Federal land claims negotiators thus viewed negotiations for the IFA, as well as those with representatives of the Dene in the Northwest Territories and Inuit outside of the western Arctic, as presenting a range of options for land tenure that were not available in the James Bay settlement.[11] Negotiators for northern Indigenous political organizations therefore had to develop convincing arguments for the provisions they wanted federal officials to accept, thus highlighting the role of knowledge production in the Arctic land claim settlement process. The negotiations for the IFA then were in some ways a precedent for negotiations on six other comprehensive land claims regarding Arctic territories that were signed after 1984.

The archival trail of the Inuvialuit Final Agreement is also more readily available than other land claims settlements. In *Taimani: At that Time,* Inuvialuit authors narrated the process of their land claim agreement, describing how Inuvialuk men and women developed bargaining positions in consultation with Inuvialuit living across the Mackenzie Delta and Beaufort Sea coast and then negotiated with the federal government. Indigenous communities in other regions of Canada, Arctic Alaska, and Greenland have created similar documents.[12] In contrast, the primary documents required for analysis of the Canadian government's roles in Arctic land claims settlements have not been available until recently, given thirty-year embargoes on sensitive documents.[13] Heretofore, the historical treatment of the federal government's roles and responsibilities in northern modern treaties has been based on private collections and personal recollections, by those intimately involved as consultants to Indigenous political groups or those part of the federal team.[14] While extraordinarily insightful, these histories are understandably incomplete.

In the analysis that follows, I draw on files from the Department of Indian Affairs and Northern Development's Office of Native Claims—which staffed negotiation teams—and that Office's Claims Policy Committee—which, based on negotiations during the 1970s, crafted updates to federal policy for comprehensive land claims in 1981 and 1986. I also consult the personal files of one of the lead federal negotiators for the Inuvialuit Final Agreement, Simon Reisman, which were opened at Library and Archives Canada in early 2016. Finally, I compare various unpublished working documents used during negotiations with the original proposal, draft agreements on various sections, the Agreement-in-Principle, and the Final Agreement.

The terms of the Inuvialuit Final Agreement

In exchange for their exclusive use of ancestral lands, Inuvialuit received through the Inuvialuit Final Agreement financial compensation of $45 million (in 1977 dollars), approximately 91,000 square kilometers (35,135 square miles) of land with a mosaic of surface and subsurface rights, and select political and economic rights guaranteed by the Crown. Describing these provisions, the Inuvialuit Final Agreement sprawls across 150 pages and includes sections on eligibility and enrollment of Inuvialuit beneficiaries, the creation of Inuvialuit corporations to manage financial transfers and other benefits, a $7.5 million social development program, procedures for arbitration should signatories disagree on the implementation of the settlement, and more. Key environmental provisions are described in Table 10.2.

Table 10.2 Key environmental provisions of the Inuvialuit Final Agreement

Lands	Inuvialuit receive title to 90,600 square kilometers of land: 13,000 square kilometers of land with surface and subsurface rights and 77,600 square kilometers of land with surface rights and with subsurface rights that exclude oil, gas, and mineral rights.
Wildlife Harvesting and Management	Inuvialuit receive exclusive rights to harvest game on Inuvialuit lands and exclusive rights to certain species throughout the Inuvialuit Settlement Region (muskox, black bear, grizzly bear, polar bear). Inuvialuit also have preference for other subsistence species in the settlement region. Inuvialuit have guaranteed participation in overall management of wildlife in Western Arctic Region through advisory bodies.
Environmental Screening and Review	All significant development proposed in the Inuvialuit Settlement Region is subject to an environmental screening and review, conducted by an advisory council with equal representation of Inuvialuit and federal/territorial representatives. This council recommends terms and conditions for proposed development to the relevant federal regulatory authority. Also, all development is subject to wildlife compensation provisions, which compensate affected Inuvialuit harvesters for losses suffered and provide for restoration of disturbed habitat.
Yukon North Slope	Area west of Babbage River will be considered no-development zone and will be withdrawn for a national park. Area east of the Babbage River will be a special management zone subject to a public land use planning process.

Since its completion, the IFA has been recognized by political scientists, Indigenous scholars, and conservation biologists as including some of the most robust and just measures for wildlife and fisheries co-management of all northern land claim settlements.[15] In 1986, the Crown revised its policy on comprehensive land claims negotiations to include these and similar environmental provisions as part of the Aboriginal rights stemming from Aboriginal title.[16]

The Final Agreement established two categories of private Indigenous land: 5,000 square miles with surface and subsurface title, adjacent to six Inuvialuit communities (referred to as 7(1)(a) lands); and 30,000 square miles with surface rights and limited subsurface rights—excluding oil, gas, and minerals, but including sand and gravel (referred to as 7(1)(b) lands). The remaining lands and waters within the settlement region—roughly 133,000 square miles or 344,400 square kilometers—were designated as Crown lands. Here, Inuvialuit had neither exclusive surface nor subsurface rights, but they retained select harvesting rights and managed wildlife, fisheries, and development proposals on advisory boards in conjunction with federal and territorial governments.

The 90,600 square kilometers of private Inuvialuit land was a much higher percentage of traditional lands than that in the two existing land claim settlements in the North American Arctic—the Alaska Native Claims Settlement Act of 1971 and the James Bay settlement of 1975.[17] Further, Inuvialuit emphasized in their 1977 proposal and subsequent negotiations that their land selections were based not only on areas of importance for traditional pursuits of hunting, trapping, and fishing, but also on areas significant for "reasons of biological productivity."[18] How did Inuvialuit and federal parties justify such quantities and selections of land? In the following sections, I interrogate negotiations in the IFA and the evidence Inuvialuit presented therein, as well as the social factors that drove scientists and Indigenous Peoples to emphasize these provisions in the settlement process.[19]

Aboriginal title and land use and occupancy studies

In a history of knowledge production the legal context of Aboriginal title is significant. Between 1968 and 1978, various courts in Canada established that Aboriginal title existed, that it was dependent on historic use and occupation of land, and that certain rights were tied to this title (Table 10.3). According to federal policy, though, Indigenous Peoples, not federal governments, were responsible for defending claims to land—and remain so to the present day. The Department of Indian Affairs and Northern Development provided funding to Indigenous associations to develop their land claims proposals. Four major Indigenous political groups formed in the Northwest Territories between 1969 and 1973. These included the Committee for Original Peoples Entitlement, The Indian Brotherhood of the Northwest Territories, and

Table 10.3 Significant decisions for treaty-making in Arctic Canada and their impacts, 1968–1978[1]

Decision	Venue	Impact
1973 *Calder*	Supreme Court of Canada	Nisga'a in British Columbia have Aboriginal rights, which stem from prior occupancy. The federal government announced a policy for negotiating land claims later in the year.
1973 *Kanatewat*	Quebec Superior Court	Provided injunction to halt hydro-electric development in James Bay because Aboriginal rights there had not been addressed. Spurred negotiations among the Province of Quebec, the Government of Canada, and the Cree and Inuit of northern Quebec.
1973 *Paulette*	Supreme Court of the NWT	Dene argue that Treaties 8 and 11 did not extinguish Aboriginal title. Court does not rule on question of continuation of Aboriginal title in NWT.
1977 Berger Inquiry	Recommendation to National Energy Board	Government of Canada enforced a moratorium on pipeline construction in the Mackenzie Valley for 10 years to settle land claims in the region.
1978 *Baker Lake*	Federal Court of Canada	Inuit in Baker Lake area hold Aboriginal rights, but permits and authorizations for development do not infringe on these rights. Development can proceed concurrently with land claims negotiations in the Yukon and Northwest Territories beyond the Mackenzie Valley.

1 Adapted in part from Alastair Campbell, "Overview of Modern Treaties," a presentation given before the 2014 Land Claims Agreement Coalition Workshop, http://www.landclaimscoalition.ca/assets/Overview-of-Modern-Treaties-Alastair-Campbelll.pdf, accessed 3 January 2017.

the Métis Association of the Northwest Territories—all representing Inuit, Dene, and Métis of western Northwest Territories—and Inuit Tapirisat Canada, representing Inuit across Canada.[20]

Court decisions also set precedents for what information would be admitted as evidence for establishing Aboriginal title and who might be authorized as experts in this arena.[21] In the 1973 *Kanatewat* case, for instance, the judge allowed testimony from both elder Indigenous residents who had hunted, fished, and trapped in northern Quebec as well as academics who were familiar with histories of trade and subsistence activities. As a result, land use and occupancy surveys became an essential research project of those Indigenous associations in Canada seeking formal ownership and recognition of their traditional territories from the Crown.[22] In this way, land claim settlements initiated a significant research program in Arctic Canada during the last three decades of the 1900s. The conditions by which studies of land use and occupancy came into being stand in contrast to the ways disciplinary interests, industry standards, defense imperatives, or regulatory requirements shaped Arctic research during the Cold War.[23]

These conditions were evident in the events that led to the proposal for the Inuvialuit Final Agreement. Initially, the Committee for Original Peoples Entitlement—the political association representing Inuvialuit—worked as part of the national Inuit organization, Inuit Tapirisat Canada (ITC), on a land claim for all Inuit in the territorial north. ITC was responsible for sponsoring the landmark Inuit Land Use and Occupancy Project, which documented through oral testimony and individualized map biographies that Inuit used roughly 2.8 million square kilometers (1.08 million square miles) of land and water in the Canadian North.[24] ITC submitted their proposal to the federal government in spring of 1976, but withdrew it in September of that year for further revision.[25] The Inuvialuit decided to separate from ITC and submitted their own proposal in May 1977, relying on the Inuit Land Use and Occupancy Project to assert title to more than 425,000 square kilometers (164,000 square miles) of land in the Western Arctic region. They also hired as negotiators two consultants who had worked for ITC, one of which, Peter Cummings, was an attorney versed in Aboriginal law.[26] The Inuvialuit proposal of 1977 thus reflected and incorporated the court's precedents on Aboriginal title and the kinds of knowledge necessary to delineate it.

The Inuit Land Use and Occupancy Project was effective in convincing federal bureaucrats of the Inuvialuit's legal claim to Aboriginal title. In June 1977, representatives of the Committee for Original Peoples Entitlement referred to the Land Use and Occupancy Study as providing "the extensive factual data" to support the Inuvialuit legal claim to land. "Before the Crown can legally interfere with Inuvialuit property rights," representatives noted, "it has to either obtain their consent or else properly compensate them for an expropriation of their lands."[27] When the Claims Policy Committee of the Office of Native Claims within the Department of Indian Affairs and Northern Development first met in June 1977 to review the Inuvialuit proposal, there was wide agreement of the study's scope and validity. Federal bureaucrats noted "a preliminary analysis of the study clearly shows that the Inuvialuit have used almost the entire area within the Western Arctic Region for the past 15–20 years."[28] This was a profound shift in outlook, given that federal policy prior to 1973 did not recognize, let alone substantiate, Indigenous claims to land in the territorial north.

That said, in their settlement with the Crown, Inuvialuit did not retain the full amount of lands they traditionally used and occupied. They also emphasized with federal negotiators that they did not only want to select areas where wildlife was harvested; they were also interested in "the areas that produce the animals."[29] How, then, did the parties reach agreement on land quantity and selection?

Territorial land management and a union of Inuvialuit with concerned scientists

The answer to this question starts with other federal policies at stake in land claims negotiations, including a commitment to northern non-renewable resource development at the expense of environmental protection and renewable resource economies. Inuvialuit disillusionment with federal commitments to land management, along with similar concerns from scientists in academia, government, and the non-profit sector, made the Inuvialuit proposal another channel for research on biological productivity of the tundra and the market value of surface and subsurface resources—as well as a mechanism for Arctic environmental policy.

To see how this could be so, we must return to the origins of the Inuvialuit proposal of 1977. When they informed ITC and the federal government of their intentions to submit their own land claim, the Inuvialuit cited the "pressure of an impending decision regarding northern pipelines."[30] In 1974, a consortium of oil companies under the name Arctic Gas had submitted an application for a right-of-way to build a pipeline connecting Prudhoe Bay, Alaska with processing facilities in northern Alberta; their proposed route would cross the North Slope of the Yukon, enter the Mackenzie Delta, and then hug the Mackenzie River on its way south.[31] Some of the area covered by this Mackenzie Valley Pipeline, especially much of the portion above the Arctic Circle, overlapped with traditional Inuvialuit lands. That same year, the federal government appointed Judge Thomas Berger to lead an inquiry into the social and ecological impacts of the potential pipeline. Suggestive of the linkages between development pressures and land claims in the north, the Inuvialuit submitted their proposal on the very same day that Berger submitted a letter to the federal government recommending a ten-year moratorium on pipeline construction until land claims could be settled in the Mackenzie River valley (Table 10.3). The Inuvialuit had drafted that proposal under the impression that the pipeline would be approved and built shortly thereafter.[32] In 1978, with the *Baker Lake* case, federal officials received the support of the Supreme Court to continue exploration and development they had already approved in the territorial north outside of the Mackenzie Valley while negotiating land claims in this same area. Within the Mackenzie Valley, including the western Arctic, other forms of exploration and development beyond pipeline construction could and did continue (Table 10.3).

Recognizing this, Indigenous organizations viewed land claims negotiations as an opportunity to create an integrated land management system, one that prioritized subsistence lifestyles and renewable resource economies and balanced ecosystem protection with natural resource development. This was a system that they felt was not in place and was not possible within the jurisdictional and decision-making structures of the Department of Indian Affairs and Northern Development. Inuvialuit representatives stated this emphatically in their 1977 proposal. "Land management in the Northwest Territories is in a deplorable state," the authors wrote, "for meaningful land management is nonexistent." They continued,

> The Department of Indian and Northern Affairs' jurisdiction over land has led to the untenable split between wildlife and land management. It has placed wildlife considerations in a subservient position to the considerations of industry. Moreover, within the structure of the Federal Government, the Northern Natural Resources and Environment Branch of the Department of Indian and Northern Affairs has a decision-making power unconstrained in any significant way by other departments. Finally, the Government itself is often the developer.[33]

To be fair, the Department had made strides in land use management in the Northwest Territories during the early 1970s. The Department promulgated Territorial Land Use Regulations under the Territorial Lands Act in 1972, which gave them statutory authority to manage the territorial land surface for the first time.[34] These changes mirrored similar expectations set forth in the Expanded Guidelines for Northern Pipelines of 1972, which compelled federal officials to collect ecosystem information ahead of pipeline construction.[35] Still, Indigenous parties found these steps unsatisfactory, especially in terms of full implementation and enforcement. In their 1977 proposal, Inuvialuit representatives noted that government and industry love to boast "how far they have come," but they seldom "admit to how inadequate land management and planning is today." The problem, they concluded, was a lack of "policies, planning, regulations, and enforcement."[36]

Inuvialuit were not alone in identifying this problem. In seeking solutions through their land claim proposal, they formed partnerships with concerned scientists in government, academia, and the non-profit sector. Many biologists had noted the lack of an effective northern conservation strategy and the Department of Indian Affairs and Northern Development's "multiple use" policy of land management in the late 1960s and early 1970s. In the Tundra Conference of 1969 and the Symposium on Arctic Development and Ecological Problems of 1970, for instance, biological scientists employed at universities and within the federal government pushed for regulations that would improve this situation.[37] According to William Pruitt, Jr., a wildlife ecologist at the University of Manitoba, these researchers were told by officials in the Department of Indian Affairs and Northern Development that they did not have "sufficient information to allow regulation of exploitation activity" in the territorial north. In response, and to prove these statements incorrect, the Canadian Wildlife Service, housed within the Department of Fisheries and Forestry in 1971, developed an Arctic Ecology map series. These maps located areas important to the maintenance of an animal population and those especially sensitive to damage by industrial activity. To compile them, the Canadian Wildlife Service relied on existing literature and personal communication with "virtually everyone with northern experience." Reviewing the project in the *Arctic Circular,* Pruitt expressed pride in the scientific accuracy of the maps, but dismay that they lacked the "sharp teeth" of regulatory authority. One means of creating this, he said, was to train and employ a crew of land use permit inspectors who were "unshakeable."[38]

Shortly after the creation of the Arctic Land Use series, the Department of Indian Affairs and Northern Development funded an Arctic Land Use Research program, which collaborated with the Department of Fisheries and Environment to develop a series of Northern Land Use Information maps. These maps, which identified sensitive ecological areas like calving and spawning grounds, were meant to "provide a convenient information base to assist regional land use planning."[39] Yet, again, the Minister deferred on adjusting regulations or staffing to prioritize ecological concerns over development interests in land use inspections or permitting of land use, much to the frustration of scientists within the Department of Fisheries and Oceans and the Department of Environment. Not coincidentally, these departments lacked statutory authority over many aspects of northern land and resource management in the early 1970s and felt that Indian Affairs and Northern Development maintained an "empire" in the Arctic.[40]

Because of the pace of non-renewable resource development in the Canadian Arctic, and the consolidation of jurisdiction within the Department of Indian Affairs and Northern Development, some researchers outside that department started to view the land claims settlement process as means to create binding agreements on conservation and environmental protection in Arctic Canada. Inuit Tapirisat Canada, which sponsored the Inuit Land Use and Occupancy Project, commissioned two other research projects directed at land management issues between 1973 and 1976: a Renewable Resources Study and a Non-Renewable Resources Study. Both of these projects involved teams of researchers in academia and produced ideas about land use planning, wildlife management, and conservation areas that appeared in the Inuvialuit proposal of 1977.[41] Doug Pimlott and Peter Usher, for instance, compiled information about government plans for development and their possible social and ecological impacts and shared these findings with Indigenous political organizations and Hunters and Trappers Committees.[42] Pimlott was a professor at the University of Toronto and a member of the environmental advocacy group the Canadian Arctic Resources Committee; he is remembered as a founder of the modern environmental movement in Canada.[43] Usher was a geographer who formerly worked in the Northern Science Research Group within the Department of Indian and Northern Affairs; he had a public falling-out with that Department's Minister in 1972 after criticizing federal policies on subsistence trapping in Inuvialuit traditional territory.[44] The Inuvialuit also hired Bob DeLury, a biologist working for the Canadian Arctic Resources Committee in the Mackenzie Delta, and Pedro van Meurs, a former consultant to ITC and former employee in the federal govern-ment's Department of Energy, Minerals, and Resources.[45] Drawing from the work of Pimlott, Usher, DeLury, van Meurs, as well as many Inuvialuit fieldworkers, the 1977 Inuvialuit land claim proposal framed land quan-tity and land selections as the essential base of power to enforce a coordi-nated land management system. Proposal authors argued that Inuvialuit ownership of land met several goals, including retaining their "land-based identity," helping control development activities, and preserving wildlife habitat.[46] In short, in their proposal for a land claim, the Inuvialuit also hoped to enshrine certain social, economic, and environmental rights.

Today, the federal government recognizes such provisions as fundamental to Indigenous self-determination and a basis for land claim settlements still in negotiation.[47] This was not the case in 1977, however (Table 10.3). When mem-bers of the Office of Native Claims received the Inuvialuit proposal, they were stunned by the scope of provisions regarding land management. They quickly prepared a memorandum to cabinet to seek guidance on what they identified as a critical problem: whereas the Crown's position had been to assert sov-ereignty to ceded Indigenous lands and compensate claimants for losses of traditional activities, the Inuvialuit viewed the land claim as an exchange of Aboriginal title for a set of rights specified in the provisions of a settlement.[48]

Arguments for land quantity and land selections: economic value and biological productivity

To convince federal representatives to agree to their proposal, Inuvialuit and their consultants presented evidence based on forecasts of the hydro-carbon potential in the Western Arctic and the biological productivity of the tundra, while linking this information with Inuvialuit perspectives on nature and Canada's national interests. This evidence was expressed in a trio of supplementary documents submitted by the Inuvialuit to federal negotiators in June 1977. These documents clarified for the federal nego-tiating team many of the provisions outlined in the proposal and thus "set the stage for detailed discussions" in the first official meetings between the parties scheduled for that month.[49]

In "Analysis of the Basic Elements of the Inuvialuit Land Rights Proposal," the author—presumably energy specialist Pedro van Meurs—quantified the economic value of lands ceded by Inuvialuit as a means of providing support for the lands they sought to retain. To begin, van Meurs differentiated the value systems of Inuvialuit and Canadian society, an important precursor to understanding the amount and location of land Inuvialuit proposed to own. Whereas in "southern perspective," land had value in relation to non-renew-able development, Inuvialuit placed value in hunting, trapping, fishing, and an overall "cultural relationship to the land."[50] Therefore, in terms of land selections, a balance of Inuvialuit-owned and Crown lands "would be opti-mal if the land given up by the Inuvialuit is the most valuable land from a southern Canadian point of view, while the land retained by the Inuvialuit represents very valuable land from a native peoples' point of view."[51]

In setting up such an optimal balance, van Meurs was well aware of the economic value of potential oil and gas deposits in the western Arctic. Drawing on a recent report released by his former employer, the Department of Energy, Mines, and Resources, van Meurs pointed out that the Mackenzie Delta and Beaufort Sea were known to be "the most promising petroleum region of the frontier." Moreover, while the particular lands selected by Inuvialuit were rich "from an Inuvialuit point of view," they were not the same lands that government-produced surveys concluded bore the highest likelihood for commercially significant petroleum. In other documents, the Inuvialuit quantified the market value of the lands they proposed to own, drawing on harvest records and nutritional surveys to develop a figure based on the total pounds of meat consumed by Inuvialuit per year.[52] This value, the Inuvialuit concluded, ranged from $2.5 million to $7 million, and paled in comparison to the value of would-be Crown lands in the western Arctic, which stored natural gas equivalent to the "anticipated production of the James Bay project for the next 40 years."[53] For these reasons, van Meurs argued the Canadian government could not object if the Inuvialuit kept lands that offered little economic value to other Canadians while surrendering lands to the Crown that "may be an order of magnitude more valuable than any other land in the frontier areas."[54] Indeed, Inuvialuit representatives noted that they, not the federal government, were "being generous" in their land claim proposal.[55]

In a separate supplemental report, Inuvialuit and their consultants presented additional evidence to support their land provisions. As van Meurs did, the author of this separate document—presumably biologist Bob DeLury—linked the "well-being of the Inuvialuit" with the "well-being of the wildlife." DeLury, however, took his line of reasoning away from the economic value of surface and subsurface resources and to the biological productivity of the Canadian Arctic. The lands and water of the western Arctic, he wrote, "are characterized by biologists as being a low productivity area if compared to other parts of Canada." This was due in part to short growing season, low temperatures, low sunlight levels, and low nutrient levels.[56] DeLury turned these biological realities into a rationale for land quantities. Because of the "generally low biological productivity of northern lands" it was necessary for both Inuvialuit and federal negotiators "to consider larger land areas to safeguard the total productivity base of each species than would be required for a similar species in the south."[57]

DeLury and Inuvialuit representatives marshaled a similar argument to identify particular land selections, a principal means of achieving the kind of land management system Inuvialuit desired. Perhaps hoping their arguments about economic value and biological productivity made a strong case for land quantities, DeLury pivoted to the quality of particular lands, especially as regards Inuvialuit use and wildlife habitat. There was a difference, he pointed out, between an "area of high harvest potential" and an "area of high biological productivity." Where the former may depend on access, stones for building food caches or other factors unrelated to biological productivity, the latter was absolutely dependent on "critical parts of the environment" like calving grounds, concentration points in migration routes, staging areas for geese, and seabird nesting colonies.[58] Rather than presume that government would properly safeguard these critical environments, Inuvialuit selected many of them for private ownership under both 7(1)(a) and 7(1)(b) classifications. In their eyes, this was a better management strategy anyway. The Inuvialuit, DeLury argued, "are closer to the resource [and] thus able to maintain the resource better by responding to fluctuations in the resource faster and in an appropriate manner."[59] To officially select these lands in 1978 and 1979, Inuvialuit had many sources of data at their disposal: the Inuit Land Use and Occupancy Project maps, and the map series produced by both the Canadian Wildlife Service and the Department of Fisheries and Environment for the western Arctic were all available at that time. However, Inuvialuit likely also relied on maps produced by the International Biological Program (1964–1974), given that six of the sixteen proposed ecological sites for the western Arctic region were included in Inuvialuit private lands.[60]

These supplementary documents and the evidence they contained were instrumental in convincing federal officials to accept key environmental provisions found in the Inuvialuit Final Agreement. Many of the figures, comparisons, and arguments made by Inuvialuit representatives were later reused by federal officials in private and in public. In preparing to seek approval for the land quantity proposal from cabinet, members of the Claims Policy Committee within the Office of Native Claims sensed the need to explain differences in land quantities retained by Indigenous claimants in the Inuvialuit claim and the existing precedent of the James Bay Agreement, as well as the precedent set by the United States in the Alaska Native Claims Settlement Act. They pointed out the particular land selections of the Inuvialuit, which did not seek total control of oil and gas deposits, and thus represented incredible value for a government grappling with an energy supply problem. Moreover, the committee members said, the biological productivity of land for surface use was "a compelling factor in the determination of land quantities."[61] Similarly, standing before the Legislative Assembly of the Northwest Territories in 1978, federal negotiator Dr. J. K. Naysmith sought to put his audience at ease with the amount of land retained by Inuvialuit and the infringements on territorial jurisdiction over natural resources posed by advisory boards. "We have an annex," he said, "which indicates that the value of land on the prairies is at least 100 times that in the Western Arctic region." Indeed, Inuvialuit negotiators had created that annex in one of their supplementary reports.[62] By compiling and presenting information on the tundra's productivity, the importance of renewable resources to Inuvialuit society, and the surrender of oil and gas rights to the Crown, Inuvialuit helped federal officials to consider the Inuvialuit claim outside "the precedent of the [historic] treaties."[63] Inuvialuit thus achieved leverage through land quantities and land selections to win agreement on provisions like wildlife and fisheries co-management, environmental screening and review boards, and a wilderness designation for the Yukon North Slope (Table 10.2).[64] Four days after Naysmith's address in Yellowknife, Inuvialuit and federal negotiators signed an Agreement-in-Principle in Sachs Harbour.[65]

Conclusions

Many subsequent negotiations for land claims in Arctic Canada attempted to replicate the success achieved by the Inuvialuit in these areas.[66] There were a variety of outcomes: indeed, both Indigenous and federal negotiators supported flexible interpretations of federal policy on land claims, agreeing that every treaty ought to be unique, while striving for parity across treaties.[67] The heterogeneity of land claim settlements—in timing, scope, and context—may resist straightforward archival research or sweeping historical narratives. However, the diversity of land claim settlements is itself a historical marker: it signals the point at which Arctic imaginaries based on Indigenous experiences began their institutionalization in national governmental agencies.

In political and military rhetoric during the Cold War, the Arctic was a vulnerable, extreme, backward, and largely empty land. These representations reflected an ideology and language of engineering, which drew from the physical sciences and the social sciences of anthropology and sociology. In seeking to claim land and other mechanisms of self-determination, the Arctic was another place altogether. One rarely reads references to "the Arctic" or "the North" in the governmental archive of land claims negotiations for northern Canada. Instead, Indigenous places and experiences assume a larger presence: the area was portrayed as a collection of regions, whether in the documentation of livelihoods in traditional use and occupancy studies or in the projection of Indigenous names for new political jurisdictions, like Inuvialuit Nunangat or Nunavut. Rather than a monolithic expanse, waiting to be modernized, the Arctic of land claims negotiations became a lived place. Its histories, ecologies, and cultures were explicitly under scrutiny and could not be easily ignored or misrepresented.

The long-term success of land claim settlements requires sustaining this attention to the Arctic's diversity. Historians have a role to play in this. In December 2015, more than 15 Indigenous organizations working together to ensure full implementation of modern treaties in Canada hosted a conference to shape the future of research on land claim settlements. This group, the Land Claims Agreement Coalition, asked academics for more nuanced historical accounts of land claim settlements.[68] Indigenous representatives desired a different history of land claim settlements specifically for civil servants charged with implementation—so these federal employees would better understand the significance and subtleties of their jobs. I attended that conference; this chapter is one response to the Coalition's request. Future research on land claims settlements ought to examine negotiations as a means of excavating the many layers of bureaucracy around land in the Arctic, especially the differential experiences of Indigenous Peoples there with procedures for managing wildlife, fisheries, and development and the knowledge gathering processes embedded within them.[69]

As advocates for Indigenous rights have argued, land claims settlements are more like a marriage than a divorce: they did not break up Indigenous and national governments, but rather bound them in a unique partnership.[70] The Inuvialuit Final Agreement reflects both Indigenous cultural values and Canada's national interests. Its language is borne of impulses for conservation, development, and for the recognition of Aboriginal title. These impulses, in turn, emerged from various sources at a particular historical moment: Inuvialuit communities, conservationists in academia, legal scholars, and former civil servants concerned about Canada's northern policies in the 1970s. In these ways, land claims settlements served as a space for science and Indigenous knowledge during the Cold War that existed outside of the military, industry, universities, and think-tanks—but was not at all separate from them.

Notes

1. In this paper, I use the terms "Native," "Indigenous," "Aboriginal." I use "Native" only in reference to government departments, publications, or decisions of the historical period under study—i.e., "Office of Native Claims." I use "Aboriginal" to highlight the ways historical actors referred to significant legal concepts, like Aboriginal title and Aboriginal rights. For every other relevant instance, I use "Indigenous."
2. Finn Breinholt Larsen, "The quiet life of a revolution: Greenlandic Home Rule 1979–1992," *Etudes/Inuit/Studies*, 1992, 16(1/2): 204–7; Hans Christian Gullov, "Home Rule in Greenland," *Etudes/Inuit/Studies*, 1979(1): 131–142. The texts of comprehensive land claim settlements in Canada are available online. See Indigenous and Northern Affairs Canada, "Comprehensive Claims," https://www.aadnc-aandc.gc.ca/eng/1100100030577/1100100030578, accessed 10 May 2018. The Alaska Native Claims Settlement Act can be found at Cornell Law School, "43 U.S. Code Chapter 33— Alaska Native Claims Settlement," https://www.law.cornell.edu/uscode/text/43/chapter-33, accessed 10 May 2018.
3. "Historic treaties" included promises by the Crown to Indigenous Peoples of reserve lands and benefits like annual payments, ammunition, clothing, and limited rights to hunt and fish. "Modern treaties," on the other hand, are nation-to-nation relationships among Indigenous Peoples, the federal and provincial Crown, and in some cases a territory. They define the land and resource rights of Indigenous signatories and include a range of provisions for education, cultural preservation, and management of wildlife, fisheries, and non-renewable resource development. See Land Claims Agreement Coalition, "Modern Treaties in Canada," http://www.moxiemedia.ca/NVision/, accessed 17 July 2017. Modern treaties are by no means complete. Treaty negotiations continue in southern Northwest Territories and several provinces. Indigenous and Northern Affairs Canada, "Comprehensive Land Claim and

Self-Government Negotiation Tables," https://d3n8a8pro7vhmx.cloudfront. net/idlenomore/pages/1140/attachments/original/1408398010/Comprehensive_ Land_Claim_and_Self-Government_Negotiation_Tables.pdf?1408398010, accessed 1 January 2019.

4. In Greenland, public opinion polls conducted a decade after the Home Rule Act of 1979 demonstrated that notions of Greenlandic autonomy hinged on the incorporation of Greenlanders and their perspectives in the growing domestic portfolio for environmental and social affairs. Larsen, 204–7. The Siumut party, which was largely responsible for getting Home Rule on the national agenda, won 13 out of 21 possible seats in the first Home Rule election and remained in power until the early 1990s. In Arctic Alaska, the failure to legislate such mechanisms of cooperation, representation, and inclusion of Indigenous knowledge in the Alaska Native Claims Settlement Act of 1971 spurred further political organizing around Inuit self-determination, which produced the powerful municipality of the North Slope Borough and the Inuit Circumpolar Council. Andrew Stuhl, *Unfreezing the Arctic: Science, Colonialism, and the Transformation of Inuit Lands,* (Chicago: The University of Chicago Press, 2016), 128–143.

5. See Graham White, "Treaty Federalism in Northern Canada: Aboriginal-Government Land Claims Boards," *Pubilus,* Summer 2002, 32(3): 89–109; Graham White, "'Not the Almighty': Evaluating Aboriginal Influence in Northern Land-Claim Boards," *Arctic,* 2008, 61(5): 71–86.; Paul Nadasdy, "The Anti-politics of TEK: The Institutionalization of Co-management Discourse and Practice," Anthropologica, 2005, 47(2): 215–232; S. Spak, "The Position of Indigenous Knowledge in Canadian Co-management Organizations," Anthropologica, 2005, 47(2): 233-246; F. Berkes, *Sacred Ecology: Traditional Ecological Knowledge and Resource Management,* (Philadelphia: Taylor and Francis, 1999).

6. Christopher Alcantara, *Negotiating the Deal: Comprehensive Land Claims in Canada,* (Toronto: University of Toronto Press, 2013). Gullov, "Home Rule in Greenland," 131–142. See also Guy Martin, "The Politics of Passage," in *The Native Land Claims,* a booklet compiled and produced jointly by Alaska Department of Education and Center for Northern Educational Research, University of Alaska-Fairbanks, June 1975, available at http://www.ankn.uaf. edu/curriculum/ANCSA/DeptEd/passage.html, accessed 10 May 2018.

7. The Inuvialuit Final Agreement went by other names from proposal stage to completion. It was also referred to as the "COPE Claim" (following the claimant, Committee for Original Peoples Entitlement) and the Western Arctic claim. I use IFA here for simplicity's sake, though I recognize it is anachronistic, given that the Final Agreement can also refer to an actual document that was not signed until 1984.

8. Milton M.R. Freeman, "Looking back—and looking ahead—35 years after the Inuit land use and occupancy project," *The Canadian Geographer,* Spring 2011, 55(1): 22–29.

9. Stephen Bocking, "Science and Spaces in the Northern Environment," *Environmental History,* October 2007, *12*: 867–94. See also Stuhl, *Unfreezing the Arctic,* 128–143.

10. A full treatment of land claim settlements, modern treaties in Arctic Canada, or even one such treaty, is beyond the scope of this chapter. Importantly, negotiations varied considerably in length. The James Bay settlement, for instance, moved from proposal to Agreement in Principle in 12 months. The Nunavut Land Claim Agreement negotiations, on the other hand, spanned nearly 20 years (1976–1993). New federal policy statements on comprehensive land claim agreements were released in 1981, 1986, and 2014. Aboriginal Affairs

and Northern Development Canada, "Renewing the Comprehensive Land Claims Policy: Towards a Framework for Addressing Section 35 Aboriginal Rights," September 2014, 6.

11. Negotiations for the James Bay settlement were also conducted without reference to the government's 1973 policy on Aboriginal title. "Memorandum to cabinet: Western Arctic Claim Final Settlement, January 27, 1984," MG31 E112, Volume 19, File 6, LAC.

12. Charles Arnold, Wendy Stephenson, Bob Simpson, and Zoe Hoe, eds. *Taimani: At That Time: Inuvialuit Timeline Visual Guide,* (Inuvialuit Regional Corporation, 2011). See also *So That You Can Stand,* directed by Ole Gjerstad (2015: Makivik Corporation), DVD. William L. Iggiagruk Hensley, *Fifty Miles from Tomorrow: A Memoir of Alaska and the Real People,* (New York: Picador Publishing, 2010). According to the author's biography, Hensley was "a founder of the Northwest Alaska Native Association and spent twenty years working for its successor, the Iñuit-owned NANA Regional Corporation. He also helped establish the Alaska Federation of Natives in 1966 and has served as its director, executive director, president, and co-chair. He spent ten years in the Alaska state legislature as a representative and senator."

13. On the thirty year embargo, see Andrew Stuhl, Bruce Uviluq, Anna Logie, and Derek Rasmussen, "Modern Treaties in Canada: A Call for Engaged, Collaborative Historical Research," ActiveHistory.ca, http://activehistory.ca/2016/09/modern-treaties-in-canada-a-call-for-engaged-collaborative-historical-research/ (accessed 3 January 2017).

14. Terry Fenge and Jim Aldridge, eds, *Keeping Promises: The Royal Proclamation of 1763, Aboriginal Rights, and Treaties in Canada* (Chicago: McGill-Queen's University Press, 2015).; Tony Penikett, *Reconciliation: First Nations Treaty Making in British Columbia* (Douglas and McIntyre, 2006).; and Terry Fenge, "Political Development and environmental management in northern Canada: the case of Nunavut agreement," *Etudies/Inuit Studies,* 1992, 16(1–2): 115–141. For an example of an in-depth study of a single land claim and the roles of environmental knowledge in its negotiation and implementation, see Robert McPherson, *New Owners in their Own Land: Minerals and Inuit Land Claims* (Calgary: University of Calgary Press, 2003). See also Christopher Alcantara, *Negotiating the Deal: Comprehensive Land Claims in Canada,* (Toronto: University of Toronto Press, 2013).

15. L.N. Binder and B. Hanbidge, "Aboriginal People and Resource Management," in J. Inglis, ed., *Traditional Ecological Knowledge: Concepts and Cases* (Ottawa: Canadian Museum of Nature, 1993), 121–132.; Derek Armitage, Fikret Berkes, Aaron Dale, Erik Kocho-Schellenberg, and Eva Patton, "Co-management and the co-production of knowledge: Learning to adapt in Canada's Arctic," *Global Environmental Change,* 2011, 21: 995–1004.

16. John Merritt, "In Search of Common Ground: Ottawa Rethinks its Approach to Comprehensive Claims," *Northern Perspectives,* January-April 1987, 15(1), http://carc.org/pubs/v15no1/1.htm, accessed 17 July 2017. This revision was largely a response to protests by Inuit from northern Quebec who had finalized a land claim settlement with the federal government before the Inuvialuit, but did not earn similar forms of representation and authority in environmental affairs. Such a turn of events underscores the differential experiences of Indigenous Peoples and national governments in land claim settlements and, thus, the potential value for historicizing the processes behind them.

17. "[90,600 square kilometers of Inuvialuit land title] was considered to be a generous land component which might well set an unacceptable precedent for other claimant groups. However, overall parity in claims settlements

must be considered. The Inuvialuit opted for protection of land and wild-life resources through ownership of land rather than for extensive financial compensation or an extensive social programs component. In addition, it is recognized that the comparative value of Inuvialuit land is very low." See "Memorandum to cabinet," 51. Inuvialuit retained about 21% of traditional lands, whereas Alaskan Natives retained about 11%. Committee for Original Peoples Entitlement, "Brief Arguments in Support of the 'Inuvialuit Nunan-gat' Land Rights Proposal," 4–5, RG 22, Accession 1992–93/208, Box 6, File B1165-C1-1, Part 2 enclosure, LAC. Notions of cooperative wildlife manage-ment and harvesting rights had been pioneered in Canada in the James Bay settlement.

18. Department of Indian Affairs and Northern Development, "The COPE/Government Working Group Joint Position Paper on the Inuvialuit Land Rights Claim, July 14, 1978," 9., Department of Indigenous and Northern Affairs Library. This criterion held priority over areas with "historic Inuvialuit sites or burial grounds.

19. While there are several important areas within the Final Agreement that were revised between the 1978 Agreement-in-Principle and the 1984 Final Agreement, parties decided on land quantity between spring 1977 and fall 1978 and Inuvi-aluit completed more than 85% of their private land selections in this time. I focus, then, on the period between May 1977 and October 1978.

20. Frances Abele, "Canadian Contradictions: Forty Years of Northern Political Development," *Arctic,* December 1987, 40(4): 314–316.

21. Keith Crowe, "A Summary of Northern Native Claims in Canada: The Pro-cess and Progress of Negotiations," *Etudes/Inuit/Studies,* 1979, 3(1): 31–37.

22. Memo from Milton Freeman to Peter Cumming, 16 August 1973. Permission to view and cite this material granted by Peter Usher. See also Arthur J. Ray, *Aboriginal Rights Claims and the Making and Remaking of History,* (Montreal: McGill-Queen's University Press, 2016).

23. See other contributions in this volume. While many Indigenous leaders main-tained knowledge about travel and subsistence activities across generations, and anthropologists, archaeologists, and geographers visiting the Canadian far north over the twentieth century had shown intermittent interest in these topics, land use and occupancy did not garner political significance or aca-demic interest until Indigenous rights movements in the late 1960s. See Peter Kulchyski and Frank Tester, including Peter Kulchyski and Frank Tester, *Kiumajut (Talking Back): Game Management and Inuit Rights, 1900–1970* (Vancouver: University of British Columbia Press, 2007).

24. Milton M.R. Freeman, "Looking back—and looking ahead," 20–31.

25. Canada. Department of Indian and Northern Affairs. Office of Native Claims. *COPE Claim: Background Information* [Ottawa]. April 1980.

26. Ibid. Cummings is listed as a technical advisor to COPE in "Minutes of a Meeting with Technical Advisors to COPE, Monday June 20, 1977, 10am to 2:30pm," RG 22, Accession 1992–93/208, Box 6, File b1165-C1-1, part 2, LAC.

27. "Brief Arguments in Support of the 'Inuvialuit Nunangat' Land Rights Proposal," 2.

28. "Notes of an Interdepartmental Meeting on June 30, 1977 to consider the COPE claim," RG22, Acc 1992–93/208, Box 6, File B1165-C1-1, Part 2, LAC.

29. "Biological Productivity in the Canadian Arctic as it relates to the Rationale for Ownership of Inuvialuit Lands as contained in the Inuvialuit Land Rights Proposal (Schedule D) 'Inuvialuit Nunangat' Presented to the Govern-ment of Canada May 13, 1977," p1, RG 22, Accession 1992–93/208, Box 6, File B1165-C1-1, Part 2 enclosure, LAC.

30. "COPE claim: Background Information," 5.

31. I have written elsewhere on the events precipitating pipeline proposals in the western Arctic. Starting in the late 1950s, Canada had leased thousands of acres of the Arctic tundra to oil and gas companies, who explored it through seismic testing. The discovery of North America's largest oil field in 1968, at Prudhoe Bay on Alaska's Arctic slope, was the fulfillment of dreams of a new domestic oil frontier—and a signal of more frenetic exploration and development activity. Stuhl, *Unfreezing the Arctic*, 105–107, 113–115. See also Timothy Mitchell, "Carbon Democracy," *Economy and Society*, 2009, 38(3): 399–432.

32. Penikett, 103. Many land claims proposals for Arctic territories were drafted in the context of pressures for natural resource development. The Alaska Native Claims Settlement Act (1971) was triggered and succeeded by a proposal for the Trans-Alaska Pipeline System (1969–1977); the James Bay settlement (1975) was similarly an artifact of the development of hydroelectric dams in James Bay (1967-1986). According to Tony Penikett, a former senior Canadian governmental official with experience in land claims negotiations, the Yukon Umbrella Agreement of 1990—which produced the Vuntut Gwich'in First Nation Final agreement—was the first modern treaty to be finalized without the pressure of resource development. Indeed, with the 1978 *Baker Lake* court decision, federal officials within the Department of Indian Affairs and Northern Development were able to legally pursue exploration and development in the territorial north in the midst of land claims negotiations (Table 10.3).

33. "'Inuvialuit Nunungat': The Proposal for an Agreement-in-Principle to achieve the Settlement of Inuvialuit Land Rights in the Western Arctic Region of the Northwest and Yukon Territories between the Government of Canada and the Committee for Original Peoples Entitlement, May 13, 1977," p. 93, Department of Indigenous and Northern Affairs Library.

34. A.R. Thompson, "A Conservation Regime for the North—What Have the Lawyers to Offer?" in "Proceedings of the Conference on Productivity and Conservation in Northern Circumpolar Lands, Edmonton, Alberta, 15 to 17 October 1969," 304.

35. These guidelines stated that the federal government will, in relation to pipeline corridors, "identify geographic areas of specific environmental and social concern or sensitivity, areas in which it will impose specific restrictions concerning route or pipeline activities, and possibly areas excluded from pipeline construction. These concerns and restrictions will pertain to fishing, hunting, and trapping areas, potential recreation areas, ecologically sensitive areas, hazardous terrain conditions, construction material sources, and other similar matters. Statements announcing the above will be released through the office of the Director, Environmental-Social Program, Northern Pipelines." Task Force on Northern Conservation, "Report of the 1984 Task Force on Northern Conservation" December 1984, Department of Indian Affairs and Northern Development, p 40, Annex A.

36. "'Inuvialuit Nunungat': The Proposal for an Agreement-in-Principle to achieve the Settlement of Inuvialuit Land Rights in the Western Arctic Region of the Northwest and Yukon Territories between the Government of Canada and the Committee for Original Peoples Entitlement, May 13, 1977," 93.

37. "Proceedings of the Conference on Productivity and Conservation in Northern Circumpolar Lands, Edmonton, Alberta, 15 to 17 October 1969." "Symposium on Arctic Development and Ecological Problems," *Arctic Circular*, 1970, 20(1): 3.

38. "Arctic Ecology Map Series," *Arctic Circular*, 1971, 21(1): 61–62.

39. "Northern Land Use Information Map Series," *Arctic Circular*, 1977, 25(3): 47–48.

40. References to the "departmental empire" of the Department of Indian Affairs and Northern Development appear throughout discussions of governmental scientists within the Department of Environment and Department of Fisheries and Oceans between 1970 and 1980. See Memorandum from EF Roots to JK Fraser and EA Balfour, 14 July 1978, RG 108, Accession 1993–1994/003, Box 7: Arctic Environmental Committee, LAC.
41. J.G. Nelson, "Arctic Renewable Resources: Summary and Recommendations," Renewable Resource Project, Inuit Tapirisat of Canada, July 1975.
42. Peter Usher to Milton Freeman, 21 February 1974. Permission to view, cite, and quote these materials granted by Peter Usher. "1974 Annual Report of Inuit Tapirisat Canada," RG 22, Accession 1992–93/208, Box 4, File: B1070/T1 Pt 1 Enclosed, LAC.
43. The Canadian Encyclopedia, "Douglas Humphreys Pimlott," http://www.thecanadianencyclopedia.ca/en/article/douglas-humphreys-pimlott/, accessed 10 May 2018.
44. "Canadian Assails Policy on Arctic," *The New York Times,* 20 February 1972.
45. "1975 Annual Report of Inuit Tapirisat Canada," p38, RG 22, Accession 1992-93/208 Box 4, File: B1070/T1 Pt 1, LAC. Pedro van Meurs was listed as a technical advisor to COPE in "Minutes of a Meeting with Technical Advisors to COPE, Monday June 20, 1977, 10 am to 2:30pm," RG 22, Accession 1992-93/208, Box 6, File b1165-C1-1, part 2, LAC. van Meurs' background with the federal government is listed at van Meurs Corporation, "Dr. Pedro van Meurs," http://www.vanmeurs.org/pedrovanmeurs.aspx, accessed 17 July 2017.
46. "Inuvialuit Nunungat': The Proposal for an Agreement-in-Principle to achieve the Settlement of Inuvialuit Land Rights in the Western Arctic Region of the Northwest and Yukon Territories between the Government of Canada and the Committee for Original Peoples Entitlement, May 13, 1977," 57, 37.
47. Indigenous and Northern Affairs Canada, "Comprehensive Claims," https://www.aadnc-aandc.gc.ca/eng/1100100030577/1100100030578, accessed 10 May 2018.
48. "Claims Policy Committee Minutes of Meeting Held June 21, 1977," RG 22, Accession 1992-93/208, Box 6, File B1165-C1-1, Part 2, LAC.
49. van Meurs and Associates Limited, "Analysis of the Basic Elements of the Inuvialuit Land Rights Proposal," June 1977, 1.
50. van Meurs and Associates Limited, "Analysis," 3–4.
51. van Meurs and Associates Limited, "Analysis," 14.
52. "Biological Productivity in the Canadian Arctic as it relates to the Rationale for Ownership of Inuvialuit Lands as contained in the Inuvialuit Land Rights Proposal (Schedule D) 'Inuvialuit Nunangat' Presented to the Government of Canada May 13, 1977," 8–11.
53. van Meurs and Associates Limited, "Analysis," 6.
54. van Meurs and Associates Limited, "Analysis," 6.
55. "Brief Arguments in Support of the 'Inuvialuit Nunangat' Land Rights Proposal," 3–4.
56. "Biological Productivity in the Canadian Arctic as it relates to the Rationale for Ownership of Inuvialuit Lands as contained in the Inuvialuit Land Rights Proposal (Schedule D) 'Inuvialuit Nunangat' Presented to the Government of Canada May 13, 1977," 1.
57. "Biological Productivity," 2.
58. "Biological Productivity," 3.
59. "Biological Productivity," 1 The Inuvialuit also selected a large tract of land in the Husky Lakes region for 7(1)(b) lands to act "as an ecological buffer against development." "Minutes of a Meeting with Technical Advisors to COPE, Monday June 20, 1977, 10 am to 2:30 pm," RG 22, Accession 1992-93/208, Box 6, File b1165-C1-1, part 2, LAC.

60. Richard D. Revel, "Conservation in Northern Canada: International biological programme conservation sites revisited," *Biological Conservation*, December 1981, 21(4): 263–287.
61. "Claims Policy Committee Minutes of Meeting Held June 21, 1977," 3 RG22, Accession 1992-93/208, Box 6, File B1165-C1-1, Part 2, LAC.
62. Legislative Assembly of the Northwest Territories, *Debates,* 27 October 1978, 71–74.
63. Ibid.
64. Legislative Assembly of the Northwest Territories, *Debates,* 27 October 1978, 70–71.
65. "Inuvialuit Land Rights Settlement: Agreement in Principle," 31 October 1978, available at Indigenous and Northern Affairs Canada library, E100.C55 1584. The Inuvialuit Final Agreement was not finalized and ratified until 1984. There was much back-and-forth in the years between 1978 and 1984, including attempts from federal negotiators to reopen the Agreement in Principle. Some of this history is documented in *Taimani: At That Time.* I hope to write about this fuller history of the IFA in future publications.
66. Tony Penikett, phone call with the author, February 17, 2016; Bob Simpson, phone call with the author, March 31, 2016.
67. C. Fairholm to NJ Faulkner, April 27, 1978, MG31 E112, Volume 19, File 9, Library and Archives Canada (LAC).
68. "Modern Treaties in Canada: A Call for Engaged, Collaborative Historical Research," ActiveHistory.ca, http://activehistory.ca/2016/09/modern-treaties-in-canada-a-call-for-engaged-collaborative-historical-research/ (accessed 3 January 2017).
69. Robert McPherson, *New Owners in their Own Land.*
70. Terry Fenge, "Negotiation and Implementation of Modern Treaties between Aboriginal Peoples and the Crown in Right of Canada," *Keeping Promises,* 109. Peter J. Usher, "Environment, race, and nation reconsidered: reflections on Aboriginal land claims in Canada," *The Canadian Geographer*, 2003, 47(4): 365–382.

Part 4

Science crossing borders

11 Knowledge base: polar explorers and the integration of science, security, and US foreign policy in Greenland, from the Great War to the Cold War

Dawn Alexandrea Berry

The largest of the bases at Thule, in Northern Greenland, only 900 miles from the Pole and 1300 air miles from the nearest Soviet Territory. Built with tremendous difficulty ... it is designed as a take-off strip for jet interceptors and a staging base for intercontinental bombers ... among other smaller installations is a weather and testing station on an ice 'island' near the pole. Thus the Arctic could be either a battleground for global aerial war or a way-stop for peaceful air traffic.[1]

In 1953 construction of the US Air Force's Thule base was completed, becoming the northern-most military installation in the world. The base was a major engineering feat, a tangible indication of Greenland's increasing geopolitical significance, and of American commitment to the security of the North American Arctic. Thule, like many Arctic military bases, embodied a new integration of science, security, and American foreign policy. The bases necessitated detailed diplomatic agreements and political impetus to be established and maintained, scientific and technical knowledge for their construction, and military expertise for their operation and staffing. The bases were also sites of scientific monitoring and experimentation, the results of which had defense and domestic policy implications.

American polar explorers played a vital role in the creation of Greenland's bases, including Thule, through their involvement in this integration of scientific, military, and political spheres.[2] They had records of distinguished military service, technical expertise, and close ties to intellectual and political elites. Through their expeditions, they not only served multiple simultaneous military and scientific functions, but they also facilitated the exchange of knowledge between Greenland and the United States. They acted as chaperones, mediating geographic space and the transfer of information between government, military, and academic interests.[3] Central to the explorer's role as chaperones connecting these spheres was the rapidly developing field of aviation. Aviation was crucial in connecting Greenland to the rest of the world, and, as Stephen Bocking has argued, aviation can itself act as a "site of exchange, enabling collaboration among disciplines as well as scientists, aviators, and institutions."[4] These explorers both directly and indirectly promoted this collaboration, resulting in American military bases in Greenland, which in turn provided physical spaces for increased direct interaction between scientists, the military, and government officials.

There is a growing body of scholarly literature on the intersections between scientific research, the military, and the environment in both the North American Arctic generally, and Greenland in particular.[5] American foreign policy with respect to Greenland, particularly in the late nineteenth and early twentieth centuries, however, has been largely overlooked in the English language literature.[6] While individual explorers, scientists, organizations, and academic networks have been beautifully treated, in the American context these have been generally divorced from discussions of the political, diplomatic, and military contexts in which they operated.[7] As a result, the complex interactions between the American government, military, and scientific communities in Greenland, along with the broader technological changes that affected the island's geopolitical position, have received little attention in the period prior to the Cold War.[8] While it is often recognized that American bases played a vital role in Cold War defensive strategy, there is limited writing on how they were established, or on the role of polar explorers in this process.

This chapter traces the evolution of the networks linking polar explorers, scientists, and US government and military officials, through the lens of technological changes wrought during the interwar period, and how these advances, particularly in aviation, changed American foreign policy with regard to Greenland. It will demonstrate the ways in which late nineteenth and early twentieth century polar explorers facilitated American base creation in Greenland, which in turn enabled the integration of American foreign policy, military, and scientific activities on the island. It will use case studies of three polar explorers—Rear Admiral Robert E. Peary, Rear Admiral Richard Byrd, and Colonel Bernt Balchen—to illustrate the roles their scientific knowledge and academic contacts, personal political connections, and mastery of emerging technologies—particularly in the context of global war, played in this process. Arguing that the explorers themselves were the critical nodes of the network linking academic, military, and foreign policy spheres, this paper examines how technological advances in the interwar period, in combination with strategic necessity during the Second World War, helped to establish and codify the structures, institutions, and policies used to advance scientific knowledge and military objectives in the Arctic during the Cold War.

Rear Admiral Robert E. Peary—to First World War

And stranger things have happened than that Greenland, in our hands, might furnish an important naval and aeronautical base...[9]

Rear Admiral Robert E. Peary was one of the United States' most renowned polar explorers (Figure 11.1). Peary was perhaps best known for his controversial claim to have been the first person to reach the North Pole, and is representative of the instrumental role that mid-nineteenth century American explorers played in the physical and intellectual exchange of knowledge between Greenland and the continental United States, and the eventual integration of Greenland into American security and foreign policy.

Chaperone of scientists and knowledge

One of the most significant challenges for Peary, and many of his contemporaries, in mounting expeditions to Greenland was the exorbitant expense of traveling to, and working in, the Arctic. In addition, at the time of Peary's expeditions, there was an ongoing debate in the United States about the

Figure 11.1 Polar explorer Rear Admiral Robert E. Peary during a Greenland expedition.

Source: Associated Press.

role of government and America's place in the world. A constant tension persisted between the desire to limit government spending on the one hand, and the United States' territorial ambitions in the Western Hemisphere on the other. Although there was some agreement on the expansion across the continental United States through Manifest Destiny, the place of non-contiguous Greenland was less straightforward, particularly given the perceived risks of travel to the island.[10] One by-product of these debates was that, unlike some government supported European polar expeditions, American explorers often faced significant challenges in raising funds for their expeditions, and American activities in Greenland were, for the most part privately conducted and funded, operating largely independent of government policy and military strategy concerns.[11]

The lack of readily accessible government funds forced Peary to become a strategic fundraiser and draw upon his expansive social and professional network in order to garner financial support for his numerous expeditions.[12] This network included academic scientists from prestigious universities, and Peary capitalized on a number of these relationships by charging scientists substantial fees for transport to Greenland to conduct summer fieldwork. These relationships not only subsidized Peary's expenses and lent scientific credibility to his work, but also promoted the exchange of knowledge and understanding about Greenland throughout the United States, through the academics' subsequent writings and public lectures concerning their work in Greenland.[13] These writings, which highlighted the facility of Arctic summer travel, helped to counter popular perceptions of the dangers of Arctic travel years before the publication of Stefansson's *The Friendly Arctic*.[14]

Peary used many additional avenues to promote his interest in and expeditions to Greenland, including speaking tours and the publication of numerous books and articles on his exploits.[15] This had obvious financial benefits for Peary, while raising public and political awareness about the potential value of the island to the United States. This was, however, a somewhat slow process. Peary was praised by many following his 1909 claim to have reached the North Pole, including former President Theodore Roosevelt, who wrote that he was

> inexpressibly rejoiced at his [Peary's] wonderful triumph; and proud beyond measure, as an American, that this, one of the greatest feats of the ages, should have been performed by a fellow countryman of ours.... We are all Captain Peary's debtors – all of us who belong to civilized mankind.[16]

In spite of the former President's enthusiasm, however, there remained a significant disconnect between the excitement regarding Peary's alleged feat and the administration's appreciation for Greenland's potential strategic utility. Peary would soon be reminded that it was not easy to convince the government of the value of the island.

After his 1908–1909 expedition, Peary had also telegraphed then-President William H. Taft and informed him of his achievement stating, "Have the honor to place the North Pole at your disposal." Taft, somewhat in contrast to his predecessor, replied:

> Thank you for your interesting and generous offer. I don't know exactly what I could do with it. I congratulate you on having achieved, after the greatest effort the object of your trip, and I sincerely hope that your efforts will contribute substantially to scientific knowledge.[17]

This continued lack of appreciation for Greenland's possible role in hemispheric defense was echoed during the First World War when the US government was considering abdicating potential American territorial claims to Greenland in order to purchase the Danish West Indies (now the United States Virgin Islands) from Denmark. American claims to Greenland were based largely on Peary's numerous expeditions to Northern Greenland, and he objected strongly to the proposed action. He was concerned not only about the status of his life's work, but also the long-term security of the hemisphere. In response, Peary launched a public and private campaign to convince President Woodrow Wilson's administration to retain its potential rights to Greenland. At the time, however, the United States remained neutral in the war, and President Wilson was growing increasingly concerned about the security of the Virgin Islands, which were situated adjacent to the recently completed Panama Canal.[18] The Canal not only drastically shortened global shipping routes and access to North America, but it also affected the physical security of the United States. St. Thomas, the largest of the three main Virgin Islands, had an excellent harbor that had been used as a coaling station during the American Civil War. Germany's explicit territorial ambitions and lack of a colony in the Caribbean heightened American concerns that the Germans would attempt to establish themselves at St. Thomas.[19] President Wilson considered the purchase of the Virgin Islands vital to American security, and the Danish government wanted to leverage the President's interest in order to solidify their legal claims to Greenland, another of their colonies. They made American recognition of Danish sovereignty over the entirety of Greenland one of the conditions of sale of the Virgin Islands to the United States.[20]

In a letter that would eventually be published in the *New York Times*, Peary argued forcefully for maintaining American claims to Greenland, particularly for future defensive purposes.[21] He noted the island's vast mineral resources, and stressed that its geographical location (in the Western Hemisphere) placed it clearly within the purview of the Monroe Doctrine, which sought to prevent, and ultimately reduce, European influence in the region. In addition, he predicted Greenland's potential strategic value as a location for American military bases given its location in the North Atlantic. He argued "in light of recent scientific and technological advances.... With the rapid shrinking of distances in this age of speed and invention, Greenland may be of crucial importance to us in the future." He also pointed out the ways in which the "present war has shown far flung may be the regions having a bearing on the struggle.... Greenland in our hands may be a valuable piece in our defensive armor. In the hands of a hostile interest it could be a serious menace."[22]

Peary was not alone in his concerns. American Secretary of State Robert Lansing wrote to the President directly regarding his objections to the terms of the agreement. In spite of Lansing's concerns, President Woodrow Wilson prioritized the tangible, immediate need to secure the Panama Canal over Greenland's future strategic potential.[23] President Wilson, aware of the pressing importance of the security of the Virgin Islands, informed his Secretary of State to "go ahead at once and take a chance on the defects, whatever they might be ... the treaty is so important that we would have to make every concession to get it through."[24] As a result of Wilson's insistence, in August 1916 Lansing issued a statement affirming Danish sovereignty over Greenland. Although public figures like Peary stressed the potential value of Greenland's strategic location in the event of a future war, contemporary national security considerations superseded Greenland's potential future utility, and the United States' purchase of the Virgin Islands was completed.

In spite of Peary's failure to convince the government of the potential value of Greenland, he continued to write publically about the need for the United States to increase its commitment to airpower.[25] Within two decades, however, geopolitical instability in combination with developments in materials science, meteorology, and aviation would force the United States to radically alter its policy with respect to Greenland.

Rear Admiral Richard Byrd—the interwar science, technology, and politics of aviation

The most significant factor for the shift in the geopolitical position of Greenland and the creation of US military infrastructure on the island were developments in aviation. Rear Admiral Richard Byrd is representative of the generation of aviators who made polar flight possible. Byrd, like Peary, was one of the United States' most decorated explorers (Figure 11.2).[26] A pioneering aviator, with connections to the highest levels of political office, Byrd was instrumental in establishing the place of aviation in the American military structure. He helped to test emerging aeronautical technology, and his contributions to transatlantic aviation during the First World War assisted with the "three stop hop" used to ferry North American-made planes across the Atlantic through Greenland en route to Europe during the Second World War. In addition, his deep connections to the wealthiest and most influential of American society facilitated the funding of his Greenland and North Pole flights, which in turn raised domestic and political awareness of the strategic significance of the polar regions.

Figure 11.2 Future American president Franklin Delano Roosevelt presents Rear Admiral Richard E. Byrd with a New York State distinguished service medal in 1930.

Source: Associated Press.

Pioneering aviator and polar promoter

Byrd's family was deeply integrated into elite political and military circles in the United States. During the First World War his father personally met with then-Assistant Secretary of the Navy Franklin Roosevelt and asked him to arrange for a desk job for his son in Washington.[27] Both Navy men with mutual interests in geography, Roosevelt and Byrd would develop a lifelong friendship that often involved the future president's political, financial, and moral support of Byrd's expeditions.[28] In addition to deepening his political connections in the capital the war eventually provided Byrd the opportunity to fulfill his ambition to fly.

Byrd was convinced that the next major feat in flying would be crossing the Atlantic Ocean. During the First World War, he came up with a plan to deliver one of the largest planes ever built, the NC-1, to Europe via the Azores. He thought that, particularly in the context of the war, the delivery of such a large plane across the ocean would improve the public's perception of aviation and have a "distinct impact on the morale of the enemy."[29] He enlisted his friend Walter Camp to assist him with his cause.[30] Camp was well connected and Byrd thought that "his large acquaintance among rich and influential men would make him a keystone in importance."[31] Byrd's instincts proved correct and, unbeknownst to him, on the week of May 19, 1918, Camp, accompanied by Rear Admiral Robert E. Peary went to the Navy Department and "urged that the flight be undertaken." Byrd notes in his memoirs that he found the effort on the part of Peary, one of Byrd's idols, particularly touching as he "had long thought a flight across the North Pole was possible."[32] Although Byrd did not make the NC-1 flight himself, the plans he developed to ferry planes across the North Atlantic via Newfoundland facilitated the later flights of the NC-4 in May 1919. The subsequent non-stop transatlantic flight by Royal Air Force (RAF) pilots John Alcock and Arthur Whitten Brown two weeks later demonstrated the possibilities of long-range transatlantic, and ultimately trans-polar, flight.[33]

Byrd's polar flights

Although Byrd proclaimed a life-long ambition to fly across the North Pole, it was his connection to William A. Moffett, the First Chief of the Navy's Bureau of Aeronautics, which ultimately spurred him north. In 1924 Byrd received orders to assist Admiral Moffett with the preparations for the proposed flight of the US Navy airship *Shenandoah* across the North Pole. The flight had the interest of President Calvin Coolidge, and the team comprised Peary's skipper, Captain Bob Bartlett, and Commander Fitzhugh Green. The preparations were well underway when the President unexpectedly called the project off. Byrd, who now had his eyes fixed on the Pole, quickly partnered with Bartlett and made arrangements for a private Arctic air expedition. Byrd's political and professional connections helped him to secure $15,000 for the Greenland mission from private donors Edsel Ford and John D. Rockefeller. It also brought him into contact with another polar explorer, Donald B. MacMillan, who was vying for the same planes from the US military. The paucity of suitable planes led the Navy Department to insist the two expeditions join forces, with MacMillan directing his expedition and Byrd commanding the Naval unit. Unlike Byrd, who had yet to spend any time in the Arctic, MacMillan was an accomplished polar explorer with decades of experience in the far North.[34] He, like Bartlett, had assisted with Peary's 1908–1909 expedition. MacMillan made numerous subsequent trips to the high North, and he became well known for supporting the work of scientists and investigating the efficacy of technological equipment in the Arctic.

Although the eventual MacMillan-Byrd Greenland expedition did not reach the North Pole, it made a number of advances in aviation and broadcast science. During the expedition, Byrd and Floyd Bennett flew across thousands of miles of Greenland's inland ice, often achieving in an afternoon what had taken Peary many months to accomplish on the ground.[35] Concurrently, MacMillan continued his work on radio communications, which were heard "by naval radio operators as far away as Tasmania."[36]

The Greenland expedition thus both embodied the integration of science, military, and private partnerships, and demonstrated the potential of Arctic flying. The press and high-profile backers of the expedition also raised public and political awareness of the place of Greenland in the future of flight. In their ongoing correspondence, future President Franklin Roosevelt wrote to Byrd stating that he was "thrilled to hear of his Polar trip," and noting that Byrd "would be the next Peary."[37] For Byrd personally, following the expedition he "was stirred with the conclusion that aviation could conquer the Arctic."[38] He and his assistant Floyd Bennett soon began preparations for a controversial North Pole flight.

In 1926, following the relative success of his Greenland flights, Byrd announced that he would attempt to reach the Pole by air. Backed once again by Ford and Rockefeller, and with new donors including Vincent Astor and J.D. Ryan, news of Byrd's well-funded flight was splashed across newspapers around the country, including the front page of the *New York Times*.[39] In addition to his high-placed financial backers, Byrd's flight also had an embedded reporter, and he planned to bring radio equipment along on his flight in order to "keep the world informed" of the flight's progress.[40] Adding to American interest in the flight were the other international competitors also attempting to reach the Pole. Byrd was one part of a heated three-way race with Australian airman Sir George Hubert Wilkins and Norwegian explorer Roald Amundsen, who were also vying to be the first to reach the North Pole by air.[41] Byrd and his co-pilot Floyd Bennett eventually took off from King's Bay, Spitzbergen, Norway around midnight on 9 May 1926 and returned at 6 pm GMT.[42] Within hours, the front page of the *New York Times* proudly exclaimed "Byrd Flies to North Pole and Back; Round trip from Kings Bay to the North Pole in 15 hours 51 minutes; circles top of the world several times."[43]

Although today the veracity of Byrd's claim to have flown over the Pole has been called into question, the flight raised awareness of the possibilities of polar flight at the highest level of the American government. Byrd's impeccable safety record during the flights also helped to convince the public that aviation was safe, including in the high North. The press and politicians drew numerous comparisons between Peary's expeditions and Byrd's flights.[44] President Coolidge noted that there could be no "more striking illustration of mechanical progress since the year 1909," and that Byrd had achieved in 15 hours what took Peary "about two-thirds of a year."[45] Coolidge also highlighted that it had taken months for the news of Peary's claim to reach the world 17 years earlier, but, as a result of advances in broadcast technology "within hours the radio had announced the triumph to the four quarters of the earth."[46] Byrd's achievement raised the global profile of aviation and the polar regions. The flight was seen as a great step in the advancement of aviation, as it proved that "airplanes are practical in the extreme cold of the polar regions, as well as tropical heat."[47] Byrd's flight was also seen to have "cinched" "America's claim to the Pole," and raised the profile of the Arctic in Washington.[48] Most importantly, Byrd's accomplishment was a powerful demonstration of the Arctic's geographic proximity to North America, and highlighted the potential geopolitical and strategic significance of the polar regions in the age of aviation. Byrd's flight, however, would likely not have been possible were it not for the assistance of polar aviator Bernt Balchen.

Colonel Bernt Balchen—establishment of bases: Second World War and the early Cold War

While Peary chaperoned scientists to and from Greenland and conveyed knowledge of the Arctic to the American public and political elite, and Byrd tested new aviation technologies while capitalizing on his personal and professional connections to raise domestic awareness about the polar regions, Colonel Bernt Balchen, perhaps the least well known of the three, had the most tangible role in the creation of American military bases in Greenland. Norwegian by birth and a naturalized American citizen, Balchen became an integral part of an international scientific military and political elite dedicated to the advancement of aviation (Figure 11.3). Drawing from decades of experience in the polar regions, Balchen, even more than Byrd, utilized, tested, and developed the systems and technology that made polar aviation

Figure 11.3 Colonel Bernt Balchen with aviator Amelia Earhart at Teterboro Airport, Hasbrouck Heights, New Jersey, May 1932.

Source: Associated Press.

Then, drawing from his scientific and technical expertise and experience, he helped to establish critical American military bases in Greenland, including Bluie West 8 (Kangerlussuaq) during the Second World War and Thule during the early Cold War. Later, he leveraged his international connections to link a network of commercial airports and military infrastructure throughout the Western Arctic.

International polar network

Bernt Balchen was born into a Norwegian family with deep ties to the military and polar expeditions. Balchen's uncle, Major General Oluf Dietrichson, was a member of Dr. Fridtjof Nansen's party that had crossed the Greenland ice cap in 1888.[49] His uncle also introduced the family to Bernt Balchen's childhood hero, Roald Amundsen. At age 12 Balchen met Amundsen for the first time and professed his desire to become a polar explorer.[50] An avid outdoorsman, Balchen's degree in forestry engineering reflected his technical interests and love of northern environments.[51] Following the completion of his degree in 1916 he served in several European militaries, and then attended the Royal Norwegian Navy Air Service's flight school. Balchen's technical abilities eventually led to him joining his idol Amundsen on his team's third attempt to reach the North Pole by air, which further broadened his connections to a growing international network of polar aviators.[52]

As mentioned, in the spring of 1926 Roald Amundsen's team was one part of the intense three-way race to reach the North Pole by air. Amundsen, along with his Italian and American partners Umberto Nobile and Lincoln Ellsworth, planned to fly over the pole from Svalbard in the airship *Norge*. Australian airman Wilkens planned to reach the Pole from Alaska, but was constantly hamstrung by unfavorable weather conditions and was not seen as an immediate threat.[53] Amundsen's team, however, was surprised by the arrival of Richard Byrd at King's Bay just days before the *Norge's* planned flight. During test flights Byrd's plane, the *Josephine Ford,* had technical issues and Amundsen encouraged Balchen to assist with Byrd's flight preparations. Byrd's subsequent flight and claim to have reached the North Pole by air remains a source of controversy and speculation, but one result was that Balchen was invited to join Byrd's technical team.[54] As there was no space for Balchen on the flight of *Norge*, Balchen signed on with Byrd's team and sailed to the United States. Once there he quickly extended his increasingly impressive social and professional network.

Balchen's initial responsibilities for Byrd did not require aviation or engineering skills, but instead involved traveling across the United States on a publicity tour with Byrd's North Pole plane, the *Josephine Ford*. Balchen and Floyd Bennett flew from city to city across the United States, answering questions at displays of the *Josephine Ford* during the day and attending dinners with local dignitaries most evenings. Although Balchen found his duties tedious, they were critical in establishing his connections in the United States. A crucial step in this process was an invitation to join a dinner for "Quiet Birdmen," a secret international group of aviators and aviation enthusiasts, most of whom were survivors from the First World War and dedicated to the future of aeronautics. The group was a virtual who's who of military, business, and government elite. The global membership base included future Field Marshall of the Luftwaffe Wolfram Freiherr Von Reichtoffen and famed German flying ace Ernst Udet among many others. Present at the first dinner Balchen attended were Hap Arnold, Carl Spaaz, Ira Eaker, Jimmy Doolittle, Harry Bruno, and General Billy Mitchell. Also present was Tony Fokker, who would later give Balchen a job at his aeronautical production facility.[55] Balchen's family connections, military service, and work with Amundsen and Byrd, along with his involvement in international secret societies, like the Quiet Birdman group, and later the Freemasons, set the stage for his role in the development of polar aviation and the establishment of Western military infrastructure in the high North.[56]

Polar technology

Balchen's impressive personal and professional network was further enhanced by his technical expertise. He enjoyed tackling seemingly impossible problems, and there was no shortage of such problems in the nascent field of polar aviation. Balchen was constantly developing innovative solutions for the challenges of flight in the far North. During his time with the Royal Norwegian Navy Air Service, he was assistant to the chief test pilot at the naval aircraft factory.[57] He spent time experimenting with various types of ski landing gear for aircraft, which "required both strength and flexibility to hold up." In one of his early encounters with Richard Byrd's team, Balchen noticed that the *Josephine Ford* was having issues with its landing gear. His suspicions were confirmed when the aircraft lost a ski on a test run.[58] Balchen used oars from one of Byrd's lifeboats to reinforce the *Josephine Ford's* landing gear, which facilitated its successful flight. Byrd, writing of Balchen, stated that

> at that time we knew almost nothing about flying a large three-engined plane from the snow with skis, and we had considerable difficulty in starting with heavy loads on the snow and from breaking the skis. Bernt Balchen saw our difficulties and showed us how the Norwegians burn into the skis a mixture of tar and resin which greatly increases their efficiency.[59]

In 1927 Balchen went to work for Tony Fokker, where he became involved closely in research and development for Fokker planes. Spurred by a gold rush in Northern Ontario, Balchen developed modified ski landing gear for an order of Fokker Universals for Western Canada Airways.[60] Balchen was dispatched to the Canadian High North, where in addition to assisting with the engineering, manufacturing, and operation of aircraft, he also helped train bush pilots on cold-weather flying and the use of the new technologies.[61]

Although most pilots in the period relied on sight navigation, Balchen also mastered "dead reckoning," a technique that relied on complicated mathematical calculations based on compass headings, speed, elapsed time, and fuel consumption in order to navigate aircraft.[62] He carried a slide rule on all of his flights to assist with his measurements, which often proved essential to the safety of the crew, as in his flights with Byrd.[63]

Following the transatlantic flight, Balchen was asked to join Byrd on his expedition to the Antarctic, which would be the first expedition to employ aircraft on that continent. Surrounded by Antarctic expedition scientists like geologist Larry Gould, Balchen continued to make his own study of aerodynamics and mathematics, and to improve and refine cold weather flying techniques.[64] He mixed his own engine oil, and experimented with the length of time required to warm aircraft engines in polar conditions before flying.[65] On November 29, 1929 at 1:14 am Balchen, accompanied by Byrd, piloted the *Floyd Bennett* over the South Pole. The transmission from the *Floyd Bennett* was picked up at the *New York Times* station and broadcast by loudspeaker into Times Square, something that would have been technologically impossible a decade earlier.[66]

German invasion of Denmark and Norway

By the early 1940s Balchen had worked in the Scandinavian Arctic, the Canadian Arctic, Greenland, and the Antarctic. He had worked in the military, for private companies, in the interests of science, and in national interests. His network of fellow aviators stretched around the world to the highest levels of multiple militaries and governments. He was granted American citizenship by a special Act of Congress, assisted by fellow Quiet Birdman Fiorello LaGuardia.

In the summer of 1941, prior to official American entry into the Second World War, Balchen was called to Washington for a meeting with the head of the Army Air Corps, General Hap Arnold. He wanted Balchen to head a team to set up American bases in Greenland "for staging aircraft from the United States to the Theater of Operations in Europe, and for patrolling the coast against enemy attack."[67] Arnold's request marked a major departure for the American military and the Army in particular. While the Navy had played a part in many polar expeditions, the Army had shown limited interest in the Arctic generally or Greenland in particular. American interest in Greenland, however, had become acute one year earlier on 9 April 1940, when Germany invaded Demark. Greenland remained a Danish colony and there were serious questions about the status of the island in the context of the Second World War.[68] The US Coast Guard secretly occupied

Greenland that summer, and on 9 April 1941 the United States signed an agreement for the defense of the island with the Danish Ambassador to the United States.[69]

The rapid developments in aviation, assisted by both Byrd and Balchen, in the 24 years since the American government abdicated its potential claims to Greenland, had radically shifted the political position of the island. Following the invasion of Denmark, President Franklin Roosevelt had raised the question of the Monroe Doctrine to justify the inclusion of Greenland in the western hemisphere and American military involvement on the island.[70] Now for the first time Greenland would play a critical role in American wartime military strategy. Balchen was dispatched to Greenland to perform an "almost impossible feat of conceiving and building the chain of small airports over the top [of the world] ... which enabled us to ferry short-range ships to England at a time when U-Boat losses were approaching the impossible."[71]

Balchen began work on establishing bases in July of 1941. He recalled at the first meeting in Washington between the Air Corps, the Army Engineers, and the contractors that they had "precious little to go by." They had some aerial photographs of potential sites and a letter from Hap Arnold on the importance of the mission. Balchen recalled, "it was a unique order and assignment, challenging and therefore attractive."[72] The eventual site selection for Bluie West 8 was based on an amalgamation of information derived from scientific, commercial, and military expeditions. These included Herbert Hobbs' 1928 pioneering University of Michigan expedition. The expedition helped to validate the Scandinavian Polar Front theory, which posited that weather in Western Europe could be predicted from weather patterns in Greenland. It also established a camp and small airstrip near Sondre Stromfjord, on the west coast of the island.[73] In the same year pilots Bert Haskell and Shorty Cramer made an emergency landing on the site, and in 1933 Charles Lindbergh scouted the area for Pan American Airways and mentioned it could be used as a potential commercial landing field.[74] The Coast Guard assisted with additional aerial photos, and it was decided that the area could be adapted to suit war-time purposes. Balchen's papers are replete with site survey notes and sketches of what would become Bluie West 8 (later Sondrestrom Air Base, now Kangerlussuaq Airport), and details of necessary modifications to the airplanes that would eventually be used there.[75]

In August of 1942, with Bluie West 8 established, Balchen made an aerial survey of Thule, Greenland. He then wrote a "lengthy recommendation" to General Arnold regarding its potential as an air base, noting its gravel flats, space for barracks, and deep-water harbor that could accommodate "any cargo ship." Eight years later, in 1950, Air Force Secretary Stuart Symington read Balchen's recommendation regarding Thule. Symington's successor, Secretary Thomas Finletter, met with Balchen to discuss the details of his proposal to establish Thule as the central search and rescue center for all northern military operations. By February of the following year, Balchen was back in Greenland with construction engineers to assist with the establishment of Thule Air Base. Balchen noted that "Thule was a strong rampart in the path of any enemy invader ... and thanks to the ingenuity and skill of American and Canadian engineers, we can use the natural resources of the icecap itself to create permanent runways anywhere above the melting line."[76] The base necessitated some of the most complicated logistics ever used in the high North. Between 1951 and 1952 more than 500,000 tons of cargo and 2,000,000 tons of petroleum products were delivered to the building site, involving more ships than any previous operation conducted by United States forces in the Arctic Region. Vice Admiral William M. Callaghan noted during the base's construction that "this may be a moment to unify the Armed Forces. For the construction of this important base is a shining example of the principle of cross-service support within our Armed Forces."[77] Nearing the completion of Thule airbase, the *New York Times* reported on the technical difficulties of the base's construction, noting "Thule's creation is a tribute to engineering and supply skill and, at the very outset, to the long and exact memory of Col. Bernt Balchen, veteran Arctic and Antarctic flier."[78]

Conclusions

In 1954, President Dwight D. Eisenhower met in the Oval Office with Matthew Henson, the last surviving member of Peary's famed 1908–1909 polar expedition. In stark contrast to Peary's challenges in communicating the strategic value of Greenland to President Taft, President Eisenhower himself stressed the increasing significance of the island to American security, pointing at Greenland's location on a globe and noting to Henson and his wife that "now we have air bases all along there."[79]

This interaction between Eisenhower and Henson encapsulated the radical transformation that had taken place in American foreign policy with respect to Greenland specifically and the Arctic in general over the course of the first half of the twentieth century. Throughout that period, the actions of American polar explorers had facilitated this transformation. American exploration of Greenland in the late nineteenth and early twentieth centuries established early links between somewhat distinct scientific, government, and military communities. It was not until the First World War, however, that the potential American territorial claims to Greenland from American exploratory and scientific activities were seriously considered as a matter of national security and, by extension, foreign policy. At the time, though, the argument for Greenland's strategic value was widely and roundly dismissed by the US government, over the objections of the explorers like Peary who had established the claims. However, during the interwar period, the interests of the government and the explorer-scientists would necessarily become more closely aligned, largely as a result of advances in aviation, which were themselves made possible by aviators like Byrd and Balchen. Whereas previously the US government had had only limited and intermittent interest in Greenland, the island would become a focal point of the United States' Second World War security strategies. During the Second World War and the early Cold War, the full integration of academic and government-funded scientific research in Greenland and the alignment of this research with the government's strategic and policy aims proved necessary in order to provide the manpower, expertise, and innovation needed to advance the US government's objectives. Central among these objectives was the establishment of a network of military bases across the Arctic—including Thule, which in 1953–1954 alone hosted scientists from Dartmouth, served as a proving ground for YH-21 cold weather helicopters, and facilitated exploration of the possibilities of several trans-polar air routes.[80] Throughout the second half of the twentieth century, Thule and other bases like it would serve as spaces where the integration of science, security, and American foreign policy could emerge in full form, and polar explorers' visions of the Arctic's value and strategic potential would at last be realized.

Notes

1. "Guarding our Northern Frontiers," *New York Times (NYT)*, 18 August 1953.
2. The body of academic literature on the politics of American bases is expansive, but few of these general studies deal with Greenland directly, if at all. See: Alexander Cooley, *Base Politics* (Ithaca: Cornell University Press, 2008), Catherine Lutz, *The Bases of Empire* (New York: New York University Press, 2009), Ruth Oldenziel, "Islands," *Entangled Geographies: Empire and Technopolitics in the Global Cold War, ed.* Gabrielle Hecht (Cambridge: MIT Press, 2011), 13–42. Recent studies relating to Greenlandic bases discuss their impact on culture, politics, and scientific research, but do not address the way in which they were established, or American policy perspectives with respect to the island prior to their establishment, Nikolaj Petersen, "The Politics of US Military Research in Greenland in the Early Cold War," *Centaurus*, 2013, 55(3): 294–318; Anthony J. Dzik, "Kangerlussuaq," *Bulletin of Geography. Socio-Economic Series*, 2014, 24(24): 57–69; Kristian H. Nielsen, "Transforming Greenland," *New Global Studies; Berlin*, 2013, 7(2): 129–54.
3. A discussion of knowledge chaperones, the people, packaging, and institutions that serve to facilitate the traditional mechanisms through which knowledge flows can be found in Mary Morgan, "Travelling facts," in *How Well Do Facts Travel?*, eds. Mary S. Morgan and Howlett, Peter (Cambridge University Press: Cambridge, UK, 2011), 24–27; Stephen Bocking has done interesting work on the mobility of scientific information generally, "Mobile Knowledge and the Media," *Public Understanding of Science*, 2012, 21(6): 705–23; and with respect to aviation in particular "A Disciplined Geography," *Technology and Culture*, 2009, 50(2): 265–290. See also Lissa Roberts, "Situating Science in Global History: Local Exchanges and Networks of Circulation," *Itinerario* 33, 2009, 33(1): 9–30; and Neil Safier "Global Knowledge on the Move: Itineraries, Amerindian Narratives, and Deep Histories of Science," *Isis*, 2010, 101: 133–145.

4. Bocking, "A Disciplined Geography," 268.
5. As evidenced by this volume and the work of many of its contributors e.g. Andrew Stuhl, *Unfreezing the Arctic* (Chicago: The University of Chicago Press, 2016), Susan A. Kaplan and Robert McCracken Peck, *North by Degree* (Philadelphia: American Philosophical Society, 2013), regarding Greenland in particular see Janet Martin-Nielsen, "The Other Cold War," *Journal of Historical Geography*, 2012, 38(1): 69–80; and *Eismitte in the Scientific Imagination* (New York, NY: Palgrave Macmillan, 2013); and Ronald Edmund Doel, *Exploring Greenland* (New York: Palgrave Macmillan, 2016).
6. There is a substantial amount of work on the diplomatic and security relationship between the United States and Denmark in Danish e.g. Bo Lidegaard, *I Kongens Navn* (København: Samleren, 1997) and *Overleveren* (København: Gyldendaal, 2006); Finn Løkkegaard, *Det danske Gesantskab i Washington* (Copenhagen: Glydendal, 1968); Poul Villaume and Thorsten Borring Olesen, *I blokopdelingens tegn* (København: Gyldendal Leksikon, 2005). Lidegaard's work has recently been translated into English: Bo Lidegaard and W. Glyn Jones, *Defiant Diplomacy* (New York: P. Lang, 2003).
7. One of the most well-known works on this subject consciously does not engage in American high politics/international relations: Michael F. Robinson, *The Coldest Crucible* (Chicago: University of Chicago Press, 2006).
8. Matthew Farish's work deals with geography, science, and the American military, but does not directly deal with American politics and international relations in Greenland in this earlier period. See Matthew Farish, "Canons and Wars: American Military Geography and the Limits of Disciplines," *Journal of Historical Geography*, 2015, 49: 39–48; Trevor J. Barnes and Matthew Farish, "Between Regions," *Annals of the Association of American Geographers*, 2006, 96(4): 807–26; Matthew Farish, "The Ordinary Cold War," *Journal of American History*, 2016, 103(3): 629–55; Matthew Farish, *The Contours of America's Cold War* (Minneapolis: University of Minnesota Press, 2010) in particular chapters 1, 2, and 4.
9. Robert E. Peary, "Greenland and the Danish West Indies," 1916, Official File (OF) 3953, Papers of Franklin Delano Roosevelt; Franklin D. Roosevelt Library, Hyde Park, New York, 3–4.
10. American Secretary of State William H. Seward and United States Minister to Denmark Maurice Francis Egan were interested in acquiring Greenland for the United States, however their proposals met with little public or political approval. See Walter Stahr, *Seward* (New York: Simon & Schuster, 2012), 516, "Denmark," *Papers Relating to the Foreign Relations of the United States 1917* (Washington: Government Printing Office, 1926), 457–706.
11. This is not to say that the American government did not sponsor any expeditions, for example the Navy provided a generous $50,000 towards Charles Francis Hall's third expedition, but these efforts were sporadic over the course of the late nineteenth and early twentieth century. Charles Francis Hall and Charles Henry Davis, *Narrative of the North Polar Expedition: U.S. Ship Polaris* (Washington: Govt. Print. Office, 1876).
12. Kelly Lankford, "Arctic Explorer Robert E. Peary's Other Quest," *American Nineteenth Century History*, 2008, 9(1): 37–60; Douglas Wamsley, "'We are Fully in the Expedition': Philadelphia's Support for the North Greenland Expeditions of Robert E. Peary, 1891–1895," *The Geographical Review*, 2017, 107(1): 207–35.
13. Wamsley, "We are fully in Expedition," 209; Ralph S. Tarr, "The Cornell Expedition to Greenland," *Science*, 1896, 4(93): 522–23.
14. Vilhjalmur Stefansson, *The Friendly Arctic* (New York: The Macmillan Company, 1921).

15. Some of his works include Robert E. Peary, *Northward over The 'Great Ice'* (New York, NY: Frederick A. Stokes Co., 1898); Robert E. Peary, *Nearest the Pole* (London: Hutchinson & Company, 1907); Robert E. Peary, *The North Pole* (New York: Frederick A. Stokes co., 1910); Robert E. Peary, "The Conquest of the North Pole," *The Independent,* 16, (September 1909): 623. His wife's account of her time in Greenland was also quite popular: Josephine Diebitsch Peary, *My Arctic Journal,* (1894).

16. Theodore Roosevelt, "Theodore Roosevelt's Tribute to Peary's Triumph" *Hampton's Magazine* (1910): 1.

17. "Taft has Faith in Peary," *NYT,* 9 September 1909.

18. Charles Callan Tansill, *The Purchase of the Danish West Indies* (London: Oxford University Press, 1932), 2.

19. Isaac Dookhan, *A History of the Virgin Islands of the United States* (Epping: Caribbean University Press, 1974), 249–58; Gordon K. Lewis, *The Virgin Islands, A Caribbean Lilliput* (Evanston: Northwestern University Press, 1972), 2.

20. "Denmark," 457–706.

21. Robert E. Peary, "Greenland as an American Naval Base," *NYT,* 11 September 1916, 8.

22. Peary, "Greenland and the Danish West Indies," 1916, OF 3953, 3–4.

23. Very little attention was paid at the time to the abdication of American rights to Greenland in order to purchase the islands. Indeed, although it is mentioned extensively in the State Department records of the time, very few studies of the history of the Virgin Islands mention this was a condition of the sale. William W. Boyer, *America's Virgin Islands* (Durham: Carolina Academic Press, 1983); Lewis, *The Virgin Islands;* Dookhan, *A History of the Virgin Islands;* and Tansill, *The Purchase of the Danish West Indies.*

24. Lansing Papers, Library of Congress, 20 no. 26 (July 1916): 1–3.

25. Robert E. Peary, "Defense Is in the Air," *The Independent* 19 February 1917, 302; Robert E. Peary, "Command of the Air," *The Annals of the American Academy of Political and Social Science,* 1916, 66: 192–99.

26. Richard Evelyn Byrd, *Skyward* (New York: Blue Ribbon Books, 1928).

27. Memorandum FDR to Bureau of Navigation re-Lieut. R.E. Byrd Jr. Interdepartmental Correspondence Box 11, FDR Papers as Assistant Secretary of the Navy. Franklin and Eleanor Roosevelt Presidential Library and Museum, Hyde Park, New York, United States of America.

28. Franklin Delano Roosevelt Papers as President: Official File (OF), Personal File (PPF), and President's Secretary's File (PSF).

29. Byrd, *Skyward,* 61.

30. "Walter Camp," *NYT,* 16 March 1925, 18.

31. Byrd, *Skyward,* 62.

32. Byrd, *Skyward,* 62.

33. "Alcock and Brown Fly Across the Atlantic," *NYT,* 16 June 1919, 1.

34. Donald B. MacMillan, *How Peary Reached the Pole* (Boston: Houghton Mifflin, 1934); *Etah and Beyond; or Life within Twelve Degrees of the Pole* (Boston: Houghton Mifflin, 1927); *Four Years in the White North* (New York: Harper & Brothers, 1918).

35. Byrd, *Skyward,* 150.

36. Genvieve Le Moine, et al., "Introduction," in *How Peary Reached the Pole* (Montreal: McGill University Press, 2008), xxxvii.

37. Letter FDR to D. Byrd 7 April 1925, FDR: Family Business and Personal Correspondence, Box 1.

38. Byrd, *Skyward,* 164.

39. "Byrd Sails North this Month to Hop 400 Miles to Pole," *NYT*, 27 March 1926.
40. "Byrd to use Radio on his Hop to the Pole," *NYT*, 26 March 1926.
41. "Times to have news of three attempts to explore Arctic," *NYT*, 3 March 1926.
42. "First News of Byrd's Great Feat as it Reached the Times," *NYT*, 10 May 1926.
43. "Byrd Flies to North Pole and Back;" *NYT*, 10 May 1926.
44. "Peary's Pole Trip at up 429 Days," *NYT* 10 May 1926.
45. Byrd, *Skyward*, 205.
46. President Calvin Coolidge quoted in Byrd, *Skyward*, 205. Members of the Radio Manufacturers Association passed a resolution congratulating Byrd at their convention. "Radio Men Praise Byrd," *NYT* 11 May 1926.
47. "After Reaching the Pole," *NYT* 11 May 1926.
48. "Byrd Flies to North Pole and Back;" *NYT* 10 May 1926; see also: "Nation's Leaders Laud Byrd's Feat," *NYT* 10 May 1926.
49. Balchen, *Come North*, 18.
50. Balchen, *Come North*, 17.
51. Index, Bernt Balchen Papers, Manuscript Division, Library of Congress, Washington, D.C., 3.
52. Balchen, *Come North*, 20.
53. Balchen, *Come North*, 37.
54. Byrd's claim to have reached the Pole was challenged from the moment it was announced, and these challenges persisted throughout the twentieth century see "Norwegian Views Differ; some authorities doubt that Byrd Reached the Pole," *NYT*, 11 May 1926.
55. Balchen, *Come North*, 60–61.
56. Irwin M. Yarry, "Masonry in Polar Expeditions, Part II," *The Masonic Philatelist* 19 no. 6 (February 1963): 3–6.
57. Balchen, *Come North*, 70.
58. William Dick, "Byrd Tests Skis on Fokker Plane," *NYT*, 4 May 1926, Balchen, *Come North*, 36.
59. Letter Richard Byrd read at reception honoring Bernt Balchen July 2, 1930, Box 9 folder 7, Bernt Balchen Papers, Manuscript Division, Library of Congress, Washington, D.C.
60. Balchen, *Come North*, 71.
61. Balchen, *Come North*, 76–77.
62. On sight navigation: B.A.S. Shantz, "Appendix 4: Weather Forecasting and Summary," in Nicholas Polunin, *Arctic Unfolding* (London: Hutchinson and Co, 1949): 288; Balchen, *Come North*.
63. Divergent accounts of this flight are found in Byrd's *Skyward* ch.13 and Balchen's *Come North* ch. 4.
64. Laurence McKinley Gould, *Cold* (New York: Brewer, Warren, & Putnam, 1931).
65. Balchen, *Come North* 190.
66. Balchen, *Come North* 181–191.
67. Balchen, *Come North*, 214.
68. Lawrence Martin, "The Geography of the Monroe Doctrine and the Limits of the Western Hemisphere." *Geographical Review*, 1940, 30(3): 525–28.
69. D.A. Berry, "Cryolite, the Canadian Aluminum Industry, and the American Occupation of Greenland during the Second World War." *The Polar Journal*, December 2012, 2(2): 218–35; D.A. Berry, "The Monroe Doctrine and the Governance of Greenland's Security," in *Governing the North American Arctic* (London: Palgrave Macmillan, 2016): 103–21.
70. Franklin D. Roosevelt, *Complete Typescript Reports of Press Conferences,* Vol. XV (Washington, 1933–1945).

71. Letter L.C. Reynolds to the Manager of the Associated Press, Box 8 Folder 7, Balchen Papers.
72. "Memorandum Re-Christmas1941 at Thule," OV II, 21–30, Balchen Papers.
73. William H. Hobbs, "Scientists to Study the Cradle of Storms", *NYT,* 14 February 1926; "The First Greenland Expedition of the University of Michigan." *Geographical Review,* January 1927, 17(1): 1–35.
74. Balchen, *Come North,* 216.
75. Technical Data, Supply, and Diagrams, Box 5 Folder 1, Balchen Papers.
76. Balchen, *Come North With Me,* 306.
77. "Military Sea Unit Cites Hazards," *NYT,* 24 September 1952.
78. Austin Stevens, "Building of Thule Aids Air Mastery," *NYT,* 22 September 1952.
79. "President Greets Last Survivor of Peary Arctic Dash," *NYT,* 7 April 1954.
80. "Dartmouth takes colors to the Arctic," *NYT,* 22 Aug. 1954; "USAF to Test 2 YH-21s in Arctic," *NYT,* 18 Aug 1953; "Flights over Pole to Europe Nearer," *NYT,* 13 March 1953.

12 Institutions and the changing nature of Arctic research during the early Cold War

Lize-Marié van der Watt, Peder Roberts, and Julia Lajus

Historians of science have long recognized that bricks-and-mortar institutions play central roles in the production of knowledge, and in the creation and maintenance of cultures that facilitate that production. Studies of early modern learned societies have emphasized their role as meeting-places where experiments were conducted and facts agreed upon, with these processes embedded within the political and cultural contexts of their times and places.[1] The flowering of these societies under royal patronage across Enlightenment-era Europe pointed to their additional function as emblems of national achievement: tributes to the vitality of a society and its monarch. In the nineteenth century, the acquisition and collation of geographical and cartographical knowledge at locations such as the British Admiralty produced new linkages between states and learned societies.[2] As Felix Driver has argued in the case of Britain's Royal Geographical Society (RGS), guidelines for geographical knowledge production helped to constitute and to manage a culture of exploration informed by the British imperial project.[3] It is no coincidence that many of the most famous polar expeditions of the nineteenth and twentieth centuries had close connections not only to organizations such as the RGS, the Norwegian Geographical Society, and the National Geographic Society, but also to organizations such as the Australasian Association for the Advancement of Science and the French Academy of Sciences.[4] In the case of the Russian Empire and the Soviet Union, scientific institutions such as the Academy of Sciences played a key role in the development of knowledge about this vast country, including its far northern regions.[5]

The aim of this chapter is to use a series of research institutes as windows into the changing nature of Arctic science and politics during the early Cold War. We draw upon materials concerning Canada, the United States, Norway, Britain, and the Soviet Union to sketch differences and note similarities. While political commitments at the national level were undeniably important, we argue that Arctic research institutes cannot be reduced to manifestations of geopolitical circumstance. Our focus also permits comment on professionalization in Arctic research, which we regard as a cultural phenomenon rather than an index of a field's maturity. Larger, long-term research projects—particularly in the geophysical sciences—demanded both material and moral support from states. Individuals like Graham Rowley (1912–2003) and Brian Roberts (1912–1978) could craft careers as Arctic specialists, moving between government departments and non-state (or quasi-state) institutes. Knowing the Arctic remained important for its own sake, while also serving as a means toward control and development.[6]

The picture that we draw is intended to be illustrative rather than definitive. We sketch how the geopolitical reconfigurations that emerged during and after the Second World War led to changes within existing Arctic research institutes, the creation of new organizations, and even the definition of Arctic spaces (notably the inclusion of Greenland in the North American Arctic). Our aim is to identify elements of institutional culture and identity that persisted after 1945 while also arguing that the institutional landscape was inextricably international, with organizations cooperating and competing with each other. We keep a close eye on how Arctic research institutes acted as both instruments of geopolitical activity and reflections of geopolitical visions. This increasingly included cooperation through formal and informal international networks. We draw attention to the interplay between logistical and geopolitical frameworks and choices in research agendas, arguing that a focus on institutions can reveal how they both exploited and were constrained by new challenges including most notably the International Geophysical Year (IGY) of 1957–1958.

Second World War and its legacies

When the Allied powers declared victory over Japan on 8 August 1945, it brought to an end a conflict that reshaped how Arctic spaces were viewed in European and North American metropolises. Some of the changes resulted directly from the need to create and operationalize knowledge of Arctic environments for warfare, as when the British Admiralty coopted both staff and office space at the Scott Polar Research Institute (SPRI) in Cambridge.[7] On a larger scale, the US military formed the Arctic, Desert and Tropic Information Center (ADTIC) in 1942 with a similar mandate.[8] The Arctic Institute of North America (AINA) emerged on the back of the wartime infrastructural and logistical expansion into the Canadian Arctic, which promised greater possibilities for northern experts to contribute to civilian development. The Norwegian Polar Institute (NP) emerged from a predecessor that could boast significant expertise in polar logistics (particularly in Svalbard) but also had a tarnished reputation thanks to the wartime collaboration of its leader and guiding spirit, Adolf Hoel (1879–1964).[9] And in the Soviet Union the Arctic Research Institute (ARI) continued to serve as the headquarters of Arctic research while other organizations devoted specific attention to the economic and logistical development of the Northern Sea Route and adjacent territories.[10]

In the case of Britain, a culture of Arctic exploration with its origins in Cambridge provided a pool of capable individuals keen to forge professional careers in Arctic research and administration–careers that could also be transferred to other countries with Arctic interests. SPRI had been founded in 1920 as a suitable memorial to Captain Robert Falcon Scott. The leader of the second party to reach the geographic South Pole, Scott and his comrades died on the return journey. The institute that emerged in his memory functioned as both a cultural center and an intellectual hub. Its rooms—and later its museum—helped to inculcate values of empire and exploration in young men, in addition to publishing a journal (the *Polar Record*), and hosting a library and archive. During the interwar years undergraduates from Cambridge, and to some extent also Oxford, led a number of expeditions to Svalbard, Greenland, and various parts of Arctic North America. For the great majority of these young men polar exploration was a stage through which they progressed, but for a small minority—most notably Brian Roberts—the polar regions became an obsession and a vocation. The latter became possible when the Admiralty began to substantially fund activities at SPRI during the war through its naval intelligence wing. James Wordie (1889–1962), a geologist and veteran of Sir Ernest Shackleton's famous 1914–16 Antarctic expedition, became a leading figure at SPRI and oversaw a program of collecting Arctic data that ranged from designing and improving equipment to providing background for military operations in Arctic spaces.[11]

By the time the war concluded, men such as Roberts who had cut their teeth during the interwar years could glimpse a future in which the state placed value upon Arctic experts. This included the natural sciences and defense-related research, but also omnivorous information-gathering about human activities in the Arctic that could function as intelligence to the state. The result was a cultural change at SPRI in which the Institute's founding Director, Frank Debenham, moved aside to be replaced by Roberts's contemporary Colin Bertram (1911–2001). Roberts divided his time between SPRI, with its unmatched library and growing collection of rare publications, and the Foreign Office, where he was an active participant in polar policy-making. The linguist-turned-geographer Terence Armstrong (1920–96) forged a career as an expert on Soviet Arctic activity, initially from a historical perspective (his doctoral thesis focused on the historical development of the Northern Sea Route) and later from a contemporary perspective, compiling a picture of Soviet activity in the Arctic from the scarce secondary material that reached Britain and later through his own personal connections.[12] His interest in sea ice research also led to contract work for the Defence Research Board of Canada, joining Wordie's former protégés Tom Manning (1911–1998), Graham Rowley, and Pat Baird (1912–1984) in a thriving Canadian Arctic milieu.

By the 1920s the Soviet government was placing considerable emphasis on studying and managing its Arctic territories.[13] In 1932 it founded the Chief Directorate of the Northern Sea Route (Glavsevmorput') as an instrument of state with power equal to that of a special ministry, fully devoted to Arctic exploration and research.[14] Otto Schmidt (1891–1956)—'the commissar of the ice,' networker, organizer, sportsman and professional mathematician—was appointed as its head.[15] The All-Union Arctic Institute, the research wing of the Northern Scientific-Commercial Expedition (established in 1920, renamed the Arctic Research Institute in 1939) that was charged with investigating the marine and geological resources of the Soviet Arctic, became the scientific hub of Glavsevmorput', with Rudolph Samoilowitsch as its director.[16] Samoilowitsch was arrested and executed in 1938 as part of the Stalin-era purges but his deputy, the meteorologist and oceanographer Vladimir Wiese (1886–1954), survived and remained in his position from 1930 to 1950.[17] Glavsevmorput', together with ARI, was active in research and data collection in addition to logistics. In contrast to SPRI, however, ARI did not gather information on the Arctic activities of other states or the rest of the Arctic, and it focused on applied polar research in geophysics, oceanography, and especially sea ice studies. Nor did ARI's research agenda include studies of Arctic peoples and their activities beyond a short-lived department dealing with reindeer herding, marine hunting and other Arctic industries.[18] But during the war it did provide environmental data services for maritime operations in the Arctic—research of a kind similar to that pursued in Britain and the United States, and a field that would prove vital during the early Cold War.

Samoilowitsch, Schmidt and Wiese were well known abroad and foreign polar scientists could visit the Soviet Union during the 1930s—albeit on a limited scale—on occasions such as during the organization of the Second International Polar Year (1932–33).[19] Even during the Great Purges, international correspondence remained possible: Wiese, for instance, received telegrams from around the world for his employment anniversary in 1937.[20] Evgenii Fedorov (1910–81), a geophysicist who served as ARI director during 1938–39 before becoming head of the USSR Hydrometeorological Service, had strong wartime contacts with Allied meteorologists. In November 1945 he participated in discussions on the "warming of the Arctic" phenomenon, held at the RGS in London and organized by British climatologist Gordon Manley. Also present at this conference was the well-connected Swedish geographer Hans Ahlmann (1889–1974),[21] who had been one of the guests invited to the jubilee celebrations of the Soviet Academy of Sciences in June 1945. In connection with the jubilee tour Ahlmann also visited ARI in Leningrad. Ahlmann left the Soviet Union impressed but also anxious about the strength of Soviet polar research,[22] an impression he shared with his Canadian fellow attendee Harold Innis (1894–1952), a political economist and influential public intellectual. Innis expressed admiration for how the USSR utilized science in developing its natural resources in the Arctic, arguing that Canada would benefit from a freer exchange of scientific knowledge with Russia pertaining to the "opening up of the Canadian Arctic."[23] The scale of activity that was apparent in 1945, combined with the known commitment of the Soviet Union to Arctic development during the 1930s (and to research in particular fields such as sea ice), led to pessimistic assessments in the West of the likely trend once the Iron Curtain was drawn. Soviet Arctic activity, it was feared, would continue to outpace the West, and it's extent would now be inscrutable.[24]

During the first years of the Cold War, information about Soviet polar research and polar data almost completely dried up, mirroring a wider trend of secretiveness and information suppression during the final years of Stalin's rule. Internationally active scientists were prevented from continuing their prewar connections. Even the influential Fedorov, who led the Soviet delegation to the International Meteorological Conference in Geneva in 1946, was accused in 1947 of making political mistakes during and after the war by providing data and instruments to his Allied colleagues. He was subsequently stripped of all his titles and positions but managed to avoid Samoilowitsch's fate, and was instead allowed to continue with some of his research from a demoted position.[25] Institutional correspondence between ARI and its foreign counterparts essentially ceased. The lack of research exchanges between the East and West turned individuals like Armstrong, who coupled excellent language skills with expert knowledge about the Soviet Arctic, into valuable commodities.

Behind this opaque veil, the Soviet Union increased its research and other activities in the Arctic. Its polar research institutions continued to function, sending large expeditions to the Arctic. These activities were kept secret— even from Soviet citizens—contrasting with past use of polar heroism for propaganda purposes in the 1930s, when the news about ice floe expeditions or flights of Soviet pilots across the Arctic Ocean had been widely promoted in the media.[26] Glavsevmorput' resumed its high-latitude vessel-based and later ice drift program along the Northern Sea Route in 1948, and organized a program of high-latitude expeditions called *Sever* (The North), focusing on oceanography and sea ice research. At least one of these expeditions, the 1950 expedition to establish the ice station (SP-2) (directed by Mikhail Somov, a leading sea ice specialist), had military tasks such as maintaining an air field for training pilots to land heavy military aircraft on ice, although its main purpose was civilian.[27] Expeditions of this kind enhanced the Soviet state's knowledge of environmental phenomena in the far north and its ability to operate in the Arctic,[28] even as Moscow concealed the military's role and positioned its own activities as antithetical to an aggressive, capitalist American military presence in the Arctic.[29]

Following his 1945 visit Ahlmann came to view the USSR as the most frightening specter in the post–1945 Arctic.[30] Ahlmann had been closely connected before the war to Norges Svalbard- og Ishavsundersøkelser (NSIU), a Norwegian state body formed in 1928 to coordinate research and logistics on Svalbard. NSIU was closely associated with its founder, Adolf Hoel, whose early geological training blossomed into a more general belief that Norway ought to exercise sovereign control over the Spitsbergen archipelago in addition to contributing to its exploration.[31] Under Hoel's leadership NSIU conducted annual expeditions to Svalbard and became associated with a controversial effort to assert Norwegian sovereignty over a section of East Greenland, leading ultimately to a decision by the International Court of Justice in April 1933 to uphold Danish sovereignty over all of Greenland. This element of nationalist political activism, which reflected Hoel's personal worldview, became closely associated with the culture of NSIU even as the organization maintained a reputation for competence in Arctic logistics, including mutually beneficial cooperation with international partners such as Ahlmann.

By the end of the war Hoel was thoroughly discredited by his collaboration with the German occupiers. The Norwegian Polar Institute was founded in 1948 to project a new image of continued Norwegian influence in both polar regions, under the leadership of Harald Ulrik Sverdrup (1888–1957). An internationally recognized geophysical scientist, Sverdrup was trained in Bergen and Leipzig before successfully building up the Scripps Institution of Oceanography in La Jolla, California. Sverdrup had connections with Soviet scientists such as Wiese before he left Norway for the US in 1936, and despite spurious concerns about his wartime loyalty he maintained strong links to the United States.[32] Sverdrup was attracted to the job in part because he felt there were opportunities to create a scientific research organization, although many NSIU employees remained, as did a culture of practical orientation. For Ahlmann and his contacts within the Norwegian political establishment, possessing a strong instrument of state engagement with the polar regions – attentive to research in addition to logistics – was essential to ensuring continued Norwegian (and indeed Nordic) presence in the Arctic at a time when the nascent superpowers seemed set to dominate.

Before 1939 the United States possessed no institution to match SPRI, ARI or the NSIU, with Arctic research largely the preserve of elite universities and to a lesser extent private bodies such as the National Geographic Society. The war provided both infrastructure to more easily access Arctic North America – notably the Alaska Highway and a network of airfields – and a presence in Greenland that began in 1941 as a temporary occupation to ward off a potential German invasion but which held continued significance after 1945 given the strategically valuable position of the island for a future conflict with the USSR.[33] With such a broad palette of both possibilities and imperatives, it is perhaps unsurprising that no dominant state body charged with Arctic research emerged. In the US, the military assumed a far more direct role in executing polar activities than in either Norway or Britain.[34] Arctic dimensions existed within bodies such as the Army Quartermaster Corps and ADTIC, and new organizations such as the Snow, Ice and Permafrost Research Establishment (SIPRE, founded in 1949) and the Arctic Research Laboratory in Barrow, Alaska (ARL, founded in 1947) soon emerged with more specialized remits.

In Canada, social scientific research in the Arctic prior to the war was mostly driven either by individual universities or by individuals like Innis and Vilhjalmur Stefansson (1879–1962).[35] With a few exceptions, such as anthropological research by Diamond Jenness (1886–1969) of the National Museum of Canada, government-initiated research tended to focus on the natural sciences, particularly in applied form like geological surveys that could lead to resource exploitation. Canadian military-led initiatives lacked the resources and reach of their American counterparts, and links across the 49th parallel—whether it be funding, research cooperation or project design—were often crucial to achieving research goals.[36] Building on cross-border links between a few scientists and well-connected social entrepreneurs with an interest in scientific research in the North American Arctic, one significant new institution *did* emerge – the Arctic Institute of North America, founded in 1944.

Unlike the other institutions discussed here, AINA was established as a bi-national organization, a private, independent institute incorporated in both Canada and the United States. Already in late 1942, separate informal discussions were held both in the US and Canada about establishing an institute focusing on Arctic research, although the groups' motivations diverged somewhat, reflecting the different national contexts and priorities. In the United States, discussions involved a narrow group, mostly scientists under military contract who were interested in continued post-war scientific access to and research in the Canadian Arctic (and who were very much aware of the strategic nature of knowledge about the North), and who also aimed to build careers as Arctic professionals. Several of the American scientists involved in the establishment of AINA had worked at ADTIC, an institution that in the words of Matthew Farish "actively engineered Arctic terrain in the name of scholarly advancement and military necessity."[37] Geologist Laurence M. Gould (1896–1995), Richard E. Byrd's second-in-command during his first Antarctic expedition in 1928–1929, served as ADTIC's scientific head, while Richard F. Flint (1902–1976) headed ADTIC's Arctic section, and Lincoln A. Washburn (1911–2007), who served as an intelligence officer in ADTIC, completed his PhD at Yale under Flint's supervision. Under Gould's direction, Washburn drew up proposals for an international institute that would help young men who gained polar experience during the war to pursue a polar-focused career, while also providing an institutional home for data gathered by ADTIC during the war, and a chance to capitalize on Stefansson's search for a permanent institutional home for his large Arctic library. Through Gould and E.M. Hopkins, the president of Dartmouth College, Washburn soon heard of Canadian plans for an international Arctic institute from the Canadian geographer Trevor Lloyd (1906–1995). At the time, Lloyd was based at Dartmouth under contract to the Wartime Information Board, where he was compiling a report on Soviet Arctic initiatives. A series of meetings organized by Washburn with support from Gould, Lloyd, Raleigh Parkin and Hugh Keenleyside (see below) were soon set up between the two groups.[38]

The Canadian discussions differed in both their general scope and their specific goals. Sovereignty over the Canadian Arctic was a key concern, especially following the intense American military activity in the Canadian north as the Pacific theater opened up. Establishing a bi-national institute could serve as a pragmatic way to draw American funding and know-how into a structure partly controlled by Canadians.[39] Major private initiatives dating back to the early 1940s aimed to gather comprehensive information on Canada's north to exercise control over resources and people. These included the Canadian Social Science Research Council Arctic Survey (initiated by Innis and funded by the Rockefeller Foundation, completed in 1947) and Lloyd's Arctic study for the Canadian Institute of International Affairs (CIIA), completed in 1946. The investment manager Raleigh Parkin (1896–1977) played a key role in both the civilian Arctic surveys and eventually also AINA. Well connected with his own country's elite and to philanthropic funders in the United States, Parkin had an abiding interest in North American international affairs and sought to promote Canadian nationality within a more continentalist context. He was an institution builder who preferred working behind the scenes to create organizations that solved specific problems: in this case the lack of knowledge needed to develop, secure, and settle the Canadian far north.[40]

Early discussions between Parkin and scientists such as Lloyd, the botanist Alf Erling Porsild (1901–77) and the anthropologists Jenness and Stefansson focused on two main concerns: government and public indifference to domestic concerns in the North, and the lack of knowledge concerning northern resource development. Like Hans Ahlmann, Lloyd and Stefansson were apprehensive about the USSR's perceived depth of logistical capacity and scientific knowledge about the Arctic.[41] Indeed, ARI was presented as a source of inspiration for AINA, with the important caveat that it should operate as a private organization:

> ... Staffed by capable specialists, the [USSR Arctic] Institute has carried out extensive research, the results of which are clearly evident in the vast development of the Soviet Arctic since the institute began to function.... [I]t is herewith proposed that an Institute similar in its scientific makeup to the USSR Arctic Institute, but smaller in size and organized privately, be established.[42]

AINA's mission was to develop research and systematically gather information to bring about "the intelligent and orderly development of the North" and advance natural and social knowledge bearing upon the attainment of that goal.[43] Notably, the North American Arctic was defined from the outset as including Greenland, to the point where AINA's constitution initially stipulated the presence of a Danish representative on its Board of Governors. Parkin was adamant that the new institution's focus should be "strictly scientific" to avoid any political tensions, and as such its members should participate in a private capacity.[44] But while Parkin and his group sought independence from government, they also wanted political goodwill and funding. Parkin lobbied for the involvement of key government officials among the blend of industry representatives (recruited for their management and fundraising expertise), military officials, and scientists.[45] This included using United States interest as a lever for mobilizing Canadian support.[46] Much like SPRI, AINA would facilitate Arctic research through grants-in-aid, coordination, logistic support and library resources, including an Arctic bibliography and a program to translate relevant articles from Russian into English. It would also come to manage military and government contract work, and became largely reliant on this income for most of the first twenty years of its existence.[47]

AINA constituted a departure from the militarized trend in United States Arctic research, even as it maintained increasingly close links with ARL and the Office of Naval Research (ONR, established in 1946). Yet it also did not neatly follow the Canadian state's concern for building an effective relationship with the United States over continental defense.[48] While they pointed vigorously to ARI to demonstrate to potential funders that an organization like AINA was needed, the Institute's early boosters drew more inspiration from SPRI in drawing up potential research goals and organizational structure.[49] AINA also profited from government contracts (often from military sources) that matched the expertise of academic researchers to the needs of governments. The relationship between AINA's Montreal office and McGill University bore parallels to that between SPRI and Cambridge. In both cases, the institute provided a means for students to access field sites through a formal structure that facilitated expeditions, through financial grants (particularly in the case of AINA) and also through a congenial institutional culture that legitimized Arctic research as an academic pursuit.[50]

Connections between AINA, SPRI, and the Canadian state were apparent at the level of personnel. Both Baird and Rowley participated in the Canadian Army's well-publicized 1946 Arctic exercise, Operation Muskox, which also featured the geophysicist and later AINA stalwart John Tuzo Wilson (1908–1993). Nevertheless, there were important differences. The bi-national character of AINA prevented any formal connections to the Canadian state akin to SPRI's close relationship with the UK Foreign Office.[51] By 1949 staff at SPRI possessed security clearances, reflecting its status as an intelligence-gathering center, whereas staff at AINA initially did not.[52] Washburn, executive director of AINA between 1945 and 1950, initially saw it as in AINA's interest to only direct projects "which can be published without restriction" to avoid political complications.[53] His concern is easy to understand given the potential difficulties that could arise from a private, bi-national organization being perceived to be an instrument of a particular state's military strategy.

Yet Washburn's position changed in 1950, as AINA's increasing dependence on government contracts in addition to a climate of international tensions led him to argue that AINA should develop along national lines to make it more attractive for classified work.[54] This was an indication of the power that funding sources could hold in determining institutional structure and direction. As the McCarthy era began AINA suffered a setback when the director of the Baltimore office, Moses C. Shelesnyak (1909–1994), was revealed to have briefly been a member of the Communist party during the 1930s—a fact the US military had apparently not deemed significant during his career to date, but which suddenly became relevant in the new political climate.[55] One of AINA's largest funders in the first decade-and-half was the ONR, but their funding was regarded as an interim measure, the ideal being a long-term budget appropriation from government. In order to be closer to American research funding sources and the National Academy of Sciences—National Research Council (NAS-NRC), Washburn moved back to the US in 1950. AINA affiliated with the NAS-NRC in 1950, after which the NAS-NRC handled contractual agreements between AINA and the US government.[56] That same year the bill creating the new National Science Foundation (NSF) passed. A new AINA office in Washington DC was officially opened in 1951.

AINA needed to navigate domestic changes in the science-policy landscapes in both Canada and the US. Washburn identified a tendency in Canadian circles to desire the direct supervision of research—especially lesser-researched parts of the Arctic—which had the potential of making it more difficult for private organizations like AINA to contribute to the wider goal of furthering northern development.[57] Among the schemes that AINA operated in the 1950s was a series of fellowships that facilitated the start of Canadian research careers for researchers from Britain and Europe.[58] While the executive director's position stayed in Montreal, the point of gravity in AINA seemed to move to Washington with Washburn. This sense persisted even after Washburn left his position at AINA to become head of the U.S. Army Snow, Ice, and Permafrost Research Establishment (SIPRE) in 1952.[59]

AINA, like its counterparts across the Atlantic and Pacific, also sought to re-orientate its priorities as space began to open for stronger connections across the Iron Curtain, while continuing to argue that Arctic research constituted a strategic national priority. Mark Adams has argued that the Cold War itself played a key part in the "liberalization" of science from politics. Both in the Soviet Union and in the West, the Cold War enabled scientists to wield influence with political leaders because scientists were willing and able to provide products such as bombs, missiles, and satellites.[60] During the last years of Stalinism, physicists who were involved in the "atomic project" gained power and ideological immunity. By 1956 they had largely taken over the Presidium of the Academy of Sciences. As Ronald E. Doel has shown, physicists in the United States who studied "the environment" as a context for military operations were able to similarly map their research onto overarching state priorities, including work in the polar regions.[61]

With Stalin's demise and the subsequent "thaw" overseen by Nikita Khrushchev, the Soviet Union became much more involved in international organizations of all kinds, including in science.[62] The Soviet leadership revealed a new interest toward "Western" science and technology in the context of so-called "peaceful co-existence" within both the industrial and academic fields.[63] Notably, the USSR decided to join the emerging plans for the IGY, an event that ultimately came to include a substantial Antarctic dimension in addition to coordinated research in the Arctic. Within the formal structure of the IGY's planning apparatus the Soviet Union played an active role in a range of theaters, particularly the Antarctic.[64]

Contacts between ARI and AINA were initially sparse, as were contacts between ARI and the NP. Susan Barr notes that although ARI director Vladimir Frolov suggested a series of Norwegian-Soviet exchanges in 1958, the NP felt it could not (or should not) function as the contact point for all Norwegian Arctic research, and the proposal came to nothing.[65] SPRI was somewhat different due to its status as an information-gathering unit, and in January 1956 SPRI director Colin Bertram wrote to Frolov proposing exchange visits between the two institutions while also expressing interest in visiting other Soviet institutions engaged in polar research.[66] A Soviet delegation arrived in London on 18 April 1956, consisting of the polar oceanologists Alexei F. Treshnikov and Igor V. Maksimov.[67] Treshnikov and Maksimov's visit to SPRI just months before leaving for Antarctica was more than a courtesy call: with the Soviet Union set to rapidly advance its activities in the far south as well as the far north, knowledge of developments in Western polar research was highly relevant.

Terence Armstrong and Brian Roberts subsequently visited ARI at the end of May the same year, similarly keen to use the official connection between institutions as a form of intelligence gathering. The report they produced upon returning to Britain confirmed the vast scale of Soviet polar research activity along with "a tendency to empire-building by rival institutions" that led to overlap in research efforts.[68] Armstrong and Roberts also noted a strong focus on practical Arctic research. ARI's main function, they reported, was to provide weather forecasts along the Northern Sea Route for its parent organization Glavsevmorput'. Staff at ARI had "an ignorance which we believe was genuine" about issues concerning Indigenous Arctic peoples and natural resource development.[69] This task was more closely associated with a body that the two Britons did not visit—and which the vagueness they noted throughout their visit concerning maps and geological information suggests was considered an inappropriate area for discussion with international visitors. This was the special Commission for Northern Problems, formed in 1955 as part of the Council for the Studies of the Productive Forces Research, the main scientific economic organization for assessing the effectiveness of resource exploration. The Commission was charged with examining the economic potential of the Soviet Arctic and preparing a plan to act upon it, thus furthering the state's commitment to colonizing and developing the Arctic. Its leader was the economic geographer Samuil Slavin (1901–89), who had previously worked for Glavsevmorput', and had been trying to persuade many official institutions of the necessity of an institution charged with the rational, effective and complex development of the region. The Commission aimed to define basic principles for effective exploration and economic exploitation of the North, with a clear focus on what would today be known as the "human dimensions" of the Arctic, including determining a balance between the economic benefits of resource exploration and establishing a good quality of life for industrial laborers in the Arctic.[70]

Roberts and Armstrong, whose home base at SPRI was increasingly concerned with Antarctic matters, noted that the Antarctic was becoming important at practically every Soviet polar institution that they visited. This dynamic soon applied to AINA also, especially as United States support of AINA—fueled in large part by contracts for Antarctic services related to the IGY—started to outweigh Canadian support. The "woefully meagre" activities on the Canadian side, compared to the "very bright" future on the US side, were starting to be regarded as a strategic burden by the late 1950s.[71] The situation even caused concern among some officials in Ottawa that US research support was starting to infringe on Canadian sovereignty.[72] Although the IGY provided tremendous opportunities for AINA to expand, it was dominated by American funding and research agenda—which largely focused on the Antarctic arena of what journalist Matthew Brzezinski aptly dubbed "a science Olympics of sorts."[73] AINA established and operated a biological laboratory at McMurdo Sound and went on to perform a range of other tasks including organizing inland traverses and establishing the US IGY program's glaciological headquarters. At one point, the Antarctic work of AINA played a sufficiently large role that the suggestion was made to add "Antarctic" to the institution's name or to change its title to the Polar Institute of North America.[74] This was a striking development for an institution that was almost entirely unconcerned with Antarctica in its early years, beyond the personal experience of some of its United States founders (notably Laurence Gould) as veterans of Antarctic expeditions.

The NP also found itself increasingly drawn into Antarctica, despite a distinct lack of enthusiasm from Sverdrup. It continued to send annual expeditions to Svalbard and to compile topographic maps of Norwegian Arctic territory in addition to maintaining a range of infrastructure, from navigation aids to cabins.[75] From the late 1940s this even included ferrying tourists to and from Svalbard—functioning in the words of Susan Barr as "a form of travel bureau."[76] Robert Marc Friedman has described how the United States pressured Norway at the governmental level to maintain an active presence in Antarctica during the IGY—driven by concern over the scale and extent of Soviet plans—and how this led to a political decision to establish Norway Station in Queen Maud Land in 1956.[77] The administrative and financial difficulties that followed the 1949–52 Norwegian-British-Swedish Antarctic Expedition, of which Sverdrup had been overall leader, must have weighed heavily on the NP's leader, who also noted a lack of enthusiasm for Antarctic research from Norwegian scientists.[78] When Sverdrup died suddenly in 1957, his successor Anders Orvin found himself having to deal not only with a set of Antarctic commitments but also with a rush of IGY expeditions to

Svalbard from a range of states, including the USSR, which at least gave him leverage to finally obtain more resources from the Norwegian Foreign Ministry.[79] It is perhaps easy to understand why the NP divested its Antarctic assets to South Africa in great haste at the end of 1959, when the imminent agreement of the Antarctic Treaty promised to diminish the political benefits of maintaining a permanent presence on the southern continent.[80]

But the body that suffered the most permanent effects from the IGY and the Antarctic rush was perhaps SPRI. Vivian Fuchs, a veteran of the Falkland Islands Dependencies Survey (predecessor of today's British Antarctic Survey), began mobilizing support in 1953 for a crossing of the Antarctic continent that would complete the journey Sir Ernest Shackleton had left unfinished nearly forty years earlier. When Colin Bertram openly expressed skepticism about Fuchs's plans he came into conflict with SPRI's Board of Management, led at the time by one of Fuchs's key backers, Sir James Wordie.[81] A disagreement over a specific expedition became a major issue in large part because SPRI's position as a quasi-official advisor to the British government on polar matters raised issues of collective loyalty to policy positions. Bertram ultimately left his post as the Director of SPRI, and when the glaciologist Gordon Robin was chosen as his replacement ahead of Brian Roberts in 1957, it symbolized a clear transition to an academic facility focused on scientific research, an identity that Robin felt would make it easier to make and maintain international connections between academic researchers without suspicions caused by close connections to the UK Foreign Ministry.[82] While SPRI's remarkable library continued to grow, and its bibliographers continued to hoover up polar information from around the globe, they were clearly now at arm's length from the British state.

Conclusions

We suggest that three key insights can be gleaned from this brief historical review of Arctic research institutes during the early Cold War. First, by investigating the reasons why particular institutes were created, or how existing institutes evolved, it is possible to understand how individual states regarded Arctic research and which priorities they had. At the same time, we can also study how those priorities were negotiated by researchers who viewed the Arctic as both field site and career path. That process of negotiation is crucial. While agreeing wholeheartedly that the role of states in post-1945 Arctic research cannot be downplayed, we reject characterizations of state priorities as rigid frames that strictly defined the possibilities for individual researchers. This is particularly true when considering the post-1945 professionalization of Arctic scientific research—at least in the Cold War West—which provided opportunities for young researchers with Arctic interests to use state interest in the polar regions as levers to further their own research and craft academic careers. Polar researchers were actors rather than simply instruments within state strategies.

Second, our analysis points to the hitherto underappreciated role of Antarctic research, particularly within the context of the IGY, as a driver for institutional change even in organizations with a primarily Arctic mandate (or in the case of AINA, an exclusively Arctic mandate). We suspect that the size and scope of the IGY Antarctic program caught many polar institutions off guard in that it demanded a capacity for action, and perhaps also a reservoir of scientific interest, in the Antarctic that was not anticipated in 1945. The funds that Antarctic contracts provided to AINA helped shift the bi-national institution's balance toward the US in the 1950s and to privilege the geophysical over the biological and the human sciences. The need to prioritize Antarctic research stretched the organizational limits of the NP, without changing its research priorities, while disputes over British Antarctic activity created a significant rupture at SPRI that helped push an institutional shift toward a more overtly academic identity. AARI obtained an extra A in its title and, although it appears to have remained primarily concerned with the Northern Sea Route and related matters, the expansion of Soviet Antarctic interest helped broaden its intellectual purview and might well have reinforced the division from social and economic development in the Soviet Arctic.[83]

Finally, attention to how institutions functioned as nodes in an international network can shed new light on how the connections between organizations and individuals reflected the possibilities afforded by wider political changes, as well as the differing functions of each institution within a national political economy. Here the glimpses of interaction between AARI and its Western counterparts are particularly instructive. Roberts and Armstrong's 1956 visit to AARI reflects the Khrushchev thaw, but a closer look at the content of their visit reveals how AARI served as a conduit through which international connections could take place without revealing the full spectrum of Soviet Arctic activities. While the scale of these interactions strikes us as modest, they invite reflection on the separation of state and institution that AINA's founders pushed so strongly—and which caught up with SPRI in the 1950s. Arctic research institutions were more than hotels housing Arctic researchers. They were organizations that faced outwards towards the world as well as inward towards governments, and they were consequently aware of their place not in splendid isolation but within an ecology of institutions.

Notes

1. See most notably Steven Shapin and Simon Schaffer, *Leviathan and the Air Pump: Hobbes, Boyle, and the Experimental Life* (Princeton: Princeton UP, 1982). Among the legions of studies of scientific societies from the early modern period, see for instance Roger Hahn, *The Anatomy of a Scientific Institution: The Paris Academy of Sciences, 1666–1803* (Oakland: University of California Press, 1971); James E. McLellan III, *Science Reorganized: Scientific Societies in the Eighteenth Century* (New York: Columbia UP, 1985); Michael Hunter, *Science and Society in Restoration England* (Cambridge: Cambridge UP, 1981); Alice Stroup, *A Company of Scientists: Botany, Patronage, and Community at the Seventeenth-Century Parisian Royal Academy of Sciences*

(Berkeley: University of California Press, 1989); and Sven Widmalm, "Instituting Science in Sweden", eds. Roy Porter and Mikulás Teich, *The Scientific Revolution in National Context* (Cambridge: Cambridge UP, 1992), 240–262. For later periods, see among many others Jim Endersby, *Imperial Nature: Joseph Hooker and the Practices of Victorian Science* (Chicago: University of Chicago Press, 2008) on Kew Gardens; Robert Marc Friedman, *Appropriating the Weather: Vilhelm Bjerknes and the Construction of a Modern Meteorology* (Ithaca: Cornell UP, 1989) on the Bergen School of meteorology; and Helen Rozwadowski, *The Sea Has No Boundaries: A Century of Marine Science Under ICES* (Seattle: University of Washington Press, 2002) on the International Council for the Exploration of the Sea.

2. See in particular Michael Reidy, *Tides of History: Ocean Science and Her Majesty's Navy* (Chicago: University of Chicago Press, 2008).

3. Felix Driver, *Geography Militant: Cultures of Exploration and Empire* (London: Wiley, 2001), especially 24–67.

4. See for example Peder Roberts, "The Politics of Early Exploration", eds. Klaus Dodds, Alan D. Hemmings, and Peder Roberts, *Handbook on the Politics of Antarctica* (Cheltenham: Edward Elgar, 2016), 318–333; Atle Næss, "De tre store", in Einar-Arne Drivenes and Harald Dag Jølle, eds, *Norsk polarhistorie 1: ekspedisjonene* (Oslo: Gyldendal, 2004), 51–171; Lisa Bloom, *Gender on Ice: Ideologies of American Polar Exploration* (Minneapolis: University of Minnesota Press, 1993), especially chapter 2; Michael Robinson, *The Coldest Crucible: Arctic Exploration and American Culture* (Chicago: University of Chicago Press, 2006), especially 107–158; Fergus Fleming, *Ninety Degrees North: The Quest for the North Pole* (New York: Grove, 2003); David Day, *Antarctica: A Biography* (Oxford: Oxford UP, 2013), especially 98–178.

5. Jonathan Oldfield and Denis B. Shaw, *The Development of Russian Environmental Thought: Scientific and Geographical Perspectives on the Natural Environment* (New York: Routledge, 2016).

6. For a comparative overview of how national interests in the Arctic impacted scientific research during the period from 1930 to 1950, see Ronald E. Doel, Robert Marc Friedman, Julia Lajus, Sverker Sörlin and Urban Wråkberg, "Strategic Arctic science: national interests in building natural knowledge – interwar era through the Cold War," *Journal of Historical Geography,* 2014, 44: 60–79.

7. Peder Roberts, *The European Antarctic: Science and Strategy in Scandinavia and the British Empire* (New York: Palgrave Macmillan, 2011), 93–95. "Environment" is understood here as the physical conditions in which people operated in the polar regions. Ronald Doel, "Constituting the postwar earth sciences: the military's influence on the environmental sciences in the USA after 1945" *Social Studies of Science,* 2003, 33(5): 635–666.

8. Shelagh Grant, *Polar Imperative: A History of Arctic Sovereignty in North America* (Vancouver: Douglas & Macintyre, 2011), 279.

9. On Hoel see for instance Einar-Arne Drivenes, "Adolf Hoel – polar ideologue and imperialist of the polar sea," *Acta Borealia,* 1994–95, 11–12: 63–72; Frode Skarstein, *Men så kom jo den 9.April i veien: Adolf Hoel, den glemte polarpioneren* (Oslo: Happy Jam Factory, 2008).

10. Between 1920 and 1939 ARI was known as the All-Union Arctic Institute (VAI), from 1939–1958, the period with which we are most concerned here, it was known as the Arctic Research Institute (ARI), and from 1958 onwards it was called the Arctic and Antarctic Research Institute, AARI. John McCannon, "The Commissariat of Ice: The Main Administration of the Northern Sea Route (GUSMP) and Stalinist Exploitation of the Arctic, 1932–1939" *Journal of Slavic Military Studies,* 2007, 20: 393–419; Paul R. Josephson, *The Conquest of the Russian Arctic,* (Cambridge: Harvard University Press, 2014).

11. On this period see Michael Smith, *Polar Crusader: A Life of Sir James Wordie,* (Edinburgh: Birlinn, 2007) and Roberts, *The European Antarctic,* Chapter 4.
12. The most notable of Armstrong's published works on the Soviet Arctic are *The Northern Sea Route: Soviet Exploitation of the North East Passage,* (Cambridge: Cambridge University Press, 1952), and *Russian Settlement in the North,* (Cambridge: Cambridge UP, 1965).
13. In 1926 the Soviet Union formally claimed all known and unknown islands within "the Soviet sector" of the Arctic. The proclamation led to the hoisting of the Soviet flag at Franz-Josef Land in 1929 and also encouraged USSR activities on Svalbard, where the USSR had the second-largest presence from 1931 onwards; Julia Lajus, "In search for instructive models: The Russian state at a crossroads to conquering the North", in Dolly Jorgensen and Sverker Sorlin, eds., *Northscapes: History, Technology, and the Making of Northern Environments,* (Vancouver: University of British Columbia Press, 2013), 110 –136. Also see Pier Horensma, *The Soviet Arctic,* (New York: Routledge, 1991).
14. Glavsevmorput' itself was founded following the first-ever successful voyage along the Arctic Siberian coasts from Murmansk to Vladivostok completed in one season without overwintering. This too was hailed as a feat of Soviet ice-breaking technology.
15. On Schmidt, see for instance John McCannon, *Red Arctic: Polar Exploration and the Myth of the North in the Soviet Union, 1932–1939,* (Chicago: University of Chicago Press, 1998).
16. Urban Wråkberg, "Science and Industry in Northern Russia from a Nordic Perspective," ed. Sverker Sörlin, *Norden without Borders: Science and Geopolitics in the Polar Region,* (Farnham: Ashgate, 2013), 111–141, 202. We use the German transliteration (not Samoilovich, Shmidt and Vise as they often are transliterated in literature in English), as the individuals used these names themselves when publishing in languages other than Russian, reflecting the fact all had German origins.
17. Samoilowitsch was rehabilitated posthumously, but not until 1988. See Vladislav Koriakin, *Rudolf Lazarevich Samoilovich,* (Moscow: Nauka, 2007).
18. Yakov Ya. Gakkel, *Za chetvert veka* (Moscow-Leningrad: Izd. Glavsevmorputi, 1945). This might be seen as consistent with a wider trend, as while studies of northern Indigenous people flourished during the 1920s in the framework of Soviet ethnography, they were much restricted during the time of Stalinism.
19. On connections related to the Second IPY see Cornelia Lüdecke and Julia Lajus, "The Second International Polar Year 1932–1933", eds. Susan Barr and Cornelia Lüdecke, *The History of the International Polar Years,* (Berlin, Heidelberg: Springer Verlag, 2010), 135–173 and on Ahlmann's in 1934 see Julia Lajus and Sverker Sörlin, "Melting the glacial curtain: the politics of Scandinavian-Soviet networks in the geophysical field sciences between two polar years, 1932/33–1957/58," *Journal of Historical Geography,* 2014, 4: 44–59.
20. St Petersburg Branch of the Archives of Russian Academy of Sciences (PFA RAN), collection 1010 (Wiese personal collection), inv. 1, f. 46. Also see Lajus and Sörlin, "Melting the glacial curtain."
21. Lord Rennell of Rodd, Evgenii Federov, A.R. (Sandy) Glenn, Noel E. Odell, Gerald Seligman, Max F. Perutz, Gordon Manley, Hans W. Ahlmann "Researches on Snow and Ice, 1918–1940: Discussion," *The Geographical Journal,* 1946, 107: 25–28.
22. For more details on Ahlmann's Soviet connections see Lajus and Sörlin, "Melting the glacial curtain."
23. William J. Buxton, "Innis's 1945 Trip to Russia," in William J. Buxton and Harold Innis, eds. *Harold Innis and the North: Appraisals and Contestations* (Montreal: McGill-Queen's University Press, 2013), 254.

24. Lajus and Sörlin, "Melting the Glacial Curtain"; Peder Roberts, "Scientists and Sea Ice Under Surveillance in the Early Cold War" in Simone Turchetti and Peder Roberts, eds. *The Surveillance Imperative: Geosciences During the Cold War and Beyond,* (New York: Palgrave Macmillan, 2014), 125–144.
25. Koriakin, *Rudolf Lazarevich Samoilovich.*
26. For a discussion of this issue see Yu.K. Burlakov, "Predislovie," in V.G. Volovich, *Poliarnye dnevniki uchastnika sekretnykh poliarnykh ekspeditsii 1949–1955 gg,* (Moscow, 2010), 7–11. The results from the SP-2 expedition had not been published prior to 1956–57; see also Z. Kanevsky, *L'dy i sud'by* (Moscow, 1980), 97 and McCannon, *Red Arctic.*
27. Somov later expanded Soviet polar research into Antarctica; on Somov, see A.F. Treshnikov, *Ikh imenami nazvany korabli nauki* (Leningrad: Publisher, 1984), 149–232 (on the SP-2 expedition, 198–207). Late in 1955, Somov led the First Complex Antarctic expedition, which utilized the research vessel *Ob* to take Soviet scientists to Antarctica for the first time.
28. John McCannon, *A History of the Arctic: Nature, Exploration, and Exploitation* (London: Reaktion Books), 247–248; Treshnikov, *Ikh imenami nazvany korabli nauki,* 195–198.
29. See Doel et. al. "Strategic Arctic Science," and references therein.
30. Robert Marc Friedman, "A spise kirsebær med de store", in Einar Arne Drivenes and Harald Dag Jølle, eds, *Norsk polarhistorie 2: vitenskapene* (Oslo: Gyldendal Norsk Forlag, 2004), 333–421; Roberts, *The European Antarctic.*
31. Einar-Arne Drivenes, "Adolf Hoel – polar ideologue and imperialist of the polar sea," *Acta Borealia, 1994–95,* 11–12: 63–72; Peder Roberts and Eric Paglia, "Science as National Belonging: The Construction of Svalbard as a Norwegian Space" *Social Studies of Science,* 2016, 46: 894–911.
32. Lajus and Sörlin, "Melting the Glacial Curtain;" Naomi Oreskes and Ronald E. Rainger, "Science and Security Before the Atomic Bomb: The Loyalty Case of Harald U. Sverdrup," *Studies in History and Philosophy of Modern Physics,* 2000, 31(3): 309–369.
33. Bo Lidegaard, *I kongens navn: Henrik Kauffmann i dansk diplomati 1919–1958,* (Copenhagen: Samleren, 1996); Dansk Udenrigspolitisk Institut *Grønland under den kolde krig: dansk og amerikansk sikkerhedspolitik, 1945–68,* (Copenhagen: DUPI, 1997).
34. This was in part due to the high importance placed upon cold weather warfare. Moreover, Britain possessed no Arctic territories to defend, and Svalbard was demilitarized under the 1925 treaty that governed it.
35. On Stefansson see for instance Barry M. Gough, *Stef: A Biography of Vilhjalmur Stefansson* (Vancouver, UBC Press, 1986); Trevor Levere, *Science and the Canadian Arctic: A Century of Exploration* (Cambridge: CUP, 1993).
36. On social science research in the Arctic just prior and during the war see Matthew D. Evenden, "Harold Innis, the Arctic Survey and the politics of social science during the Second World War," *The Canadian Historical Review,* March 1998, 79(1): 36–67. Also see Andrew Stuhl, "The Experimental State of Nature: Science and the Canadian Reindeer Project in the Interwar North", eds. Stephen Bocking and Brad Martin. *Ice Blink: Navigating Northern Environmental History* (Calgary: University of Calgary Press, 2017), 63–102.
37. Matthew Farish, "Frontier Engineering: from globe to the body in the Cold War Arctic," *The Canadian Geographer,* 2006, 50: 177–196, 177. On the history and purpose of ADTIC, see Paul H. Nesbitt, "A Brief History of the Arctic, Desert, and Tropic Information Center and Its Arctic Research Activities", eds. Herman R. Friis and Shelby G. Bale, Jr.. *United States Polar Exploration,* 1 (National Archives Conference September 8, 1967, Athens, Ohio: Ohio University Press, 1970), 134.

214 *van der Watt, Roberts, and Lajus*

38. Libraries and Archives Canada, Ottawa, Canada (LAC), MG 28 I 79 (Arctic Institute of North America fonds) Vol. 1. "Meeting held at Lisger Building, Ottawa, March 31st, 1944, 1:00pm." March 31, 1944. Washburn went on to become the first full-time director of AINA in 1945.
39. Shelagh Grant, *Sovereignty or Security? Government Policy in the Canadian North, 1936–1950*, (Vancouver: UBC Press, 1988). Political sensitivities did however play a role in choosing Montreal as AINA's headquarters, in addition to more practical reasons such as access to universities and accessibility. See Raleigh Parkin, "The origin of the institute" *Arctic*, 1966, 9: 5–18.
40. Evenden, "Harold Innis, the Arctic Survey and the politics of social science" 44; Grant, *Sovereignty or Security?*, 124.
41. Stefansson regularly corresponded with researchers from ARI, including Vladimir Wiese. Wiese also wrote an introduction to the Russian transla- tion of Stefansson's most popular book *The Friendly Arctic* (first published in the USSR in 1935 and reprinted in 1948). Their correspondence continued in 1939–1940 and may have resumed after the war. Lloyd reported on Soviet Arctic institutes for the Wartime Information Board.
42. LAC, MG 28 I 79, Vol. 1 "Proposal for an Arctic Institute of North America" final form, approved 18 September 1944. Lincoln Washburn, the first director of AINA publicly referred to ARI as inspiration, stating in an interview that "A realization of the advances made in the Soviet Arctic through the activities of the Arctic Institute of Leningrad and the desirability of similarly advancing our knowledge of the North American Arctic, led to the founding of our institute. Elizabeth Norrie, "Arctic Institute opens doors here" *Gazette* 21 November 1945.
43. Act to Incorporate The Arctic Institute of North America, Chapter 45, 9–10 George VI, 1945 Copies can also be found in LAC, MG 28 I 79 Vol. 1.
44. LAC, MG 28 I 79 Vol.1 G.R. Parkin to Trevor Lloyd, July 20, 1944. Letter marked "personal". The Canadian group included several senior government officials, amongst others Hugh L. Keenleyside (assistant Under-Secretary of State for External Affairs, from 1950 Deputy Minister of Resources and Development), ADP Heeney (Clerk of the Privy Council, 1940–49) and Charles Camsell. (Deputy Minister for Mines and Resources, 1936–47). Also see Grant, *Sovereignty or Security*, chapter 6.
45. LAC, MG 28 I 79 Vol. 1 G.R. Parkin to L. Washburn April 6, 1944. Also see the correspondence around AINA's constitution in LAC MG 28 I 79 Vol. 1, where those involved in the founding debated to what extent membership is needed to be restricted in order to promote strict scientific interest. A sum- mary of the opinions can be found in LAC MG 28 I 79 Vol. 1 H. L. Keenleyside to R.F. Flint "Comments on the constitution" August 4, 1944.
46. Grant, *Sovereignty or Security?*
47. Robert MacDonald, "Challenges and Accomplishments: A Celebration of the Arctic Institute of North America" *InfoNorth*, December 2005, 4: 440–451, especially 440–442.
48. For a summary of the historiography of this issue, see Peter Kikkert, "1946: The year Canada chose its path in the Arctic," in P. Whitney Lackenbauer, ed. *Canadian Arctic sovereignty and security: historical perspectives.* Calgary Papers in Military and Strategic Studies Occasional Paper No. 4 (Calgary: Centre for Military and Strategic Studies, 2011), 69–109.
49. LAC MG 28 I 79 Vol. 1 "An organisation for research and information about Northern Canada. Draft notes as a basis for discussion. July 1944" and LAC MG 28 I 79 Vol. 3 "The Arctic Institute of North America Information Letter 6, 1 August 1946." These information letters were compiled by the executive director and addressed to the board of governors.

50. See for example Matthew L. Wallace, "Reimagining the Arctic Atmosphere: McGill University and Cold War Politics, 1945–1970," *The Polar Journal*, 2016: 1–21 on the role AINA played in establishing the Arctic Meteorology Research Group at McGill, through channeling funding, organizing expeditions and connecting researchers from different fields.
51. The US had its own military structures to fulfill this role, while the Canadian government preferred to directly engage individual scientists or research groups. See Washburn's assessment of this in LAC MG 28 I 79 Vol. 4 Information letter 25 "The position of the Arctic Institute" February 14, 1950.
52. Scott Polar Research Institute Archives (SPRI), Brian Roberts Correspondence, Brian Roberts diary entry 12 September 1949.
53. LAC MG 28 I 79 Vol. 4 Report of the Director of the Arctic Institute of North America, October1, 1945– October 1, 1946. Also see Farish, "Frontier Engineering," p. 183. Washburn left the position of executive director of AINA to become director of the Washington office.
54. LAC MG 28 I 79 Vol. 4 Information letter 25 "The position of the Arctic Institute" 14 February 1950.
55. See the correspondence LAC MG 28 I 79 Vol. 11.
56. Correspondence LAC MG 28 I 79 Vol. 11.
57. Correspondence LAC MG 28 I 79 Vol. 11.
58. Jack Ives, *The Land Beyond: A Memoir,* (Fairbanks, AK: University of Alaska Press, 2010) p. 11.
59. Edmund Wright, "CRREL's First 25 Years 1961–1986," Hanover, 1986, p. 14. http://www.dtic.mil/docs/citations/ADA637200 He was replaced by L.O. Colbert, formerly Director of the U.S Coast and Geodetic Survey. John C. Reed, "Yesterday and Today" *Arctic* (1966), 9, 19–31, p. 21.
60. Mark B. Adams, "Networks in Action: the Khrushchev era, the Cold War and the transformation of Soviet science", eds. G.E. Allen and R.M. MacLeod. *Science, History and Social Activism: A Tribute to Everett Mendelsohn.* (Dordrecht: Kluwer Academic Publishers, 2001), 269–270.
61. Ronald Doel, "Constituting the postwar earth sciences: the military's influence on the environmental sciences in the USA after 1945," *Social Studies of Science*, 2003, 33(5): 635–66.
62. Konstantin Ivanov, "Science after Stalin: Forging a New Image of Soviet Science," *Science in Context*, 2002, 15: 317–338.
63. V. Mastny "Soviet Foreign Policy, 1953–1962", eds. Melvyn P. Leffler and Odd Arne Westad, *The Cambridge History of the Cold War, Vol 1: Origins,* (Cambridge: CUP, 2010), 313.
64. While plans for the IGY were conceived in the US, the planning apparatus was known by its French name Comité Spéciale de l'Année Géophysique Internationale (CSAGI). For an overview of the roots of the IGY, including the late entry of the USSR and its role in the Space Race, see Allan A. Needell, "Lloyd Berkner and the International Geophysical Year in Context: With some comments on the implications for the Comité Spéciale de l'Année Géophysique Internationale, CSAGI, Request for Launching Earth Orbiting Satellites", eds. Roger D. Launius, James Rodger Fleming, and David H. DeVorkin. *Globalizing Polar Science: Reconsidering the International Polar and Geophysical Years* (New York: Palgrave Macmillan, 2010), 205–224 and Fae Korsmo, "The genesis of the International Geophysical Year", *Physics Today*, July 2007: 38–43.
65. Susan Barr, *Norway, a consistent polar nation?: analysis of an image seen through the history of the Norwegian Polar Institute*, (Oslo: Kolofon, 2003), 345.

66. Central State Archive of Scientific and Technical Documentation in St.-Petersburg (TsGANTD SPb) coll. 369, inv. 1-1, f. 1097, l. 3. The earlier letter dated by 2 December 1955 was mentioned but it did not survive in the archive. A note in the margins of the reply from AARI states that "… the text was agreed with the Ministry of Foreign Affairs," indicating it was checked and edited not only in the Institute's security department (so-called First department – *Pervyi otdel*) but also in the Ministry of Foreign Affairs. TsGANTD SPb coll. 369, inv. 11, f. 1094, l. 4.

67. Thirtieth Annual Report of the Committee of Management of the Scott Polar Research Institute. *Polar Record* 1957, vol. 8, 56: 493. The term oceanology was preferred to oceanography by Soviet researchers, and is still in fact used widely across the former Eastern Bloc today.

68. Copy accessed at the Scripps Institution of Oceanography Archives, Robert S. Dietz Papers MC 28, 26/4 "ONR Technical Memoranda, 1954–56". Roberts and Armstrong, "A Visit to the USSR on behalf of the Scott Polar Research Institute, May–June 1956", unpublished report, p. 32.

69. Roberts and Armstrong, "Visit to the USSR," 15.

70. Ekaterina Kalemeneva, "Smena modelei osvoenia Sovetskogo Severa v 1950-e gg. Sluchai Komissii po Problemam Severa," Sibirskie istoricheskie issledovaniia 2018, vol. 2: 181-200. Also see the Russian State Archives for Economy, coll. 746 (personal collection of Samuil Slavin).

71. LAC MG 28 I 79 Vol. Report of the Executive Director to the Board of Governors, June 1958.

72. Ives, *The land beyond*, p.12. Ives quotes F.K (Ken) Hare, Chair of Geography at McGill and member of the AINA board of governors on the matter.

73. Matthew Brzezinski, *Red Moon Rising: Sputnik and the Hidden Rivalries that Ignited the Space Age.* (New York: Bloomsbury, 2008).

74. MacDonald, "Challenges and Accomplishments," 445.

75. Barr, *A Consistent Polar Nation?* 340.

76. Barr, *A Consistent Polar Nation?* 321.

77. Friedman, "Å spise kirsebær med de store," 372–403.

78. See for instance Roberts, *The European Antarctic*, 138.

79. Barr, *A Consistent Polar Nation?* 345.

80. On this handover see Peder Roberts, Klaus Dodds and Lize-Marié van der Watt "But Why Do You Go There? Norway and South Africa in the Antarctic during the 1950s" in Sörlin, *Norden without Borders*, 79–110.

81. Details of this episode are recorded in SPRI Management Committee Papers from 1955. Roberts is grateful to Professor Julian Dowdeswell, head of SPRI at the time, for granting permission to view these documents.

82. On this episode see Roberts, *The European Antarctic*.

83. A study of Danish Arctic research at this time might reveal a different story given that Denmark does not appear to have wavered from its narrow definition of Arctic research as almost exclusively concerned with Greenland. Indeed, a future study might profitably ask how the connections between bodies such as the Arctic Institute in Copenhagen and Danish administrators in Greenland formed a fundamentally different institutional arrangement.

13 Rockets over Thule? American hegemony, ionosphere research and the politics of rockets in the wake of the 1968 Thule B-52 accident

Henrik Knudsen

On US Independence Day 1968, the front page of Denmark's premier newspaper, *Politiken*, reported the beginning of a joint Danish-American "grand rocket program" on Thule Air Base located in Greenland's desolate North-western corner. The project would investigate the "splendid natural phenomena" of sunspots, solar particle emissions, and in particular polar cap absorption (PCA), which frequently hindered radio communication in the Arctic. Enthusiasm and national pride ran high in Denmark. Participation in the large PCA project, which *Politiken* reported would include no less than 50 rocket launches (actually 35), meant that Denmark's fledgling space research activities would reach a new level. Soviet scientists would participate from the Antarctic, the article claimed, leaving no doubt that this was a truly international scientific project consistent with current Danish détente politics and scientific internationalism. Yet the next day *Politiken* reported that the Danish government had called off the American part of the effort (at least for 1968), in effect ending the entire project.[1]

Over the following weeks, there was widespread newspaper coverage in Denmark of what most observers perceived as a major diplomatic mistake by the Danish government. Despite intense public debate, the decision was not reversed. This remarkable turn of events raises several questions. What was the nature and purpose of this project? What role did Danish scientists play in the project? Why cancel this joint venture after announcing it?

This analysis draws inspiration from the notion of American hegemony. In *American Hegemony and the Postwar Reconstruction of Science in Europe*, John Krige explored the American efforts to reconstruct European science in the early Cold War. A massive flow of resources from American foundations and organisations helped rebuild European scientific institutions and co-produce cadres of European elites who supported American global leadership. American aid (backed by other political measures) effectively constructed a transatlantic empire founded on "consensual hegemony," on shared values and preferences rather than hard power.[2] At the same time transatlantic relations in the late 1960s were shaped by the rise of détente politics in Europe and mounting hostility to the Vietnam War—a period when "transatlantic drift" became visible and when, for a particular set of reasons, a significant crisis occurred in the U.S.-Danish relationship.[3]

From this analytical viewpoint the paper looks at a group of Danish ionosphere scientists and their involvement in a large Pentagon-led project with rocket-borne probes that could only collect the data they required in the vicinity of one of the geomagnetic poles. Operating from Greenland, Danish ionosphere scientists possessed the leverage to negotiate partnerships with American scientists, gaining resources, authority and academic status. In fact, American scientific and military interests in the Arctic ionosphere were crucial for the formation and growth of Danish ionospheric research in the post-war period.[4] Such reciprocal exchanges of knowledge and resources between resourceful and dominant centers in the U.S. and junior partners at the periphery are a salient feature of Krige's notion of techno-scientific networks in the Cold War period.[5]

Operation PCA 68—using the Arctic atmosphere as a natural nuclear laboratory[6]

In November 1967, the US Embassy provided the Danish government with an outline of military scientific activities planned in Greenland during 1968, including two large rocket projects—later collectively known as Operation PCA 68—to be carried out from Thule during the autumn. The embassy specifically requested early approval for these projects, the aim of which was to elucidate "problems of radar and communication systems operating through a nuclear disturbed atmosphere. Certain natural disturbances in the ionosphere can be used to study problems of a nuclear disturbed atmosphere. A polar cap absorption (PCA) event is an ideal opportunity to study the lowest part of the ionosphere."[7] James Ulwick, of the Air Force Cambridge Research Laboratories (AFCRL), an institution with world leading expertise in upper atmosphere physics and extensive research activities at Thule, was in charge of the operation.[8] The justification for the $10 million price tag: "to better understand nuclear weapons' effects in the upper atmosphere more complete knowledge is needed of the atmospheric chemistry, coupling mechanisms between the expanding fireball and the ambient air, causes of debris distribution and high altitude plasma-magnetics. [...] These data will be used to develop a model of the disturbed ionosphere. Extrapolation can be made to higher energy levels to develop a more accurate model of the nuclear fireball and the highly disturbed ionosphere."[9]

The project was funded by the Defense Atomic Support Agency (DASA), a Pentagon agency dedicated exclusively to nuclear weapons research and testing of the effects of nuclear explosions on weapon systems.[10] Several spectacular high-altitude nuclear tests (e.g., TEAK and STARFISH PRIME) conducted in 1958 and in 1962 had demonstrated severe impacts on electromagnetic systems, including the blackout of shortwave radio and the degradation of radar. Information on such effects was critically important for assessing the feasibility of the anti-ballistic missile (ABM) systems under development by both superpowers in the 1960s. The military wanted a communication system that could survive a nuclear attack and continue to operate in a nuclear blackout environment of fission fragments and ionizing radiation.

The available toolkit for such R&D work had been radically reduced in 1963 when the USA, UK, and USSR agreed to the Limited Test Ban Treaty, which banned all but underground nuclear detonations. Starting in 1962, DASA's research program gave priority to studies of natural phenomena such as aurorae and polar cap absorption (PCA), whose effect on the ionosphere resembled that of nuclear bursts. PCA's are caused by energetic protons emitted from active sunspots that penetrate deep into the atmosphere in the polar regions. In lieu of actual testing, PCA-events can be used to gain information about nuclear effects on the upper atmosphere. Military researchers were interested in solar particle-induced disturbances on radar and very low frequency (VLF) systems. DASA had been developing plans for PCA-studies since the early 1960s.[11] However, they had to wait until the late 1960s for the next peak in solar activity and they also had to develop specialised probes.

Operation PCA 68 would become a comprehensive cluster of inter-related projects that used rocket measurements, satellite data, flying laboratories aboard two Boeing KC-135 aircraft, and additional ground-based measurements—an advanced research infrastructure bearing testimony to the Pentagon's deep pockets and ability to increase the level of technological sophistication in polar science. With the project making use of several Defense Department contractors who had taken part in high-altitude nuclear testing in the Pacific, its organisation and structure closely resembled that of a nuclear test.[12] In fact, Operation PCA would be AFCRL's biggest field operation since these Pacific tests.[13] The Pentagon was preparing to use the upper atmosphere above Greenland as a nuclear laboratory.

The only real problem was that rocket launches had to take place in the vicinity of the geomagnetic North Pole and the one favored location, Thule Air Base (Figure 13.1), was located on Danish territory. This meant that approval had to be obtained from the Danish government. At this time, U.S. military research projects had to be presented at an annual meeting with the Danish Commission on Scientific Investigations in Greenland (KVUG) where projects were subjected to closer scrutiny by Danish scientists and administrators. This approval procedure usually looked a lot like rubber-stamping, and was described as such in 1968 by Canadian Arctic expert Trevor Lloyd, a close and reliable observer of Danish Arctic policy with firsthand knowledge as he was writing from Denmark.[14]

An early outline of the planned PCA study was presented by Air Force (USAF) scientists in KVUG in May 1967. The intention at this time was to shoot rockets off in an eastern direction with impact on the inland icecap.

Figure 13.1 Established in 1951 at the height of the Cold War, Thule Air Base in the North Western corner of Greenland is the US Armed Forces' northern-most installation.

Source: Wikimedia Commons, courtesy TSGT Lee E. Schading/U.S. Air Force.

During the summer AFCRL personnel surveyed the facilities at Thule and decided on utilising A-Launch, a decommissioned rocket launch battery located northwest of the base. The eastern trajectory was now relinquished in favor of a westward firing direction. According to the Danish scientific adviser Jørgen Taagholt this was because the Strategic Air Command found it too risky to launch in the direction of the Ballistic Missile Early Warning System radar.[15] But this meant that all the risk of damage was placed on the local Inughuit (North Greenlandic Inuit) population. Due to the timing, which was going to be late in the year, the Americans expected the level of Inughuit hunting and traffic to be low. During 1967 Danish contractors were hired and began preparing the launch site.[16]

It took some time, however, before precise information regarding the number of rockets, timing, direction, and location of impact areas was passed on to the authorities in Denmark. In order to secure early approval of the project a short, and as it turned out, rather imprecise, description was presented to the Danish authorities in November 1967.[17] The vague and contradictory information raised a number of questions and made it difficult for the Danish authorities to assess the risks involved and hence to give immediate approval. Only at the beginning of March, following pressure from the Ministry of Foreign Affairs, did the U.S. specify these points in a detailed table. The number of rockets was now fixed at 35 and the perimeter

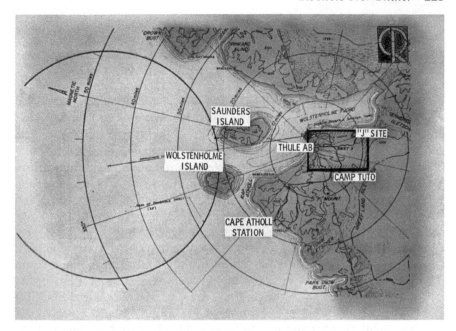

Figure 13.2 Estimated area of probable impact for rockets launched during Operation PCA 68.

Source: J.C. Ulwick: *Operation PCA 68.*

of the impact area was specified as 40-50 miles from the shore to the west of Thule (Figure 13.2).[18]

Enrolling allies

Since the early 1950s, Danish scientists had conducted ionospheric research from a network of stations in Greenland, established and operated in cooperation with the U.S. National Bureau of Standards. Beginning from the days of the IGY the USAF supported the registration and mapping of very low frequency electromagnetic noise by Danish scientists in Greenland, with the contract renewed several times up through the 1960s.[19] Due to the establishment of the European Space Research Organisation (ESRO) and Danish participation in rocket-borne ionospheric research from Andoya (Norway) the Ionosphere Laboratory underwent a period of rapid and massive growth from 1962, followed by a dramatic split up in two separate entities. A Danish Space Research Institute was established between 1966–68 with its own allocation in the State Budget—in effect taking the lion's share of the national ESRO approbations. However, in 1968, activities related to the ground-based ionosphere observations in Greenland were transferred to the Danish Meteorological Institute. Still carrying the name of the Ionosphere Laboratory (Ionlab) this group was led by Jens K. Olesen and Jørgen Taagholt, radio engineers and veterans of the ionosphere operations in Greenland. It took some time to secure the necessary funding for these activities, and as a consequence a prolonged battle about funding took place between these two groups.[20]

In the midst of the reorganisation process the somewhat financially amputated Ionlab formulated its new strategic program for the period 1967–1971 in which Greenland was pushed forward as the ultimate trump card. Research in the Arctic ionosphere was an obligation that "ought" to be carried out in Denmark for "scientific, technical and political reasons." The strategy focussed on research problems in relation to communications systems, the effect of solar particle emissions on the Arctic ionosphere, and in particular "solar flares" and polar cap absorption. It argued that an urgent strengthening of activities was needed if Denmark was going to take part in and be able to benefit from the large American PCA program proposed for 1968.[21] The strong focus on the upcoming USAF campaign must be seen in relation to the unstable funding situation faced on the home front. Tapping into the rich U.S. military funding seemed a viable alternative, and one that had succeeded in the past and present. Since 1960 the establishment of rocket research in Greenland had been a long-term objective for Danish scientists. This aim was fulfilled in 1968 by collaborating with the USAF's Air Weather Service's meteorological rocket program at Thule.[22] When Danish ionosphere researchers were invited to join this program in 1967, with Danish instruments piggybacked on the Air Weather Service's small Arcas rockets launched from Thule, a new model for US-Danish collaboration was established just around the time when plans for the large PCA-project were taking form.

With another solar activity peak approaching in 1968-69 USAF research activities in Greenland were quickly picking up momentum. In a letter to the Ministry for Greenland in November 1967 Jørgen Taagholt called for stricter measures to ensure better coordination between Danish and American geophysical research programs. Large projects should be announced in detail one year in advance in order to secure Danish participation in relevant American projects. The following day he requested the Ministry to press the Americans to release ionosphere data from on-going projects and from a decommissioned U.S. ionosphere station at Camp Tuto (near Thule) from 1955 to 1966.[23] These demands were forwarded in a verbal note in mid-November and reached the Actually, the AFCRL through the official diplomatic channels.[24] The Ionlab was playing its trump card to ensure tighter cooperation.

The tactics worked. The requested information was released and in December 1967 James Ulwick from AFCRL contacted Jens K. Olesen, head of the Ionlab, with an invitation to join forces and coordinate efforts. He added that the AFCRL would be grateful if Danish researchers could help push the PCA-program through the formal channels.[25] At this time Danish plans for a PCA-study (i.e. the aforementioned collaboration with the Air Weather Service) had reached an advanced stage. Since rocket experiments are usually a time-consuming affair requiring years of planning it is clear that the AFCRL's interest in collaboration was first and foremost motivated by a need to secure goodwill in the Danish administration. Actually, the AFCRL had little or no technical and scientific need for Danish assistance since it was 100% self-sufficient regarding instrumentation.[26]

In his reply Olesen admitted that the short notice made it "very difficult" for the Danes to coordinate their planning with the AFCRL project. He did, however, assure his transatlantic colleagues that the Danes would do their best to support U.S. activities with ground-based equipment. In reality, this meant that the Danish contribution to the PCA-project was reduced to supplying data from the three ionosphere stations. He also assured the Americans that Danish researchers would do whatever they could to make the application "go fast through the official channels here."[27] The cooperative arrangement with AFCRL was of a very limited and pragmatic nature. And yet enrolment of Danish support and cooperation even at this late time went smoothly.

Danish researchers had every reason to support their colleagues from AFCRL. Building strong ties to USAF activities in Greenland was an integral part of the strategy of the Ionlab to gain resources and authority in and outside of the field. The Ionlab was consciously seeking to take advantage of the American interest in the polar ionosphere. This strategy reached its apex in spring 1968 when the Ionlab pressed for a more direct role in the PCA-project and access to military research contracts on an equal footing with American research contractors. The request reached the AFCRL through official diplomatic channels and with reference to the Danish-American defense agreement on Greenland.[28] The request was supported both by the Danish ministries and KVUG.[29] But this time the request was met with a polite but firm rejection.

When the leaders of the Ionlab met with representatives of the AFCRL in Copenhagen in mid-May to coordinate the Danish participation in the project, discuss future collaboration and the delicate issue of economic support, Sam Silverman of the AFCRL advised the Danes to forward an ordinary research application through the Air Force Research and Development Command, European Office. Pointing to all the troubles generated by official diplomatic requests, he urged the Danes to consider what kind of collaboration they were aiming at: "one based on legal claims or one based on common interests [?]."[30] AFCRL was interested in and needed to cooperate with the Danes, but it preferred transnational cooperation to take place on the institutional and personal level rather than through diplomatic channels. The Danish researchers were trying to maximise and monopolise the gains from their access to unique northern locations. The reserved attitude of the AFCRL representatives suggests that the Danes had been playing their cards too aggressively for this time. The fact that the Danish researchers resorted to the official diplomatic channels when presenting their demands to the AFCRL points to a significant observation about the nature of Cold War scientific collaboration. Such research networks were not symmetric and flat, but rather lumpy and asymmetrical,[31] with the AFCRL holding a dominant position even in a situation as this where they needed goodwill from a foreign government and cooperation from Danish researchers.

Voices from the north

A slight tension and unease can be felt in the negotiations just mentioned, which reflected a much deeper crisis that plagued the American-Danish relationship in early 1968. On January 21 a nuclear armed USAF B-52 bomber crashed into the sea ice of the North Star Bay about 7.5 miles to the west of Thule AB scattering plutonium over a large area. The B-52 accident was a delicate matter for both countries and caused a serious diplomatic crisis between Denmark and the US during the winter and spring of 1968.[32] Another complication occurred in early June 1968, when articles in the *Washington Post* previewed the coming book *Chemical and Biological Warfare* by the military journalist Seymour Hersh. This usually well-informed writer claimed that the US was conducting experiments with chemical and biological warfare agents in Greenland. Regardless of the veracity of these claims, the controversial story added to the diplomatic tension around Greenland.[33] It looked as if Denmark had lost control of its Arctic territories. In this charged political atmosphere the fate of the PCA-program was going to be decided in Denmark.

At the beginning of January 1968, the Ministry of Commerce confirmed that Operation PCA 68 would not cause problems for shipping transports provided that information about the time and place of the experiments could be provided "well in advance." By February, the Ministry of Public Works had suggested a number of warning measures in order to safeguard flight traffic in the area. A heightened sensibility on security issues can be noted after the date of the Thule crash, but nothing in fact to suggest the later development. On 14 May KVUG received the annual briefing from the U.S. researchers engaged in activities in Greenland. Ulwick and other leading USAF scientists made presentations on Operation PCA 68 and other AFCRL activities. In his summary Ulwick now chose to omit all references to the military agenda, perhaps in response to the radically changed political environment. Yet the American visitors must have left the meeting confident since no objections were raised by KVUG-members. According to the minutes the Danish authorities were in the process of evaluating safety aspects for the local Inughuit population. Approval would rest on this evaluation.[34]

Three days after the meeting in KVUG the Ministry for Greenland initiated this evaluation by sending telexes to the Danish police, the liaison officer at Thule and to the Danish inspector Orla Sandborg in Qaanaaq (the principal Inughuit settlement 65 miles north of the base). The ministry asked 1) if any local traffic could be expected in the area during the period (September and October); 2) if cordoning off the area for limited times during launches would have economic consequences for the local population; 3) how best to secure effective broadcasting of warning signals? The liaison officer at Thule AB had the opportunity to discuss these questions with the local police, the U.S. base commander and, incidentally, also inspector Sandborg who happened to be present at the base. In the opinion of these local authorities the program would not cause any economic loss for the Inughuits since only "very limited" traffic from "small vessels" could be expected due to icy and harsh weather conditions at this time of the year.[35] On this basis the ministry prepared to approve the project.

Two days later on 22 May the city treasurer, second in line of command in Qaanaaq, replied to the telexed questions in the absence of inspector Sandberg. The treasurer had talked things over with the local member of the national council of Greenland and the newly founded municipal council of Qaanaaq which "strongly protested against the rocket launches," stating that the local population "had already taken a hard blow due to the accidental crash 21 January." According to the treasurer, a "lively traffic" took place in the area during the months in question, "since the hunt for whales among other things takes place from small vessels." An ecologist who visited the area in 1968 noted that the waters to the west of the base were a "favourite hunting ground" for the locals, especially in the autumn when "the straits of the coast are the last to freeze solid."[36] Moreover, unrestricted freedom to follow the game wherever it went was a crucial matter for the local population, which almost exclusively relied on hunting.[37] Cordoning off the area, the treasurer contended, would "absolutely have economic consequences for the local population" since these hunting activities were critical for winter supplies.[38] It was "not a matter of economic consequences but a matter of losing vital catch" and as the message frantically pointed out, "the existence of the whole population will be endangered by these rocket launches."[39] In order to stop the rockets the municipal council would now seek to involve the chairman of the Greenland Provincial Council Erling Høegh.

Faced with two contrary assessments, the Ministry for Greenland decided to advise against the project.[40] Leading figures from the Ionlab were briefed about these developments in the Ministry for Greenland on May 27. Reacting against the assessment of the city treasurer, J.K. Olesen vigorously denied the existence of an "objective" link between the B-52 crash and the rocket campaign.[41] He also voiced his concern about possible consequences of a last minute rejection "on a meager basis such as the present case", stating that:

> A rejection will reverberate in all parts of the international circle of scientific groups within space science, it will seriously harm the reputation of Denmark in general and in particular concerning future cooperation with the USA especially given the motivation for the rejection, i.e. prohibition against a very small number of North Greenlandic hunters in a certain sea area in total less than 36 hours during two months, possibly against compensation...[42]

Olesen suspected that the city treasurer had wrongly understood the safety area to be closed for a full two-month period and urged the ministry to send up Jørgen Taagholt to clarify possible points of misinterpretation. The ministry official rejected this suggestion stating that it could be interpreted as the government putting the Greenlanders under "moral pressure." He promised however to ensure that the municipal council had correctly understood the details regarding time and date. But this was not enough for Olesen, who "in all friendliness" announced his intention to use his own chain of command to mobilize an objection from the Ministry of Defense.[43] Olesen's reaction gives a clear indication of his alignment with American interests and how far he was prepared to go politically to protect these joint interests. The perspective of Inughuit people counted very little when he weighed the different interests.

In any case the pressure worked. The next day the permanent secretary Erik Hesselbjerg telexed inspector Sandborg asking him to check if the reply from the city treasurer was in fact based on misunderstandings, to give precise information on all safety measures and American guaranties, and to secretly report back on the matter. Sandborg was also requested to pass along his judgment on what kind of "guaranties and/or payments" might change the minds of the Inughuits.[44] So, at this point the ministry was prepared to buy its way to a peaceful solution.

On 8 June Kaj Bech (the chairman of the Qaanaaq municipal council) telexed Erling Høegh a status report from negotiations with inspector Sandborg, reiterating his resistance and pointing to "the psychological consequences" of the rocket campaign and the B-52 crash. The accident, he insisted, had had a much stronger impact on the local population than generally recognized. Høegh also informed the ministry of this correspondence. Having no personal objections to the project, he "above all attached importance to the judgment of the [Inughuit] hunters" who now was said to support the project.[45] Greenlandic authorities continued to give widely differing assessments of the situation. In the absence of a clear and univocal assessment from local authorities, the ministry backed by Erling Høegh decided to give the project its support.

A majority of domestic newspapers supported the project. Lacking clear statements from the government, media coverage initially gave considerable attention to the protesting voices from Greenland. A particularly insensitive approach was adopted by the social democratic newspaper *Aktuelt*, which found it outrageous that the government had chosen to give in to "emotional voices", ridiculing the Greenlandic reactions as being caused by "a primitive fear for the unknown." According to this newspaper, "it would have been more flexible if the hunters, which in theory could have been out, had stayed indoors during this period. A million dollar project of invaluable significance to science must not, literally speaking, vanish in thin air for the sake of a few seals."[46]

The politics of rockets and military research

The Ministry of Foreign Affairs ultimately decided the fate of Operation PCA 68. Its well-informed officers suffered no illusions about the nature of the project and correctly designated the operation as a military scientific investigation aiming to assess "atmospheric nuclear weapon bursts effects on radar and communication systems" by means of naturally occurring PCA-events. This aim however, did not in itself give rise to any "hesitations in principle," especially given the fact that ground-based investigations with similar aims had been going on in Greenland for some time.[47] Consequently, the ministry focused on safety issues and on backing the interests of Danish scientists.

In the final presentation (14 June) to the Foreign Minister the case officers recommended the project after a detailed discussion of safety issues, the response from local authorities, and Danish scientific interests. The PCA-project fell within the boundaries of what was normally accepted and the military aim of the project altered nothing, since, "it is also a Danish interest that western defensive radar and communication systems operates efficiently." After all, Denmark supported a strong and effective nuclear deterrent. They finally stressed the significance of the project pointing to the total U.S. costs of 10 million dollars and the strong interests in PCA-events among Danish scientists.[48]

In contrast Paul Fischer, the director of Ministry of Foreign Affairs, was much more cautious. He tended to look at the overall political context—the B-52 accident and rumors of bacteriological and chemical weapons tests in Greenland—and cautiously advised, "we must avoid activities in Thule that once more can bring Thule into the limelight."[49] He recommended rejection, to be reviewed in 1969, expecting and hoping perhaps that a more tempered atmosphere would prevail at this time.

When the PM Hilmar Baunsgaard, the Minister of Foreign Affairs Poul Hartling, and the Minister of Greenland met on 27 June to settle the issue they decided to suspend the program and leave a door open for reconsideration in 1969. The political rationale, given in a short note, followed closely the political assessment by Fischer, in as much as it stated a wish to avoid "any possible new uneasiness about Thule *even* if it appears unmistakably that such uneasiness is not *factually* justified."[50] The Danish rejection was conferred to the U.S. in a verbal note on 1 July. In case of an American request for further information the Danish embassy in Washington was instructed to reply that "carrying out the project shortly after the American air crash in Thule and the denied allegations about American testing of chemical and bacteriological weapons in Greenland could give rise to public debate on the American activity in Greenland, which is neither in accord with Danish nor American interests."[51] This indicates that the Danish government acted from a wish to avoid a public picture of offensive military scientific actions taking place in Greenland.

An official public statement from the Ministry of Foreign Affairs appeared on 5 July. It provided as explanation a governmental "wish to avoid any risk of public anxiety." The project was neutrally described as measurements of "phenomena related to solar eruptions," which in itself confirmed the government's wish to avoid public debate about the military scientific issues at hand. Such wishes were in line with the overall Danish policy to avoid public debate on the American military activities in Greenland. One of the persistent features of Danish society throughout the Cold War was a persistent public opposition to nuclear weapons and by proxy all things related to nuclear issues. The "nukes in Greenland" issue was particularly sensitive and the Thule accident had brought this to politicians' attention. The public statement of 5 July did not, however, have the intended impact and instead opened the way for journalistic speculation and critical attacks from political enemies. Since the public only knew the project as a scientific effort, the government was in a vulnerable position, facing a serious challenge in explaining its policy. Why, after all, ban a scientific experiment in such a remote area if it was not simply a matter of security for the local population?

A clue to the internal political deliberations came when in an interview on 14 July Poul Hartling explained that his "friends" would have objected if he had decided in favor of the rocket project.[52] This cryptic remark points to the Social Liberal Party (or radical liberals) as *the* determining factor. Under the leadership of Hilmar Baunsgaard this party was one of the winners of the January 1968 election and Baunsgaard had gained leadership in a center-right coalition that took office after the election. Placed in the center of the political spectrum this party had strong historic roots in pre-war antimilitarism and pacifism, and still harbored neutralist and NATO critical sentiments.

Incidentally at this time the new Danish Minister of culture and disarmament, Kristen Helveg Petersen (Social Liberal) delivered the opening address at the third Pugwash conference on "Implications of the Deployment of ABM Systems" held at Krogerup Højskole. Addressing the need for scientists to acknowledge and take responsibility for military applications of scientific research he used the occasion to take pride in the recent decision to ban the rocket program in Greenland. That decision, he explained, was grounded in a feeling of "mistrust and doubt" towards certain areas where a "confusion of civil and military missions is taking place," issuing a strong statement against what he saw as a dangerous tendency of scientific involvement in military projects. Helveg Petersen preferred that research money should go to conflict and peace research and hinted at the need to stop certain "tainted" research areas.[53] The speech gave a clear indication of the pacifist and antimilitary sentiments within the governing social liberal party from which it followed that Danish researchers ought not take part in research of direct military interest.

Despite its strong ties to the Social Liberal Party the leading daily newspaper *Politiken* initially gave full support to the project and criticised the government. Digging deeper into the matter the paper eventually discovered the military involvement in the project and informed its readers that the actual aim of the program was to "create a defense against nuclear attack" and thus linked the experiment with ongoing concerns and discussions about ABM-systems.[54] This discovery immediately prompted a much more positive assessment of the government's policy. Rejecting a project of this calibre must also be seen as an effective act to assert full Danish sovereignty over Greenland. The social liberal mouthpiece *Skive Folkeblad* found it "instrumental" that it had now once and for all been settled who "exercised sovereignty over Greenland."[55]

Public relations—classified science

During the spring of 1968 the Ionlab had prepared PR material about the PCA-program, framing it as a cooperative Danish-American effort, explicitly stating its "purely scientific character" and its possible value for improvement of radio communication and traffic in the Arctic.[56] The press material did not mention AFCRL, DASA or other DoD organisations, and gave no hint of the military aims of the project. Since the U.S. material had outlined all these military details, the Ionlab's PR exercises could be characterised as spin.

On the symbolic date of 4 July 1968 *Politiken* put the story on its front page—a full week after the government had issued its ban. The whole project was framed as a Danish-American cooperative effort, as one big joint effort and not two programs running in parallel, and as a truly international exercise since Soviet scientists in the Antarctic were claimed to participate in the effort. Readers were assured that the rockets were "totally harmless" since bits and pieces would rain down over uninhabited land. All safety measures had been worked out in cooperation with the Greenlanders, it was said.[57] The article was written by Niels Blædel, the well-informed doyen of Danish scientific journalism. His source was Peter Stauning, project leader of the Danish part of the rocket program.[58] This carefully framed PR-effort seems well advised, since the recent B-52 accident had heightened public and political sensitivities towards military and nuclear issues in Greenland.

Now, because the PCA-program was classified as "confidential" by the U.S., such public efforts were highly problematic in the eyes of the Ministry of Foreign Affairs. The ministry immediately set up a meticulous investigation to clear up the sources of what it regarded as an inappropriate leak and a transgression of the limited freedom of speech for civil servants. In a defensive move Stauning pointed to the fact that U.S. scientists had circulated knowledge about the PCA-program at international conferences.[59] In any case, the investigation lost its impetus when the government a few weeks later was compelled to release information about the project.[60]

As mentioned the Danish government had left the door open for re-evaluation of the project in 1969. Angry and frustrated Stauning instantly launched a vigorous rebuttal of the Danish government, claiming the American project to be of a purely scientific character and to have the exact same "scientific aim" as ongoing rocket activities conducted by the ESRO from Esrange, Kiruna, which also had Danish participation. He stated that 1969 would be too late and that the world would now have to wait 11 years for the next solar maximum. In his mind Denmark simply owed too much to say no, since all earlier Danish rocket-borne space investigations had been conducted with rockets received free of charge from the U.S. He thought it "inconceivable" that the decision was final, since it could only rest on "misunderstandings."[61] The notion that earlier help and support from the U.S. obligated a "yes" found support among many newspaper commentaries in the following days.[62] Such arguments points to the strong moral alignment felt not only by the involved scientists but also among the media and the public in general.

Like his colleagues Stauning choose to present the project as "pure science" though he actually had quite substantial knowledge of the military aspects of the project. A few days later in a cryptic letter directed to Hilmar Baunsgaard, the PM, Stauning hinted at these military aspects and indicated that he reckoned that this was the real reasoning behind the rejection. Speaking directly to the PM Stauning stressed that he did not find it "correct" that the Danish government would use a full year to evaluate the project, nor did he find the official justification given so far to be "worthy for the Danish Government."[63] Stauning had been a member of the social liberal party and knew the PM through his affiliations with this party.

The political ban on the PCA campaign and the intense media coverage created an aura of suspicion around ionospheric research adding to the feeling of insecurity in the community. The Ionlab's staff expected the Americans to rapidly withdraw their planned experiments and reduce activities in Thule. In that situation, they argued, there would be no economic basis for Danish ionospheric research from Thule. As one rank and file engineer explained: "We fear that the politicians will cut us down to the ground if they find us troublesome—and we soon expect that to be the case."[64] Ionosphere scientists felt rejected by the politicians and feared losing their jobs due to political repercussion.

Top-level scientists on the other hand worried about the possibility of losing international prestige and of possible repercussions for Danish-American cooperation in science. In a four-page personal letter the director of the Danish Geodetic Survey Einar Andersen told Niels Boel of the Ministry of Foreign Affairs that the decision in his opinion was "unwise" and that he "as a Danish scientist felt ashamed." Andersen, a prominent veteran in international scientific organisations, asserted that the "rocket affair" would be "very unpleasant" for him when he took part in future international meetings. In the light of the incident he would even consider withdrawing all together from his many international positions.[65] He went on to reflect upon the moral economy of science in an age of Cold War scientific cooperation with the resourceful hegemonic power on the other side of the Atlantic. "Our American friends are of course quite different from us, and we often become rather annoyed by them, but we must acknowledge that they have done a very great deal for us and made a great many things possible here. Science does not exactly have the best conditions in Denmark and this is largely independent of who is ruling the country."[66]

When referring to Helweg Petersen's fear that some research fields had been "tainted" because of military patronage, Andersen found these concerns ridiculous and equated them with a young girl's fear of small bugs. The rocket campaign, he explained, had "no direct military significance" but on the other hand telephones, food, and hospitals were of military significance in the right contexts. Andersen thus minimized the military aspects of the project, and in this way justified a scientific project under scrutiny. On the other hand his argument made it clear that US military patronage was a necessary precondition for Danish geophysical scientists working in Greenland. Domestic fiscal reluctance towards scientific research would continue to push researchers into a borderland that politicians and the public might find questionable.

Reconciliation—American hegemony at work

The outcome of the "rocket affair" was that the main part of the experiment was relocated to Fort Churchill in Canada (less suitable from a geophysical point of view) and carried out in November 1969 (redubbed Operation PCA 69). The expected repercussion and decline in U.S. activities in Greenland, however, failed to materialise. Reconciliation came almost immediately and was effected through channels at a lower level. No official diplomatic response appeared from the U.S., but in mid-July 1968 Major General Otto Glasser of the USAF R&D staff met up with representatives of the Greenland administration and KVUG. Instead of adding to the disagreement the U.S. side chose a smooth pragmatic response.

Expressing the deep concerns felt by USAF scientists, Glasser pressed hard for an indication of the future political line of the Danish government— that is "whether a negative attitude might as a whole be expected from the Danish side towards the American activity?" Could any lessons be learned to avoid future failures, he asked, requesting a "firm Danish viewpoint." In response, the representative of the Ministry of Foreign Affairs explained the Danish need "to take all necessary precaution" and stated that "after the B-52 incident and the rumours about American test of biological weapons in Thule the government was bent on avoiding any more problems around this area within the same year." Reassured by this statement the general concluded that it was indeed a singular event, that Operation PCA 68 was canceled for unusual reasons, and that no major change in the American-Danish cooperation had taken place! Nobody objected and clearly to the relief of the Danes, Glasser then made it plain that the USAF was prepared to continue cooperation with Danish scientists, "irrespective of the present circumstances."[67] The crucial point here is to note how careful the general acted to solve or circumvent the diplomatic issue in a pragmatic manner and thus manage the American hegemonic relations.

The US did not withdraw or reduce their research activities in the Thule area; in fact the exact opposite happened. Regarding future cooperation the participants in the meeting reached an agreement on three points: (1) that more openness about research programs was needed; (2) active measures to promote public understanding had to be taken; (3) long term planning should be considered to improve the possibilities for Danish scientists to partici-pate.[68] Transatlantic cooperation between the Ionlab and the AFCRL and Air Weather Service continued and a new active information procedure was adopted by the AFCRL.[69] The USAF took measures to accommodate the specific needs and peculiarities of the reluctant partner in order to stabi-lise the political boundary conditions around the Arctic activities and secure future access to the area. Hegemony was not something that could be taken for granted but a fragile arrangement that had to be continuously renegotiated.

Conclusions

Operation PCA 68 was not rejected because of its military origin *per se*, but rather because the Danish government feared that heated public discussion about the project would endanger future transatlantic cooperation. The rejection must be viewed as a singular diplomatic act to restore balance and an image of full Danish sovereignty in Greenland. The B-52 accident made it clear that the U.S. had stepped on (nuclear) sensitive toes of a close and loyal ally, and the Danish government used the first opportunity to react. After this event it was business as usual. Moreover the intense diplomatic activity in the first part of 1968 is better depicted as a temporary distur-bance in consensual hegemony than as a clear sign of "transatlantic drift."

Initially public media favored the project. Reactions became mixed when the public finally got a glimpse of the military agenda. Faced with rising political dispute and unease about the military-academic nexus Danish sci-entists tended to belittle and even reject the military significance of scientific investigations in Greenland and generally succeeded in doing so because of the wall of secrecy that surrounded the American activities in Greenland.

Cooperation between Danish civilian scientists and U.S. military insti-tutions was a significant feature in ionospheric science in Greenland. For Danish scientists, such joint efforts provided additional funding opportu-nities and access to world centers of ionosphere research. Such cooperation fostered robust ties and strong loyalties towards American military research institutions. When these activities were challenged Danish scientists actively sought to stabilise the boundary conditions, for instance through outreach and PR efforts that framed military scientific investigations as international and purely scientific in scope, and by acting as information agencies for U.S. military organisations and (for lack of a better word) as a hegemonic force within the Danish administration. The scheme of "consensual hegemony" and consensual elites fits well the ionosphere scientists engaged in Danish-American research cooperation in Greenland. But this study has also supplied ample evidence to conclude that scientists were not alone in this respect. Many media actors could also be fitted into this scheme.

In this case it is clear how American military institutions pragmatically included Danish scientists in their projects and adapted in response to demands from their counterparts in Denmark. American hegemony rested on reciprocal relations and exchanges of "gifts" and services that had to be carefully managed and balanced on many levels.

Notes

1. "50 raketter skal sendes op fra Thule," *Politiken*, 4 July 1968; "Hartling aflyser raket-program," *Politiken*, 5 July 1968.
2. J. Krige, *American Hegemony and the Postwar Reconstruction of Science in Europe*. (Cambridge, Mass. & London: MIT Press, 2006), 5.
3. G. Lundestad, The United States and Europe after 1945: From "Empire" by Invitation to Transatlantic Drift (Oxford University Press, 2003).
4. H. Knudsen, "Battling the Aurora Borealis: The Transnational Coproduction of Ionospheric Research in Early Cold War Greenland," in: R. Doel, K. Harper and M. Heymann eds., *Exploring Greenland: Cold War Science on Ice,* (New York: Palgrave Macmillan, 2016), 143–165.
5. J. van Dongen, *Cold War Science and the Transatlantic Circulation of Knowledge*, (Leyden: Brill, 2015), 3–4.
6. A part of this section has been adapted from H. Knudsen, "Cold War Greenland as a Space for International Scientific Collaboration," in: R. Doel, K. Harper and M. Heymann, eds., *Exploring Greenland: Cold War Science on Ice.* New York: Palgrave Macmillan, 2016), 228–230.
7. Confidential note no. 96. US Embassy to Ministry of Foreign Affairs, 21 November 1967 (Rigsarkivet, Udenrigsministeriet 1945–1972, 105 F.9.a.) [Hereafter: 105.F.9.a.]. Elaborate descriptions became available later, e.g., "Operation PCA 1968" and James C. Ulwick, *Operation PCA 68* (Rigsarkivet, Grønlandsministeriet, Departementschefs arkiv (fortroligt), journalsager 1950–78, XLI B V, PCA Raket Program) [Hereafter PCA-Program].
8. R. Liebowitz, *Air Force Geophysics: Contribution to Defense and to the Nation, 1945–1995*. PL-TR-97-2034, Special Reports, No. 280, 8 April 1997. (Hanscom AFB, Mass.: Geophysics Directorate, 1997), 30–32.
9. Office of Aerospace Research, *Air Force Research Resumes 1968*, 484.
10. Defense Threat Reduction Agency, *Defense's Nuclear Agency 1947–1997*, (Washington: U.S. Department of Defense, 2002). DASA's interest in the Thule project is treated in N. Davis, *Rockets over Alaska: The Genesis of the Poker Flat Research Range*, (Fairbanks: Alaska-Yukon Press, 2006), 80–102.
11. "Program Change Proposal for DASA," 15 May 1962, see Knudsen, "Cold War Greenland as a Space for International Scientific Collaboration," 229.
12. "Operation PCA 1968", PCA-Program.

13. D. Flinders, "The Space Forecasting System: Confluence of Military and Scientific Interests," in *Air University Review* Nov-Dec 1969. http://www.au.af.mil/au/afri/aspj/airchronicles/aureview/1969/nov-dec/flinders.html (accessed 7 May 2017).
14. Lloyd to Executive Director, AINA, 11 July 1968. Arctic Institute of North America, File: Seismic—"Blue Ice."
15. J. Taagholt, "Dansk Videnskabelig Forbindelsesofficer for Grønland," *Forskning i Grønland*, 1992, 4, 27.
16. Annual Report of 1967 from the scientific advisor to FOTAB. 105.F.9.a.
17. Confidential note no. 96. US embassy to Ministry of Foreign Affairs, 21 November 1967. 105.F.9.a.
18. Blankenhip to Boel, 4 March 1968. 105 F.9.a.
19. Knudsen, "Battling the Aurora Borealis," 155–6.
20. G. Gudmandsen, *ESRO/ESA and Denmark. Participation by Research and Industry*, (Noordwijk: ESA Publications Division, 2003), 9.
21. Undated and untitled strategy and working plan for the Ionosphere Laboratory 1967-1971. 105.F.9.a. The strategy of collaboration with USAF's activities at Thule Air Base was also presented in KVUG in May 1967.
22. Evidence of this strategy can be found in "Memo for Records", July 7 1964, James Giraytys, (Captain, USAF, Assistant Chief, Meteorological Rocket Planning Group). (National Archives and Records Administration [NARA], RG 59, Entry A-1(5590) Lot#73D224, B1, F: Greenland Area Research Progress. Meteorological Rocket Project: SCI 11-1 1964); Supplementary comments on "Proposal of Solar Particle Rocket Experiments launched from Thule Air Base, Greenland 1968." PCA-Program; J. Taagholt, "Danish Arctic Ionospheric Research," *Arctic*, 1972, 25(4): 259.
23. Taagholt to Ministry for Greenland, 1 November 1967; Taagholt to Ministry for Greenland, 2 November, 1967. 105.F.9.a.
24. Note Verbale, 16 November 1967. 105.F.9.a.
25. Ulwick to Olesen, 6 December 1967. PCA-Program.
26. "Notat angående det amerikanske Thule projekt", Bernard Peters, 1 August 1968. 105.F.9.a.
27. Olesen to Ulwick, 19 December 1967. PCA-Program.
28. Otto Jensen to Ministry of Foreign Affairs, 4 April 1968; Boel to Brady Barr, 10 April 1968. 105.F.9.a.
29. *Summary report of discussions at the Ionosphere Laboratory in meetings May 13–15, 1968 with representatives of the U.S.A.F. and the Ionosphere Laboratory and the Geophysical Division of the Danish Meteorological Institute*, 25 June 1968, 4. 105.F.9.a.
30. *Ibid.*, 8.
31. Here I am indebted to a talk given by John Krige at the 4th Int. European Society for the History of Science Conference, Barcelona, 18–20 November 2010.
32. N. Petersen, "The H.C. Hansen Paper and Nuclear Weapons in Greenland," *Scandinavian Journal of History*, 1998, 23(1): 21–44; DUPI, *Grønland under den kolde krig. Dansk og amerikansk sikkerhedspolitik 1945-68*, (Copenhagen: DUPI, 1997), 451 ff.
33. DUPI, *Grønland under den kolde krig*, 524.
34. Minutes of meeting in KVUG 14 May, 1968. Rigsarkivet, KVUG, journalsager 1939–1971, box 4.
35. Military liaison officer, Thule to Ministry for Greenland, 20 May 1968 (telex). PCA-Program.
36. H.D Bruner (US Atomic Energy Commission): "Trip to Thule, Greenland, January 23 to February 3, 1968 in connection with B-52 Crash, January 20, 1968" March 18, 1969, 10. NARA, RG 59, Entry A-1(5587), B5, F: "Thule Crash—Internal Memos, 1968."

37. K. Hastrup, *Thule på tidens rand*, (Copenhagen: Lindhardt og Ringhof, 2015), 213–218, 412–413.
38. City treasurer to the Ministry for Greenland, 22 May 1968 (telex). PCA-Program.
39. City treasurer to the Ministry for Greenland, 22 May 1968.
40. Ministry for Greenland to Sandborg, 24 May 1968 (telex). PCA-Program.
41. Olesen to Andersen, 28 May 1968. PCA-Program.
42. Olesen to Andersen, 28 May 1968.
43. Olesen to Andersen, 28 May 1968.
44. Ministry for Greenland to Sandborg, 28 May 1968.
45. Bech to Høegh, 8 June 1968 (telex), with ministerial remarks. PCA-Program.
46. "Kæmpeafbrænder for videnskaben—", *Aktuelt*, 6 July 1968.
47. Quoted from Notits, signed by Niels Boel, 9 January 1968. 105.F.9.a.
48. Quoted from Notits, signed by Gunnar Blæhr (14 June) and Niels Boel (19 June). 105.F.9.a.
49. *Ibid.* Additional note of 25 June 1968.
50. *Ibid.* Underscoring as in original document.
51. Ministry of Foreign Affairs to Danish Embassy, Washington, 28 June, 1968 (telex). 105.F.9.a.
52. "Nu vil regeringen give befolkningen store kendskab til udenrigspolitik", *Jyllands-Posten*, 14 July 1968.
53. "Dansk minister appellerer til videnskabsmænd", *Politiken*, 15 July 1968.
54. "Thule-raketterne", *Politiken*, 22 July 1968.
55. "Et nej til Amerika", *Skive Folkeblad*, 6 July 1968.
56. Taagholt to Otto Jensen, 26 March 1968. PCA-Program.
57. "50 raketter skal sendes op fra Thule", *Politiken*, 4 July 1968.
58. Notits, "De amerikanske raketprojekter i Thule", 4 July 1968; Notits ("Det amerikanske raketprojekt i Thule. Ionosfærelaboratoriets rolle i presse-kampagnen"), 8 July 1968. 105.F.9.a.
59. Notits, "Det amerikanske raketprojekt i Thule. Civilingeniør Peter Staunings offentlige udtalelser om udenrigsministeriets afgørelse", 9 July 1968. 105.F.9.a.
60. Interview with Peter Stauning, 25 November 2010.
61. "Hartling aflyser raket-program", *Politiken*, 5 July 1968. See also Memorandum vedrørende de Amerikanske Thule-Eksperimenter, Haldor Topsøe, 31 July 1968. 105.F.9.a.
62. "Formørkede Solpletter", *Politiken*, 6 July 1968; "Kæmpeafbrænder for videnskaben—", *Aktuelt*, 6 July 1968.
63. Stauning to Baunsgaard, received 8 July 1968. 105.F.9.a. Stauning also pointed to such knowledge in an interview conducted by the author. Interview with Peter Stauning, 25 November 2010.
64. "Hele laboratoriet går imod Karl Andersen!", *Aktuelt*, 17 July 1968.
65. Andersen to Boel, 17 July 1968. PCA-Program.
66. Andersen to Boel, 17 July 1968.
67. Meeting in the Ministry for Greenland, 19 July 1968. 105.F.9.a.
68. Meeting in the Ministry for Greenland, 19 July 1968.
69. A series of annual reports from the AFCRL during 1968–1971 is still available through the Danish library system.

14 Applied science and practical cooperation: Operation Morning Light and the recovery of cosmos 954 in the Northwest Territories, 1978

P. Whitney Lackenbauer and Ryan Dean[1]

When Cosmos 954, a Soviet Radar Ocean Reconnaissance Satellite (RORSAT), tumbled across the early morning sky over Yellowknife on 24 January 1978, witnesses reported an object that "looked like a jet on fire ... with flaming jet stream." Yellowknife resident Marie Ruman observed the event on her way to work, later recounting that the main body of Cosmos was followed by "dozens of little pieces" with smaller but similar flaming tails that glowed a bright red.[2] Cosmos 954's power-plant, a nuclear reactor fuelled with approximately 45.5 kg (100 lbs) of enriched uranium-235 to operate the large radar array, had failed to eject from the stricken craft and boost itself into a higher disposal orbit as per its design. As the satellite broke up plunging through the atmosphere, its reactor compartment failed, spewing out radioactive fuel 55–60 kilometres (34–37 miles) above the Earth's surface. Scientists were concerned that the uranium-235 and other long-lived reactor by-products, such as the isotopes strontium-90 and cesium-137, would contaminate the area.[3]

During its three-minute burn through the upper atmosphere, Cosmos scattered debris from the western edge of Great Slave Lake, tracking east-north-easterly over an 800-km (500 mile) stretch along the Thelon River, through the barrens, to the region just north of Baker Lake. Authorities were quick to respond. A 22-person Canadian Forces Nuclear Accident Support Team (NAST) had been sent from Canadian Forces Base (CFB) Edmonton to Yellowknife that morning to begin their preliminary assessment of radioactive contamination and recovery of satellite debris.[4] "The normally easy-going citizens of Yellowknife were startled by the sight of yellow-garbed troops walking the streets, reading radiation meters and taking air samples," Major W.R. Aikman observed.[5] The immediate response to a potential nuclear disaster required a concerted scientific effort, with military personnel supporting Canadian and American scientists who combed the projected debris area for radioactive wreckage.

Operation Morning Light unfolded over the next 84 days. The recovery effort ultimately spanned 24,000 square kilometres (9,266 square miles) of sub-arctic wilderness in conditions that dipped below −40°C (−40°F).[6] Much of this activity took place from the hastily constructed military "tent city" known as Camp Garland or Cosmos Lake in the Thelon Game Sanctuary where the first large piece of debris was discovered.[7] Specially-equipped CC-130 Hercules aircraft flew search patterns to detect "hits" of radiation, while helicopters carried recovery teams to find and secure the radioactive debris. Encasing the remnants of the fallen satellite in specially constructed lead-lined containers, these materials were then transported to Baker Lake or Yellowknife and then shuttled south for analysis. In this manner, Operation Morning Light successfully recovered 66 kg (145.5 lbs) of wreckage, with all but one 17.7 kg (39 lb) piece proving to be radioactive.[8]

Shortly after the operation concluded, author Leo Heaps and official military historian Captain Colin A. Morrison produced books examining the subsequent challenges of Operation Morning Light. The cover of Heap's popular history *Operation Morning Light* boasts the hyperbolic tag line: "it was a science-fiction nightmare come true!"[9] Filled with anecdotes from those involved in the operation, the book preys on popular fears about lingering radioactivity for dramatic effect. Morrison's more technical history *Voyage into the Unknown* emphasizes the difficulties that searchers had to overcome in their recovery operations.[10] Morrison treats Morning Light as a prototypical case study—the first mission of its kind to locate and remove radioactive debris from one country that had fallen onto another country's territory from space—and a resounding success.[11]

This chapter tells the story of how a mixed group of civilian and military personnel covered tremendous distances and endured frigid temperatures to survey a wide area for radioactive contamination. After-action reports that critically evaluated the methods, equipment, and personnel employed during Morning Light elucidated how a combination of civilian scientific expertise and military capabilities succeeded in effectively locating and recovering the remnants of the downed nuclear-powered satellite scattered across a frigid, sub-arctic environment. As Morrison insisted in 1983, "those involved in the planning and execution of the search for and recovery of *Cosmos* 954 were venturing into a new field of operations—a voyage into the unknown—a process that entailed much trial-and-error." The potential danger to humans, fish, and wildlife in the region gave the operation its driving imperative and demanded a "crisis-management" approach.[12] Crisis was averted, however, through tight bi-national cooperation and systematic scientific monitoring. "The mere entry of the satellite into Canada was a breach of Canadian sovereignty and, therefore, constituted an international wrong," a legal appraisal later observed in seeking damages from the USSR.[13] Interestingly, a dominant historiographical theme emphasizing Canadian sensitivity over US "threats" to Arctic sovereignty from the Second World War onwards is conspicuously absent in the case of Operation Morning Light. In the face of a tangible Cold War nuclear threat, the search, recovery, removal, testing and cleanup of radioactive fragments that had fallen from the sky necessitated a joint effort.

Nuclear histories, historian Itty Abraham notes, are dominated by a discourse of control that has narrowed the focus to national efforts at non-proliferation and less on the implications of nuclear programs more broadly. More attention should be paid to the scientific-technological underpinnings of these programs,[14] and to the national (and in the case of Morning Light, bi-national) response systems set up to deal with nuclear disasters and accidents. While accidents at the SL-1 (1961), Three Mile Island (1979), Chernobyl (1986), and Fukushima Dalichi (2011) power plants, as well as those associated with nuclear-powered submarines, have been subjected to significant analysis and debate, the application of science and technology to detect and cleanup small nuclear incidents has received less attention. Adapting responses that had been developed in southern laboratories and offices and devised for global application to an austere Arctic environment, Operation Morning Light demonstrated the transferability and application of Cold War applied science in the Canadian North. While civilian scientists and military operators had to render the Arctic scientificially "legible"[15] to identify and cleanup nuclear debris, this knowledge and concomitant use of technology was not used to (re)create Arctic environments during the operation and in its aftermath.[16] Instead, an immediate joint Canadian-American effort, involving multiple government agencies, was coordinated to protect the landscape and Northern peoples from radionucleide contamination. While a modest story of bilateral scientific cooperation, it is also an important case study in how government action during the Cold War could be channeled constructively to prevent toxic legacies—either environmental, diplomatic (in terms of Canada-US relations), or between Northern residents and the Canadian government.

Setting the context

The Americans had good reason to track the launch and orbit of Cosmos 954, which lifted off on 18 September 1977. Fourteen metres (46 feet) long with a mass of 3500 kilograms (7716 lbs), the nuclear-powered RORSAT was built around a powerful X-band radar that could look through thick cloud layers to scan the world's oceans for naval vessels (primarily American aircraft carriers, as Cosmos was able to track smaller warships like destroyers only in clear weather conditions). The satellite had the ability to send its observations back to Moscow or directly to Soviet naval units and possibly even Tu-22M "Backfire" bombers.[17]

The U.S. became aware that Cosmos 954 was in trouble in late October 1977. North American Air Defense Command (NORAD)[18] noted Cosmos 954's slowly decaying orbit and began updating plots of when and where the satellite would re-enter the atmosphere. Most of these calculations were done at the Lawrence Livermore National Laboratory by engineer Milo Bell and mathematician Ira Morrison, supported by engineer Robert Kelley. The trio had access to the highly sophisticated Control Data Corporation 7600 supercomputer, with its C-shaped frame stretching twenty feet and filling an entire room at the laboratory.[19] The problem was clear: "what does one *do* about a live nuclear reactor reentering the earth's atmosphere aboard a Soviet surveillance satellite?" Gus Weiss, a special assistant to the Secretary of Defense, explained how "a quick scan of literature showed no textbook answer, nor even a textbook question. It remained for the National Security Council [NSC] Staff to put together a group to cope with the problem."[20]

On 19 December, the NSC formed a working group (the *Ad Hoc* Committee on Space Debris) to prepare contingency plans and prepare to mount a quick search and recovery operation of Cosmos 954 if needed, thus birthing Operation Morning Light. Contributing agencies included the Central Intelligence Agency (CIA), the Department of Defense (DoD), the Department of Energy (DoE), the State Department, the Environmental Protection Agency (EPA), the Federal Preparedness Agency, and the Office of Attorney General. The NSC placed the DoE's nuclear emergency response capabilities on alert "to assist in the protection of public health and safety should radioactive debris from Cosmos 954 come to earth in the United States." This included organizations such as the Accident Response Group (ARG) and the Nuclear Emergency Search Team (NEST), which had the experts and equipment necessary to find and recover radioactive materials. Due to the "uncertainty in determining when or where (in the world) Cosmos 954 would reenter," experts anticipated "that there was no preventative or preparatory action that could be taken by the public." Subsequently, both the American public and the US's allies were kept in the dark until experts could plot a more accurate projection of Cosmo 954's return.[21]

In early January, Cosmos 954's orbit decayed precipitiously. Updated calculations estimated a reentry date of the 24th of that month, but *where* the satellite would crash remained hazy. American authorities summoned the Soviet ambassador to secure information on the radioactive hazard that Cosmos 954 posed. The USSR's response was rather sparse, noting that the power plant on the satellite was "explosive-proof" and was designed to burn up when it entered denser layers of the atmosphere. Nevertheless, the depressurization (for unclear reasons) that had caused the satellite to lose control meant that some destroyed parts of the plant could still reach the earth's surface, and "in that case an insignificant local contamination may occur in the places of impact with earth which would require limited usual measures of cleaning up." One US official remarked that he was not sure what "'usual measures of cleaning up' a reactor crashing in from outer space might be, and there was also some ambiguity in the meaning of 'explosive-proof.'" Nevertheless, the firm knowledge that the nuclear reactor had been designed to burn up during re-entry offered a measure of comfort to American planners.[22]

By this time, computer modeling discerned that the wave-like orbital path of the doomed satellite overflew Australia, Britain, Canada, Japan, and New Zealand, and the US notified its allies accordingly.[23] Canada first learned that Cosmos 954 could crash in its territory on 19 January, and the Department of National Defence (DND) alerted all regional commanders and its NAST of the impending threat the following day. Air Command Headquarters alerted CFB Edmonton Base Commander Colonel D.F. Garland on 23 January that Cosmos would be entering Edmonton's Search and Rescue Region the following day, and the NAST was informed and placed on two-hour standby. At this time, the Prime Minister's Office (PMO) notified several of the civilian departments of the threat that the satellite posed to the country and their responsibilities in the response effort. This meant that many of the key agencies and actors who became involved had less than twenty-four hours notice, and some did not receive notification until after it had crashed.[24]

As soon as American experts confirmed Cosmos 954's reentry over the NWT on the morning of 24 January, President Jimmy Carter contacted Prime Minister Pierre Trudeau and offered American assistance. Trudeau immediately accepted. The principal mission for the U.S. NESTs was to help the Canadian government locate radioactive debris. Accordingly, they enlisted American experts to provide technical assistance in calculating the reentry of the Cosmos and the ballistics properties that various pieces of it would likely exhibit in their fiery plunge back to Earth. This involved sophisticated reentry calculations and computer modeling, establishing the perimeters of the search area, and estimations of where larger pieces of debris would land.[25] NEST also operated aerial measuring equipment and assisted with ground recovery activities.[26] At the request of DND, the DoE provided two gamma ray spectrometers and operating personnel, who arrived in Edmonton on 24 January to install their equipment on Canadian Hercules aircraft. Canada provided the technical assistance to mount the detection equipment onto the aircraft, as well as on-site logistics support such as providing the NEST with military clothing for sub-arctic operations.[27]

Despite having received little to no warning, Canadian civilian scientists responded immediately and began arriving in Edmonton on mid-morning of 24 January—at roughly the same time as American NESTs, which had seven weeks of forewarning and preparation. The first of these scientists was Dr. Bob Grasty of the Geological Survey of Canada (GSC), whose expertise in aerial surveying for naturally occurring uranium was mobilized to detect Cosmo 954's highly enriched uranium-235 core. A GSC gamma ray spectrometer designed for uranium exploration and mapping was quickly shipped, along with Grasty, from Ottawa to Edmonton to enable the search.[28]

NORAD had provided Operation Morning Light with projections of Cosmo 954's probable debris field between Great Slave and Baker Lakes, delineated as an area 800 km (500 miles) long and 50 km (31 miles) wide

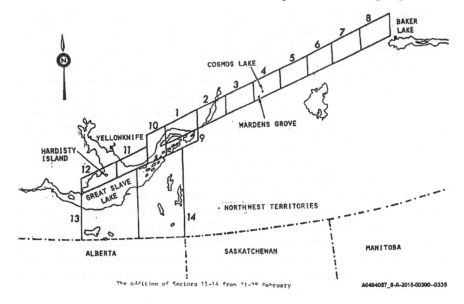

Figure 14.1 COSMOS 954 search area.

Source: ATIP 2015-00300.

(see Figure 14.1). The first phase of the operation called for CC-130s, specially equipped with gamma ray spectrometers to detect radiation emitted from the surface, to fly a grid pattern 1000 feet above ground level (AGL) over the suspected satellite crash area.[29] While the pilots focused on carefully flying their intended search tracks under difficult conditions, "back in the cargo compartment, the [NEST] scientists took turns watching several needles as they slowly swayed up and down across a piece of graph paper, waiting for the telltale swing that would indicate a hit."[30] NEST members operating these devices quickly began registering "hits" along the search area, which were recorded on data tapes and then fed into NEST computer vans at Yellowknife and Baker Lake for analysis. "Each hour of search flight time for each of the C-130s created four hours of computer analysis time, creating a major assessment backlog," the DoE's official report recounted.[31] "Hits" would then be located on navigation charts[32] and helicopters fitted with detection equipment sent to these sites to precisely locate the radioactive source. One helicopter would drop a brightly colored streamer on the suspect site, and a second helicopter carrying a three-person recovery team would follow to inspect the area on the ground and recover any radioactive materials (Figure 14.2).[33]

The largest piece of debris, found through aerial search on 1 February, became known colloquially as the "stovepipe." The head of the search team that recovered this piece recounted to excited reporters that it was evident "something [had] really gone through the ice at high speed." Paul Murda, the leader of a five-man American scientific team that analyzed the object, described it as "sort of like a cylinder that got smashed," with what "looks like structural tubing" sticking out the ends.[34] Fortunately it was not radioactive, which made its detection from the air a stroke of luck. Three days later, another recovery team—wearing their trademark thick yellow coveralls, parkas, and Arctic boots, with radiation detectors hanging their waists[35]—found some of the most radioactive material: a clutch of beryllium rods and cylinders partially embedded in the snow and ice.[36] When the recovery team, led by Atomic Energy Control Board (AECB) members Tom Robertson and Wick J. Courneya, cautiously approached the debris, their "Geiger-counter readings exceeded 100 roentgens per hour."[37] Courneya, a health physicist, put this level of radiation into perspective in a later interview. "If a person held [an object measuring some 200 roentgens] for one hour, he would probably get ill," he explained. "If a person held it for two hours, he probably would die."[38] Accordingly, it was standard operating procedure after every mission to check recovery teams and aircrew for radiation, and "any item of clothing which produced a reaction on the meters was immediately removed."[39]

The most famous piece of recovered debris, known as "the antlers," was initially discovered by adventurers during a fifteen-month trip across northern Canada. Traveling from the Yukon into the Northwest Territories (NWT) along the Mackenzie River, the party was wintering over in the Thelon Game Sanctuary when Cosmos 954 broke-up in the skies above them.[40] At 2:30 in the morning on 26 January, Christopher Norment recorded that "we are awakened by a large, four-engined plane passing low over the cabin; it appears to be flying a grid pattern, as if it is conducting a search."[41] It was one of the CC-130 Hercules from Edmonton's 435 Transport Squadron searching for radioactive debris.[42] Although the aerial search failed to turn up a "hit" in this area, John Mordhorst and Mike Mobley (both part of Norment's expedition) had undertaken a two-day dog sled from their camp at Warden's Cove to visit cabins erected by the English naturalist John Hornby and his party during their 1926–7 expedition (which Norment's group was retracing).[43] Crossing the frozen Thelon River en route back to their camp on 28 January, they came upon "an odd-looking metal object" which they stopped to examine.[44] When the pair returned to Warden's Cove they reported their discovery to Norment. "We found a strange object in the river ice just below Grassy Island, not more than a hundred yards from where we'd passed three days ago," he recalled. "There was this crater, six or seven feet across, where something hot had hit and melted into the ice, and several charred metal struts were visible.[45] By then, news of the satellite's crash had been broadcasted across Canada and the wider world. Norment had heard about it on his radio and had been visited two days earlier by land use officers interested in any signs of Cosmos. "You guys have found a goddamn (etc., etc.) Russian satellite!" exclaimed Norment.[46] The party radioed in their discovery.

The next day, a military CC-138 Twin Otter arrived with a recovery team to inspect and remove the adventurers and secure the debris. Norment recounted how:

> Two boffins in full protective regalia emerge with radiation detection devices and begin measuring contamination levels; the other passengers remain on board, awaiting word that it is safe to deplane. I can see faces peering out of the plane's windows, and I feel like an animal on display in a drive through wildlife park. The scientists run their instruments over the dogs and sled, and take readings of John and Mike. One of Mike's mitts is found to be very faintly radioactive, and it's confiscated.... An hour later, a twin-engined Chinook helicopter comes whopping into camp with a Hercules flying support and making large circles overhead. The Chinook disgorges twenty people ... scientists, military types, photographers. Everyone is outfitted in what will become familiar attire: white insulated boots, regulation green overpants and parkas, and white-and-tan mitts with synthetic fur backing.[47]

The NAST team pushed through the snow and scrub to the river and found twisted metal protruding from the ice like a pair of antlers. Their Geiger-counters produced readings of 10-1000 milliroentgens per hour, "not the several hundreds of roentgens per hour that the solid core would produce."[48]

Due to time constraints brought on by extreme cold and impending darkness, the recovery teams were flown out to Baker Laker by Chinook helicopters later that day. John, Mike, Christopher, and the rest of their party were loaded onto the Twin Otter and flown to Yellowknife before being sent to a hospital in Edmonton for further radiation testing. Considering the growing concerns about members of the news media arriving at the unsecured site, as well as the welfare of the dogsled team that had to be left behind, the military dropped four paratroopers into Warden's Grove on the morning of 31 January to secure the area. The adventurers, meanwhile, were found to have absorbed the equivalent of one or two X-rays of radiation from their experience. After a press conference, the men were released from hospital and hired by the military to act as guides in the Thelon Game Sanctuary to assist scientists before returning to their dogs and expedition.[49]

Over the course of Operation Morning Light, scientists used two general methods of data collection to determine if the uranium-235 core (or a part of it) survived re-entry and posed a risk to people and the surrounding environment (see Figure 14.2). The first involved collecting ground samples along the debris field. The second method was aerial survey by helicopter-mounted gamma detectors flying search patterns. Surveyors found that particles were randomly distributed and far apart, and scientists concluded by early March that people living in the affected area had little to be concerned about.[50]

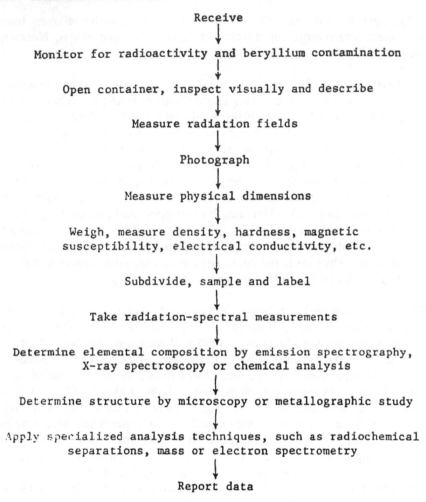

Receive
↓
Monitor for radioactivity and beryllium contamination
↓
Open container, inspect visually and describe
↓
Measure radiation fields
↓
Photograph
↓
Measure physical dimensions
↓
Weigh, measure density, hardness, magnetic
susceptibility, electrical conductivity, etc.
↓
Subdivide, sample and label
↓
Take radiation-spectral measurements
↓
Determine elemental composition by emission spectrography,
X-ray spectroscopy or chemical analysis
↓
Determine structure by microscopy or metallographic study
↓
Apply specialized analysis techniques, such as radiochemical
separations, mass or electron spectrometry
↓
Report data

Figure 14.2 Examination sequence of satellite debris.

Source: ATIP 2015-00300.

While the debris area was large, barren, and sparsely populated, searchers were highly aware that it was also a homeland for people.[51] "The inhabitants of the Northwest Territories in the path of the Cosmos 954 Satellite were concerned about their safety and it was necessary to undertake search and recovery operations so that the inhabitants could be assured that all debris dangerous to their health had been recovered," an official summary noted. Dan Billing, the Chief of Emergency Services for the Government of the Northwest Territories, explained that about 10,000 people lived in the "hit-zone," leading municipal officials and citizens' committees in Fort Smith, Hay River, Snowdrift, Fort Resolution and Pine Point to express tremendous concern about radio-active debris that might threaten their safety. Signs alerted residents to report any unusual debris (and to avoid touching it), and some people "restricted their normal use of of the territory for fear of contact with radio-active material." People worried that drinking water, fish, and caribou might be unsafe to consume. In this context, Morning Light crews completed foot searches for radioactive material in the municipalities and around hunting and fishing lodges, finding radioactive debris in several of these locations.[52]

In addition to checking settlements, authorities were dispatched across the affected area to contact Northerners who were out on the land. Although distance and sub-arctic operating conditions complicated logistics, authorities were confident they had found all the civilians in the search area by early 28 January and advised them of the possible hazards.[53] Authorities were particularly anxious about how they would explain the situation to many of the region's Indigenous inhabitants. "There was a common concern and generally not enough known about this strange element ... translated from English to Chipewyan [as] 'poisonous,'" reporter Robert Blake explained. "There are no words in Chipewyan to adequately describe radioactivity, gamma ray sweepers and the like."[54] When a NAST flew to the Chipewyan (Dene) community of Snowdrift (now Łutsel K'e), its inhabitants scattered. Canadian Northern Region Headquarters commander Brigadier General (BGen) Ken Thorneycroft flew to the village the next day to explain what was happening and to reassure local residents that no radiation had been detected near their community.[55] Although concerns never entirely dissipated, Morning Light's coordinated response offered credible reassurance to Northerners that their safety was of paramount importance, and that the search and recovery operations for debris appropriately addressed the most serious threats to human and environmental health.

Bi-national cooperation: crisis management and sovereignty as non-issue

The American contribution reached its zenith two weeks into the operation, when 120 specialists in various fields were participating. Author Leo Heaps, in his dramatic account, observed that:

> When the Americans went into full gear with their immense back-up resources, there was very little in the world that would be able to equal them. The motive of competition, of sensitive pride where the Americans were concerned, was all one-sided. Canadians are traditionally apt to have some acute feelings in these matters. However, this was an emergency and the clear-headed Garland and his team appreciated the assistance. The American scientists and technicians stayed out of

sight in spite of the urgings of their public relations man, allowing the Canadian scientists and military to make all the announcements. They would have their turn when they arrived home.[56]

Importantly, the sovereignty concerns that dominated most discussions of American involvement in the Canadian North since the Second World War—as other chapters in this book make clear—were not in play.[57]

When the United States eventually published its official "non-technical" summary of the operation detailing its agencies' roles, it highlighted Morning Light as an "example of international cooperation for the protection of the health and safety of the population of North America."[58] An internally-directed Canadian report affirmed that the two countries' intimate cooperation during the operation proved seamless and effective. "The American agencies provided excellent technical support ... plus the all important scientific expertise for re-entry, health physics and radioactive material recovery advice and support," it notes. From an organizational perspective, this technical support "melded well into an efficiently functioning team that preformed the job safely."[59] Furthermore, as more Canadians arrived on the scene, the Americans drew down their assistance as planned.[60] The first NEST left on 8 March, and two weeks later the last Americans left for Las Vegas with the remaining US equipment.[61] NEST expertise proved to be tailor-made for the Cosmos 954 search. "The much smaller resource base in Canada did force some adjustments on the American time accomplishment expectations," an official Canadian reported noted. "Beyond this ... without reservations, this was an excellent, productive exercise in international cooperation."[62] In the end, the Canadians were saddened to see their American counterparts go.[63]

"There was no historical precedent for Operation *Morning Light*," Lieutenant General (retired) William Carr noted afterwards. "From my vantage point as Commander, Air Command during the events recorded here, I was privileged to see the spontaneous cooperation which invariably surfaces when Americans and Canadians, under pressure, work toward a common goal."[64] Supporting this assessment, the CFB Edmonton report explained that individual responsibilities assigned to Canadian and American participants became well defined from the onset. "The two national teams of the Task Force worked extremely well together ... [in] a common purpose easily and productively with amazingly few problems," it extolled. "The blend of skills each side brought to the task was essential to the other side's requirement and success, which is an exceedingly important factor. Without reservation, this was an excellent, productive exercise in international cooperation."[65] Accordingly, Prime Minister Trudeau expressed Canada's appreciation for American assistance in a message to President Carter on 22 March.[66]

For their part, the Americans participating in Operation Morning Light concluded that "the Canadians were outstanding hosts, both in technical support and personnel consideration. This likely represents the best of international assistance conditions that we could ever expect to encounter; many other situations could be far from ideal."[67] While the Canadians provided the bulk of personnel and logistics, the "previous specialized experience of the U.S. team with nuclear radiation search and measurement over large areas was a key Morning Light resource; the operation could not have been completed as expeditiously without it."[68] In addition to helping a close ally, the U.S. was able to glean intelligence about Soviet reactor design from recovered fuel samples.[69] Perhaps the greatest benefit for both countries was the "genuine emergency response" that it entailed, "much larger than any simulation that would have been reasonable to fund. To some degree the costs of participation in Morning Light represent an investment for invaluable experience."[70]

Morning Light as a "Whole of Government" scientific operation

The gradual draw down and departure of the Americans necessitated further cooperative measures between Canadian federal deparments and agencies involved in Morning Light. The AECB, responsible for the recovery, transportation, and storage of debris, expanded the pool of civilian scientists that it could draw upon to take over from the departing NESTs. The AECB also provided additional scientific instrumentation to the Department of Energy, Mines, and Resources (EMR), itself responsible for the management of the airborne search for radioactive wreckage. As the American drawdown continued, DND, responsible for overall operations, increased logistical support to the AECB and the EMR to help in the achievement of their objectives.[71]

The Canadian response in Morning Light ultimately embodied a nascent Whole of Government (WoG) approach to interdepartmental/agency cooperation. In a northern context this approach means sharing information, assets, facilities, supplies, and even personnel between agencies and departments—all of which operate in the region with extremely limited resources. WoG operations are essential in the North, not only to leverage capabilities, but to ensure that departmental mandates are fulfilled properly. It is in this supporting role that the military normally operates in a domestic emergency, providing transport, ships, and human resources that enable other government departments to react to contingencies in a rapid, coordinated manner.[72] Although the need for an integrated framework to guide northern operations was well established, at least in theory, prior to Morning Light, WoG approaches were never incorporated into CAF northern training during the 1970s. Nevertheless, the military was uniquely equipped and positioned to provide rapid, coordinated support to other departments and agencies in an emergency or crisis, thus enabling them to fulfill their mandates.[73]

During Operation Morning Light, planners and operators confronted two challenges that hindered (but did not prevent) multi-departmental cooperation. The first problem was the lack of a clear and unified line of authority and communication. This resulted in decision-making lags between the military (which was in overall command of the operation) and civilian participants,[74] with last-minute notification hampering some departments' efforts to mobilize the right personnel and meld them into a cohesive search and recovery effort.[75] The second challenge came after the American departed and a shortage of civilian scientists hindered field operations.[76] The 120 Americans that had participated in Morning Light were mostly specialists with their NESTs and had been specifically trained for such an operation. "Canadian personnel were identified to take over search planning, scientific photography, search operation, health physics, computer support, and other functions,"[77] the US report noted, but many of these Canadians did not have the levels of specific expertise that their American counterparts possessed. Instead, they were scientists and technicians with skills amendable to the detection and recovery of radioactive material. A shortage and rapid turnover of AECB scientific personnel (to the point that "rarely did a familiar face return") and the rotation of their departmental leadership out of the field compounded this challenge. The military's NAST teams had to compensate with their own personnel. Thus, although the Canadian military-civilian team accomplished the task, it did so with less efficiency than in the early stages when Canada drew upon American technical support.[78]

About 250 CAF personnel worked alongside thirty scientists from AECB, GSC, and EMR to complete the mission.[79] Given the austere conditions in which they had to operate, Operation Morning Light demonstrated the adverse effects that cold had on human effectiveness in the field. The official US report emphasized that the cold demands "more people, more reserves, to withstand the additional environmental fatigue and to carry out tasks which take longer to accomplish in extreme physical stress." Given that personnel could only work for a few minutes at time in sub-arctic conditions, meeting a fixed deadline would require two to three times the people necessary to respond to such an event in more temperate climates.[80]

In terms of generating experiential knowledge, Operation Morning Light also represented a prolonged test of techniques and equipment in Northern conditions. Cold temperatures, extremely limited infrastructure (such as roads, airfields, and communications networks), and great distances from main logistical support hubs dramatically increase the resource intensity of northern operations. Because the operation had depended so heavily on airlift, Canadian aircraft were subjected to scrutiny—and fared quite well. Furthermore, the exceptional flying skills of the pilots and aircrews throughout the operation allowed them to manage a steep operational tempo, poor weather, and navigational challenges.[81] Other equipment and techniques did not fare as well. The reliance on the American-provided microwave ranging system (MRS) underscored the unsuitability for

Northern operations of the navigation systems deployed on CAF aircraft (such as the Omega Navigation System) and the flying of precise tracks over long distances.[82] These navigational challenges were compounded by the lack of 1:50,000 scale maps east of 108°W longitude in the NWT and the inability to obtain replacement photos quickly.[83] Furthermore, military shelter left much to be desired. Tents were "poorly insulated" and there was a "lack of a safe, reliable method of heating field shelters." Shelters and electrical equipment—most importantly radios—also suffered from the cold draining battery power.[84]

Mission accomplishment

Operation Morning Light successfully met its objective to find, secure, and define the radioactive risk to civilians in the affected area.[85] In the end, scientific crews recovered about 65 kg (143 lbs) of satellite material, including 3500 particles that appeared to be the remnants of the enriched uranium fuel used in the satellite's nuclear reactor.[86] The lead effort in recovering, storing, and disposing of the radioactive debris fell to the AECB,[87] which contracted the Whiteshell Nuclear Research Establishment (WNRE) of AECL to analyze and store recovered debris. By the time the project was completed in the summer of 1978, scientists at Whiteshell had examined hundreds of specimens and conducted more than 4700 analyses.[88] Studies of the radioactive fragments quickly yielded debris of particular interest, including a highly-radioactive steel "hotplate" determined to be part of the reactor container, beryllium "slugs" that were thought to be part of Cosmos 954's reactor core, and a series of small cylinders in pristine condition that may have been part of the reactor control device. WNRE staff quickly determined that the reactor core had broken up and pieces of it were distributed across the search area. By analyzing the recovered fuel, staff determined the approximate size and power of the *Romashka*-type reactor, discerning that the powerplant produced an output of 132 kilowatts and would "have left in excess of 13,000 Curies of radioactivity 90 days after re-entry." WNRE concluded that "much of this [radioactivity] may never have reached the ground."[89]

Scientific reports based on investigations of radioactive contamination of people and the environment confirmed that much of Cosmos's reactor core had vaporized in the upper atmosphere, thus limiting irradiation of the terrestrial environment.[90] Most of the uranium-235 fuel remained in the atmosphere. The major effort placed into searching the debris area and measuring radiation levels gave scientists the confidence to conclude that dangerous radioactive debris which had fallen to the ground had been cleaned up, with smaller particles deteriorating rapidly and thus posing no serious threat to human health.[91]

All told, Morning Light represented a success story in bi-national and WoG cooperation—a significant outcome given that the crash of a nuclear-powered satellite was uncharted terrain for both the United States and Canada. Amidst tremendous uncertainty, officials had to decide how to address a scenario that could pose an acute threat to public safety. The mere potential of human impacts demanded that governments prepare for serious contingencies. This was far from a simple mission, given the vast areas that had to be covered by scientists and search and recovery teams. "Morning Light was an unmitigated success when measured against the Operational constraints," the Canadian military report concluded. "In the main these were cold, weather, distance, aircraft shortages, aircraft under-equipped for navigation, trained manpower, the pitting of resources against the cold, and the general lack of deployable equipment to support such an operation." There had been no major aircraft accident, over-exposure to radiation, or serious injuries.[92]

A professional, systematic scientific effort was also essential to securing compensation from the Soviet Union for scattering radioactive satellite debris across Canadian territory. Although Cosmos 954 was the seventh nuclear-powered vehicle to return to Earth, it was the first example of one state's space asset inadvertently crashing onto another state's territory.[93] Accordingly, the response to Cosmos 954 served as a precedent-setting case in international law,[94] focusing on the issue of liability pursuant to the *1972 Convention of International Liability for Damage Caused by Space Objects* and Soviet compensation of $3 million to Canada.[95] From an international legal standpoint, this settlement ended the Cosmos 954 affair. For Northern Canadians, however, the sweep of a diplomatic pen did not eradicate all concerns about residual legacies. "Despite soothing official words" about the clean up of the satellite debris, residents remained apprehensive about the longterm effects of low-level radiation on their health, *Globe and Mail* journalist Mitchell Beer reported in October 1980. Brian Wainwright, a union representative at the Pine Point lead-zinc mine, claimed that a nagging fear lurks below the surface. "None of us know for sure whether we've been got," he told Beer, "or if our lifespan as individuals has been shortened. I don't know how much longer I've got to live, or what my kids have got to look forward to - this is the kind of cloud that's left hanging over everybody."[96] The Cold War nuclear threat hung over all of North American society, but satellite Cosmos 954 caused particular stress for Northerners over whose communities and homelands radioactive particulates fell to earth in early 1978.

Notes

1. A previous version of this chapter was published as: Ryan Dean and P. Whitney Lackenbauer, "Introduction: 'The Satellite that Came into the Cold'" Ryan Dean and P. Whitney Lackenbauer, eds., *Operation Morning Light: An Operational History*, Arctic Operational Histories, no. 3. (Antigonish: Mulroney Institute, 2018), v–xlviii.

2. Roland Semjanovs, "Satellite looked 'like a jet on fire'; witness" *Yellowknifer, 13* January 1978.
3. U.S. Department of Energy (DoE), *Operation Morning Light: Canadian Northwest Territories, 1978—A Non–Technical Summary of United States Participation* (September 1978), 67, 71; "The Unscheduled Return of Cosmos 954," *Science News*, 2 April 1978, 113(5): 69; Gus W. Weiss, "The Satellite that Came into the Cold: The Life and Death of Cosmos 954," *CIA Historical Review Program*, 1978, 22: 1.
4. DoE, *Operation Morning Light*, 8, 11.
5. Major W.R Aikman, "Operation Morning Light," *Sentinel*, 1978, 14(2): 6. He noted that "tension dropped when negative results were announced." While Yellowknife was spared from contamination, analysts projected that massive radioactive objects could survive re–entry and reach the Earth's surface further down range towards Baker Lake. DoE, *Operation Morning Light*, 12n.
6. "Canada Wants Cash for Cosmos 954 Cleanup," Science, 16 February 1979, 203: 632–3; W.K. Gummer, "Summary of Cosmos 954 Search and Recovery Operation" (Ottawa: Atomic Energy Control Board, January 1979), 1.
7. Nancy Cooper, "Forces end search for Soviet satellite parts," *Globe and Mail,* April 8, 1978; and Major Bill Aikman, "Ice Strip on the Thelon: The Story of Camp Garland," *Sentinel,* 1978, 14(2): 17–19.
8. W.K. Gummer, F.R. Campbell, G.B. Knight, and J.L Ricard, *Cosmos 954: The Occurrence and Nature of Recovered Debris* (Ottawa: Minister of Supply and Services Canada, 1980), iii.
9. Leo Heaps, *Operation Morning Light: The True Story of Canada's Nuclear Nightmare* (Toronto: Random House, 1978).
10. C.A Morrison, *Voyage into the Unknown: The Search and Recovery of Cosmos 954* (Stittsville, ON: Canada's Wings, 1983), 120.
11. Morrison, *Voyage into the Unknown*, 120–1. For more on satellite failures in general, see Les Johnson, *Sky Alert! When Satellites Fail* (Chichester, UK: Springer–Praxis Books, 2013).
12. Morrison, *Voyage into the Unknown*, 4.
13. Annex A: Legal Basis of Canada's Claim: Nature of Damage, to A.E. Gotlieb, Memorandum to Ministers, Cosmos 954–Claim Against the USSR, 18 October 1978, AECB file 15–200–24–12–0 pt.2, ATIP 2016–000082.
14. Itty Abraham, "The Ambivalence of Nuclear Histories," *Osiris*, 2006, 21(1): 49–65.
15. On legibility and Arctic environments during the Cold War, see Trevor J. Barnes and Matthew Farish, "Between Regions: Science, Militarism, and American Geography from World War to Cold War," *Annals of the Association of American Geographers*, 2006, 96(4): 807–826; and Farish, *The Contours of America's Cold War*, (Minneapolis: University of Minnesota Press, 2010).
16. On this theme, see for example Liza Piper, *The Industrial Transformation of Subarctic Canada,* (Vancouver: UBC Press, 2010); Dolly Jorgensen and Sverker Sörlin, eds., *Northscapes: History, Technology, and the Making of Northern Environments,* (Vancouver: UBC Press, 2013); P. Whitney Lackenbauer and Matthew Farish, "The Cold War on Canadian Soil: Militarizing a Northern Environment," *Environmental History*, 2007, 12: 921–950; and Farish, "The Lab and the Land: Overcoming the Arctic in Cold War Alaska," *Isis*, 2013, 104(1): 1–29.
17. Jeffrey T. Richelson, *Defusing Armageddon* (New York: W. W. Norton & Company Inc., 2009), 48–50; Weiss, "The Satellite that Came into the Cold," 1.
18. NORAD was renamed North American *Aerospace* Defense Command in 1981.

19. "Operation Morning Light: Department of National Defence Final Report," 1, DND Directorate of History, ATIP A–2015–00308; DoE, *Operation Morning Light*, 66; Richelson, *Defusing Armageddon*, 53. On NORAD, see Joseph Jockel, *Canada in NORAD, 1957–2007*, (Montreal: McGill–Queen's University Press, 2007).
20. Weiss, "Satellite that Came into the Cold," 1.
21. DoE, *Operation Morning Light*, 2.
22. Weiss, "Satellite that Came into the Cold," 3–4. Heaps noted that Anatoli Dobrynin, a former aerodynamic engineer, would only inform the US National Security Advisor that "Cosmos 954 was not an atomic bomb." *Operation Morning Light*, 27. On the uncertainty over the depressurization, see L.I. Sedov, *Isvestia*, 5 February 1978, 3, quoted in Alexander Cohen, "Cosmos 954 and the International Law of Satellite Accidents," *Yale Journal of International Law, 1984*, 10: 80.
23. Weiss, "The Satellite that Came into the Cold," 4. While the countries who received information initiated their own preparations to deal with Cosmos 954, the U.S. DoE's field units were ready for deployment by 22 January, with all personnel on a two–hour alert and NEST equipment loaded onto four Air Force C–141 Starlifter aircraft in Washington D.C., California, and Nevada. DoE, *Operation Morning Light*, 2–3, 5; Aikman, "Operation Morning Light," 6.
24. Gummer, "Summary of Cosmos 954 Search and Recovery Operation," 2; CFB Edmonton, "Post Operation Report," 3.
25. DoE, *Operation Morning Light*, 8, 15–7. Canada's first contribution to this bilateral, collaborative effort was meteorological reports to enhance reentry modelling.
26. DoE, *Operation Morning Light*, 62.
27. Aikman, "Operation Morning Light," 6; DoE, *Operation Morning Light*, 8–9, 14.
28. Aikman, "Operation Morning Light," 6. On the spectrometer, see Barb Livingstone, "In Search of Radiation in Barren Land," *Edmonton Journal*, 2 February 1978.
29. By 28 January the whole search area had been overflown at least once by CC–130 aircraft. Gummer et al., *Cosmos 954,* 3, 8; Aikman, "Operation Morning Light," 5–6; DoE, *Operation Morning Light*, 25, 39. Ironically, the gamma ray spectrometers aboard the CC–130 Hercules proved much more effective in detecting Cosmos debris than specially–equipped U.S aircraft desgined to measure for radioactivity in the atmosphere. See DoE, *Operation Morning Light*, 14, 42; Richelson, *Defusing Armageddon*, 55–56.
30. Aikman, "Operation Morning Light," 7.
31. DoE, *Operation Morning Light*, 22. Data was also sent on to Los Alamos and Livermore for further study. Richelson, *Defusing Armageddon*, 64.
32. Gummer et al., *Cosmos 954,* 8.
33. DoE, *Operation Morning Light*, 53–4.
34. Canadian Press "Searchers find satellite debris," *Fort McMurray Today*, 30 January 1978.
35. John Noble Wilford, "Canadians Pick Up 'Hottest' Satellite Fragment Yet," *New York Times*, 6 February 1978.
36. Aikman, "Operation Morning Light," 10.
37. Wilford, "Canadians Pick Up 'Hottest' Satellite Fragment Yet."
38. Canadian Press "Satellite fragment is radioactive" *Fort McMurray Today*, 2 February 1978.
39. Aikman, "Operation Morning Light," 12.

40. DoE, *Operation Morning Light*, 33. For a complete record of their travels based on diary entries, see Christopher Norment, *In the North of Our Lives* (Renssalaer, N.Y.: Hamilton Printing Co., 1989).
41. Norment, *In the North of Our Lives*, 158.
42. Aikman, "Operation Morning Light," 6.
43. See Malcolm Waldron, *Snow Man: John Hornby in the Barren Lands* (Montreal & Kingston: McGill–Queen's University Press, 1997).
44. DoE, *Operation Morning Light*, 33.
45. Norment, *In the North of Our Lives*, 159.
46. Norment, *In the North of Our Lives*, 158–9.
47. Norment, *In the North of Our Lives*, 160.
48. Aikman, "Operation Morning Light," 9.
49. Aikman, "Operation Morning Light," 9; DoE, *Operation Morning Light*, 33.
50. DoE, *Operation Morning Light*, 56, 58; Gummer et al., *Cosmos 954*, iii.
51. CFB Edmonton, "Post Operation Report," 22.
52. Witness Dan Billing, Chief of Emergency Services, Government of the Northwest Territories, AECB 15–200–24–12–2 vol.2, ATIP A–2016–00082. On impacts on specific communities, see the reports on AECB file 15–200–24–12–2 vol.2, ATIP A–2016–00082.
53. DoE, *Operation Morning Light*, 25.
54. Robert Blake "Snowdrift safer than most places, General tells residents" *Yellowknifer*, 2 February 1978.
55. Aikman, "Operation Morning Light," 9. See also Heaps's description of the visit (replete with offensive racial stereotypes) in Heaps, *Operation Morning Light*, 117–22. While the reported results were reassuring, mixed official messaging left some observers skeptical. See, for example, Mitchell Beer, "Aftermath of Cosmos Crash," *Globe and Mail*, 25 October 1980.
56. Heaps, *Operation Morning Light*, 76.
57. On these concerns, see for example Ken Coates, P. Whitney Lackenbauer, Bill Morrison, and Greg Poelzer, *Arctic Front: Defending Canada in the Far North,* (Toronto: Thomas Allen, 2008); Shelagh Grant, *Polar Imperative: A History of Arctic Sovereignty in North America,* (Vancouver: Douglas & McIntyre, 2011); and Adam Lajeunesse, *Lock, Stock, and Icebergs: A History of Canada's Arctic Maritime Sovereignty,* (Vancouver: UBC Press, 2016).
58. DoE, *Operation Morning Light*, iv.
59. CFB Edmonton, "Post Operation Report," 11.
60. Beer, "Aftermath of Cosmos Crash," 2; DoE, *Operation Morning Light*, 62.
61. Aikman, "Operation Morning Light," 16; DoE, *Operation Morning Light*, 62.
62. CFB Edmonton, "Operation Morning Light Post Operation Report,"33.
63. Aikman, "Operation Morning Light," 6.
64. Lieutenant General (ret'd) W.K. Carr, "Foreword," in Morrison, *Voyage into the Unknown*, 1.
65. CFB Edmonton, "Operation Morning Light Post Operation Report," 33.
66. Aikman, "Operation Morning Light," 16; DoE, *Operation Morning Light*, 62.
67. DoE, *Operation Morning Light*, 73.
68. DoE, *Operation Morning Light*, 22.
69. Richelson, *Defusing Armageddon*, 70.
70. DoE, *Operation Morning Light*, 73.
71. Gummer et al., *Cosmos 954*, 2, 4.
72. P. Whitney Lackenbauer and Adam Lajeunesse, "The Emerging Arctic Security Environment: Putting the Military in its (Whole of Government) Place," in *Whole of Government through an Arctic Lens,* P. Whitney Lackenbauer and Heather Nicol eds. (Antigonish: Mulroney Institute on Government, 2017), 1–36.

73. P. Whitney Lackenbauer and Daniel Heidt, eds., *The Advisory Committee on Northern Development: Context and Meeting Minutes, 1948–67*, Documents on Canadian Arctic Sovereignty and Security (DCASS) No. 4 (Calgary and Waterloo: Centre for Military and Strategic Studies/Centre on Foreign Policy and Federalism, 2015); Department of National Defence, *Defence in the 70's* (Ottawa: Queen's Printer, 1971), 11. See also Lackenbauer and Kikkert, eds., *Canadian Forces and Arctic Sovereignty.*

74. Military personnel would report to Air Command Headquarters in Winnipeg while DND civilians would report to National Defence Headquarters in Ottawa (NDHQ). Similarly, AECB, Atomic Energy of Canada Ltd. (AECL), and EMR personnel reported back to their respective National Headquarters in Ottawa as well. Consequently, the Edmonton Report notes "that all military communications had to be reassessed at Air Command while the civilian agencies went directly to their senior headquarters with real time and detailed information." CFB Edmonton, "Post Operation Report," 33.

75. Morrison, *Voyage into the Unknown*, 120. The CFB Edmonton report notes that the participation of American personnel helped to compensate for this initial multi–departmental coordination shortcoming. CFB Edmonton, "Post Operation Report," 3; DoE, *Operation Morning Light*, 22.

76. CFB Edmonton, "Post Operation Report," 33.

77. DoE, *Operation Morning Light*, 62.

78. CFB Edmonton, "Operation Morning Light Post Operation Report," 12, 33.

79. Gummer et al., *Cosmos* 954, 2.

80. DoE, *Operation Morning Light*, 74.

81. See CFB Edmonton, "Post Operation Report," 28, 31, 33, 34, A/1/A–1, Annex A "LAPES Operation Report," Part 1 B (4).

82. CFB Edmonton, "Post Operation Report," 28. On the MRS, see Aikman, "Operation Morning Light," 13; DoE, *Operation Morning Light*, 46.

83. CFB Edmonton, "Operation Morning Light Post Operation Report," 27–8.

84. CFB Edmonton, "Operation Morning Light Post Operation Report," 36, 37.

85. CFB Edmonton, "Operation Morning Light Post Operation Report,"4.

86. H.W Taylor, E.A. Hutchison, K.L. McInnes, and J. Svoboda, "Cosmos 954: Search for Airborne Radioactivity on Lichens in the Crash Area, Northwest Territories, Canada," *Science* 205:4413 (28 Sept 1979): 1383–5; Gummer et al., *Cosmos 954*, 2–5. Small particles measured as low as a few thousandths or millionths of a roentgen per hour, and steadily decayed to below natural background levels. September measurements found radiation levels to be one–fifth of what they were in January. For a more in–depth breakdown of the recovered materials and the nature of their radioactivity, see Gummer et al., *Cosmos 954*, 10–32.

87. CFB Edmonton, "Operation Morning Light Post Operation Report,"4. Much of the material collected by Operation Morning Light, after being flown to CFB Edmonton and then for further testing and storage at the Whiteshell Nuclear Research Establishment at Pinawa, Manitoba, was later sent to the Chalk River Laboratories at Deep River, Ontario, for disposal. Gummer et al., *Cosmos 954*, iii; CFB Edmonton, "Post Operation Report," 7; and Morrison, *Voyage into the Unknown.*

88. R.B. Stewart, "Russian Satellite Debris: Examination of COSMOS 954 Fragments at the Whiteshell Nuclear Research Establishment," May 1979, DHH 79/528, acquired through ATIP A–2015–00298.

89. Gummer et al., *Cosmos 954*, 9.

90. See, for example, Taylor et al., "Cosmos 954: Search for Airborne Radioactivity on Lichens in the Crash Area".

91. Gummer et al., *Cosmos 954*, iii.
92. CFB Edmonton, Operation Morning Light Final Report, 32
93. Morrison, *Voyage into the Unknown,* 3–4.
94. See, for example, Edward R. Finch, Jr., and Amanda Lee Moore, "The Cosmos 954 Incident and International Space Law," *American Bar Association Journal*, January 1979, 65: 56–9.
95. R. I. R. Abeyratne, "Environmental Protection and the Use of Nuclear Power Sources in Outer Space," *Environmental Policy and Law*, 1996, 26(6): 255–60. After Canada presented a bill of $6.1 million to the Soviet Union in 1979, the countries eventually agreed to a lump sum settlement of $3 million in April 1981. See Edward G. Lee and D.W. Sproule, "Liability for Damage Caused by Space Debris: The Cosmos 954 Claim," *Canadian Yearbook of International Law* (1988): 273–80.
96. Mitchell Beer, "Aftermath of Cosmos Crash," *Globe and Mail*, 25 October 1980.

15 Melting the ice curtain: indigeneity and the Alaska Siberia Medical Research Program, 1982–1988

Tess Lanzarotta

In May 1987, some sixty people, many of them scientists and physicians, gathered to attend a film premiere at the University of Alaska Anchorage (UAA).[1] They were there to watch a 30-minute production—made in a style that fell somewhere between a documentary and a news special—titled *Breaking the Ice: The Alaska Siberia Medical Research Program*.[2] The film's subject was a formal collaborative medical research agreement that the UAA and the Siberian Branch of the Russian Academy of Medical Sciences (RAMS) had signed in November 1987. The Alaska Siberia Medical Research Program (the Program), as the agreement was called, would generate shared research projects, lead to joint publications, and facilitate the exchange of students and medical delegations between the two institutions.[3] But, it also had a specific political mission. As *Breaking the Ice* explained, the founding of the Program was "the story of some Alaskans and Siberians who decided to work together" not just to improve health in the Far North, but to create a "model of U.S.-Soviet Relations" that would ultimately benefit both nations. In the words of the film's narrator: "These heavily guarded back doors to the United States and the Soviet Union have become the front doors to improved relations. The threat of nuclear exchange is countered by the promise of peaceful scientific exchange."[4] *Breaking the Ice* gestured toward a possible new future for Alaska and Siberia as diplomatic, economic, and scientific partners.

The Program's founders were committed to making this imagined future into a reality. They formulated promotional strategies and forged alliances in order to garner support for the Program and to achieve the political goals associated with it. In this chapter, I track these tactics and explore how the Program's architects made indigeneity, as a conceptual category, and Indigenous Peoples, as a study population and as a group whose healthcare needs demanded attention, central to the Program's scientific and peace-building efforts. However, I do not mean to argue that the Program's rhetoric was necessarily exploitive or disingenuous. Instead, I seek to emphasize that it was conscious and intentional. The researchers involved in the Program instrumentalized indigeneity, using Alaskan Native history and culture to send a message about the moral urgency of their scientific work.

The Program's founders insisted that health science could and should be an important facet of a growing diplomatic effort directed towards ending the Cold War in the Far North. Alaskan Natives Peoples, and circumpolar Indigenous Peoples more broadly, supported this diplomatic effort and argued that the very existence of circumpolar indigeneity as an identity category provided sufficient justification to melt the "ice curtain." For instance, in 1985, Hans Pavia-Rosing, the president of the Inuit Circumpolar Council (ICC), a prominent non-governmental Indigenous rights organization, wrote to both Ronald Reagan and Mikhail Gorbachev on behalf of "the world's 100,000 Inuit...." He called upon Reagan and Gorbachev to demilitarize the Far North and to end the isolation of "our brothers and sisters in Siberia from the rest of the Inuit world."[5] The Program's founders would make the same argument, but as my first two sections show, had their own reasons for insisting that Indigenous Peoples living in Alaska and Siberia should not be separated by political boundaries.

In my third section, I explain how Program researchers engaged with Alaskan Native culture. I frame this, too, as a conscious effort to demonstrate a certain set of political beliefs that had salience both in Alaska and internationally. In the 1960s, a powerful Alaskan Native rights movement had emerged, inspired by the Red Power movement and global pan-Indigenous activism.[6] This culminated in the largest land claim in American history, the 1971 Alaska Native Claims Settlement Act (ANCSA), which transferred some 45 million acres of territory to regional for-profit Indigenous Alaskan corporations. ANCSA is widely considered to be an ambiguous victory at best, but it nonetheless signified a radical transformation in the political power of Alaskan Native Peoples.[7] ANCSA helped to grant legitimacy and stature to Alaskan Native political leadership and, Stephen Haycox has argued, it effectively delegitimized the overt racism that had sometimes characterized Alaskan political discourse in prior decades.[8]

These changes ran parallel to the rise of transnational Indigenous organizations, like the ICC, that were an increasingly powerful force in international politics. Their presence had a particular impact at the United Nations, which had made a commitment to address issues of diversity, discrimination, and human rights, and felt compelled to consider the grievances of Indigenous Peoples worldwide. Indigenous Peoples began to hold the power to expose a state or government that ignored their concerns to international shame.[9] Liberal (or neoliberal) states, meanwhile, embraced various versions of multiculturalism and the "politics of recognition," which encouraged celebrating the diverse peoples who lived within their borders.[10] These ideas were still nascent as the Program was taking shape, but they nonetheless had an influence over the way that Alaskan health researchers chose to explain the benefits of their work. Rather than claim that scientific investigations could enable or uphold militarization, as many Arctic bio-medical scientists before them had done, Program researchers embraced Alaskan Native cultures and claimed that they worked on behalf of Alaskan Native Peoples.[11]

The founders of the Program emphasized that their work would directly benefit Alaskan Native healthcare by improving the systems of rural primary care that served Alaskan Native villages. This stated goal spoke to Alaskan concerns, but it also helped to align the Program with dominant trends in international health. In the wake of the 1978 Alma-Ata Declaration, the World Health Organization (WHO) had moved sharply away from programs that emphasized technical expertise and top-down disease eradication, and had become more interested in supporting programs and approaches that addressed the basic needs of remote communities.[12] Although emergent neoliberal politics had begun to dilute the radical potential of this change by the mid–1980s, the circumpolar health research community was still influenced by these ideas and saw emphasizing primary care as an opportunity to bolster their own form of expertise.

The Program achieved its first goal: by all accounts, it helped to broker peace and to build a stronger relationship between the Alaskan medical research community and their Siberian counterparts. However, the Program's promised effort to improve primary care was never borne out. In my final section, I explain why this was the case by tracking the Program's transformation from a wide-ranging—and perhaps overly optimistic—set of ideas about what circumpolar collaboration could achieve into a narrowly focused research agenda. I conclude by reflecting on the limitations of the Program's engagement with Alaskan Native Peoples, and by considering the Program's role within the longer history of the Alaskan circumpolar health movement.

"A unique gift": founding the Alaska Siberia Medical Research Program

After its premiere, *Breaking the Ice* aired on local television in Alaska and was broadcast nationwide by PBS; later that same year, it was picked up by the BBC and televised across Europe. Dr. Theodore (Ted) A. Mala, one of the architects of the Program and a main character in the film itself, traveled to London to premiere the film and to participate in a promotional tour.[13] Mala was widely lauded for his efforts. Alaska Governor Steve Cowper described him as "the key player in the unfolding relations between the Soviet Far East and Alaska" and the *Los Angeles Times* explained that Mala was "credited by many with melting the 'ice curtain.'"[14] He was the face of the research program and was never shy about expressing his sense of its significance. "This is a unique gift," Mala would explain "that the North can give to the world."[15] Mala's statement, or at least the sentiment behind it, was reprinted many times in newspapers and in the promotional material for *Breaking the Ice*.

Mala developed the idea for the Program in 1982, after his first visit to the Soviet Union. For Mala, the trip was the realization of a lifelong dream. He was the only child of a Russian mother, Galina Liss, and an Iñupiaq father, Ray Mala, who was one of the first Indigenous American movie stars.[16] After the untimely death of his parents, Mala spent most of his youth in California boarding schools, an unpleasant experience that left him with a strong desire to reconnect with his Alaskan and Russian roots.[17] After graduating medical school in 1977, he moved to Anchorage and began working for the Alaska Federation of Natives. In 1982, he accepted an Assistant Professorship in the Health Science Department at the UAA. Through connections he made at the Alaska Public Health Association annual meeting, Mala traveled to the Soviet Union later that year, as a guest of the Soviet Medical Workers Union.[18]

When Mala arrived in Leningrad, he toured Russian alcoholism rehabilitation centers, because he was curious about Soviet approaches to treatment.[19] However, it was the conversations he had with his Russian friends that truly convinced him of the value of medical and scientific cooperation between Alaska and Siberia. One of his Russian hosts asked: "Do you (Americans) really want a nuclear war with Russia?" The question, Mala recalled, "stunned" him.[20] Mala began to doubt the Cold War propaganda he had been exposed to: "I found when I watched TV in the United States, I was afraid of Russia," Mala explained, "and I'd go to Russia and do the same thing and found I was afraid of people in the United States. It was all baloney, so that strengthened my ... [resolve] to go forward."[21] When Mala returned to Alaska, he built a coalition of likeminded colleagues, university officials, and politicians, and dedicated himself to building a formal research relationship with Soviet health scientists.[22]

There was a growing precedent for the kind of political work that Mala hoped to do. In 1980, after the Soviet invasion of Afghanistan and the election of Ronald Reagan, Bernard Lown, a cardiologist at Harvard University, reached out to Eugene Chazov, a Soviet cardiologist with whom he had a close friendship, to consider what they might do to advocate for peace. The organization they founded, International Physicians for the Prevention of Nuclear War (IPPNW), became a prominent voice for the cause of nuclear disarmament. It went on to win a UNESCO Peace Education Prize in 1984 and the Nobel Peace Prize in 1985 for its efforts.[23] In 1987, Mala spoke about the Program at a meeting with nearly 3000 physicians from around the world sponsored by the IPPNW on the subject of "Drafting Proposals for East-West-Cooperation."[24] From the very beginning, Mala and his collaborators emphasized the humanitarian potential of their project. When an opportunity to gain federal support for collaborative scientific efforts appeared, Mala was ready to take advantage.

In 1985, Gorbachev's implementation of glasnost (openness) and perestroika (restructuring) ushered in a new era of government transparency and economic revitalization in the Soviet Union. That same year, Gorbachev and Reagan met at the Geneva Summit to begin a dialogue about ending the arms race. While their conversations about defense issues garnered the most media attention, their meeting also generated a General Exchanges Agreement intended to facilitate scientific and technological collaboration. Through this agreement, the UAA and the Siberian Branch of RAMS were able to gain the support of the National Institutes of Health and the USSR Ministry of Health for a joint project focusing on "fundamental problems of adaptation to the North."[25] By October 1986, the remaining bureaucratic obstacles had been cleared, and Mala returned to Russia to draft a preliminary research agenda with his Soviet colleagues.[26]

They agreed on a set of broad priorities, which included nutrition (particularly the effects of the changing dietary patterns of Alaskan Native Peoples on their health); the physiology of cold adaptation; the genetics of alcoholism; and the training of Indigenous healthcare workers. Program researchers also identified four main study populations—transient workers, newcomers, long-term settlers, and Indigenous Peoples—who were assumed to experience different sets of health problems.[27] This list, from which a more detailed protocol would later be developed, represented the issues that Program researchers felt were most urgent.[28] It also divided study populations into racial and temporal categories whose bodies were assumed to interact with the Alaskan environment in different ways.

Studies of racial difference, particularly in studies of human adaptation and acclimatization, had been characteristic of health research in Alaska for decades.[29] When the Cold War began, the Arctic had taken on a new strategic importance for national defense, resulting in a dramatic increase in American military presence and industrial expansion into Alaska. In this geopolitical context, Matthew Farish has argued, "fears of a Soviet assault led to an alternate invasion of Arctic landscapes by research teams, administrators and troops."[30] These researchers, driven by a collective military and scientific preoccupation with human adaptation and acclimatization to "hostile" climates, framed Alaskan Native peoples' capacity to thrive in the Arctic as a "medical mystery that if deciphered would aid those troops who were in Alaska, it seemed, out of necessity."[31] This was a scientific approach that Alaskan Native peoples had contested.[32] The Program relied upon alliances with Indigenous Peoples; however, it nonetheless framed circumpolar Indigenous Peoples as a racialized "baseline" of perfect adaptation from which the gradual physical adjustments that white Northerners' bodies made over time could be measured.[33] For organizations like the ICC, circumpolar indigeneity might involve some degree of direct biological relatedness, but was primarily based on shared cultures, languages, environments, and political concerns.[34] But the Program's approach flattened this complex identity category into a singular biological population.

"In the blood of our Native Peoples": appealing to a common history

Breaking the Ice began with a scene that would have been chilling for Alaskan viewers: American Air Force F-15 Eagle fighters soaring through the clear skies off the coast of Alaska, racing to intercept a Soviet bomber as it crossed the Bering Strait from Siberia.[35] The footage was a stark reminder of the tensions that had shaped the relationship between Alaska and Siberia for decades. "Since the Cold War began," the film's narrator explained, "the people on both sides of the Bering Sea have viewed each other through the distorted lenses of rocket and bomb sights."[36] However, there were signs of political thaw; just months after the premiere of *Breaking the Ice*, Gorbachev made a speech in the Arctic city of Murmansk that called upon Warsaw Pact and NATO countries to reduce military activity in the Far North and turn the Arctic into a "zone of peace."[37] *Breaking the Ice* situated its subject at the center of this pivotal political moment.

During the early Cold War, Alaska and Siberia's strategic locations had made them ideal military outposts. But, *Breaking the Ice* employed a specific view of the past to present the argument that Alaska and Siberia would also make ideal sites for diplomatic exchange. This message, spoken over footage of Indigenous Peoples happily greeting one another, was conveyed by the film's narrator: "Long separated from each other by politics, eskimos from Siberia and Alaska found they still have common interests ... and a common language."[38] As Andrew Stuhl points out, historians must "take seriously both the overt and covert exercises of power inherent in continually imagining the shape of Arctic futures by reading its past."[39] Program founders, then, realized that foregrounding historical ties rooted in biology offered a more seemingly moral and authentic basis for peaceful relations than the transitory imperatives of Cold War geopolitics could provide.

In 1986, when Mala was preparing to leave Siberia, he was surprised to encounter a group called the "Alaskan Performing Artists for Peace" on tour in the USSR. At the time, Mala was collecting footage for *Breaking the Ice* and made sure to film the group's music and dance routines and the cheering Russian audiences who watched them perform.[40] The tour was organized by Dixie Belcher, a Juneau-based activist who felt that ordinary Alaskans needed to take charge of peace-building efforts.[41] Hers was but one in a series of contemporaneous efforts that drew attention to the situation of Indigenous Peoples in Alaska and Siberia who had experienced the "ice curtain" as an abrupt and unfair separation from friends and relatives.[42]

For instance, Little Diomede Island on the Alaskan side of the International Date Line stands only 2.4 miles from Big Diomede Island, on the Russian side. The Indigenous residents of the Diomedes, many of whom were directly related to one another, had crossed the ice bridge between the islands for centuries. After 1948, however, such crossings were forbidden. The Russian government relocated Big Diomede's population, and both islands were subsequently used as sites for military surveillance.[43] Occasionally, residents from Little Diomede still covertly made the trip across the ice to trade small goods and to learn news about their relatives from the Soviet troops. One Little Diomede resident explained: "Most of the time, they tell us that our people—our uncles, our friends, you know—have died over there.... Then we feel the separation."[44] The possibility of rectifying this situation added moral weight to the actions of those who sought to improve relations between Alaska and Siberia.

Ted Mala and his collaborators grafted their work onto this broader movement, but added a biological inflection to a discussion about relatedness that had been based on familial kinship. "Siberia and Alaska will always have a great bond not only in history, but also in the blood of our Native Peoples," Mala argued. "It is more than a romantic notion but rather a physical reality."[45] For Mala and his colleagues, this "physical reality" was a crucial facet of the Program's scientific potential. In the late 1980s, Alaska's population was still only half a million, 90,000 of whom identified as Alaskan Native. Mala insisted that "statistically significant medical studies" would only be possible if members of Siberia's population, then around 30 million, could also be enrolled as research subjects.[46] The Program was thus organized around the supposition that Indigenous Peoples who lived in Alaska and Siberia could and should be studied as a single population. The introductory pamphlet circulated at official signing ceremony of the Program was even titled "A New Bridge For Health, History and Humanity," referring to the Bering Land Bridge that had once connected Alaska and Siberia.[47] The Bering Land Bridge theory was and is controversial amongst Alaskan Native peoples.[48] However, many Alaskan Native activists had still found the notion that the Indigenous Peoples of Alaska and Siberia were constituents in a larger circumpolar Indigenous population a meaningful way to organize politically and articulate shared critiques about the effects of colonialism in the Far North.

The Inuit rarely identified themselves as a distinct ethnic group prior to the 1970s, but a regional circumpolar Inuit identity began to emerge in response to a widespread feeling that Arctic Indigenous Peoples needed to oppose, or at least set the terms of, industrial development in the Arctic.[49] This collective identity solidified into a political movement after the 1977 Inuit Circumpolar Congress meeting in Barrow (now Utqiagvik), Alaska.[50] Three years later, Eben Hopson, then the mayor of Alaska's

North Slope Borough, founded the ICC, which quickly became a power-ful force in international politics and Arctic policy-making, and pushed back against militarization and the view of the Arctic as "solely a security region."[51] The ICC cohered around a multi-faceted definition of circum-polar indigeneity and united peoples who shared a common relationship to lands and waters that were being threatened by increasing environ-mental degradation.[52]

Though their definition of circumpolar indigeneity was different than that of Program researchers, the ICC still saw biomedical scientists as potential allies. The ICC passed a resolution in 1986 supporting the Program based on its expressed aims: to improve the health of Indigenous Peoples living in Siberia and Alaska and to help ease restrictions on cross-border travel.[53] Mala and his collaborators were far from naive about the political land-scape of the Far North and likely knew that the goals they had chosen to focus on would help them to curry favor with Alaskan Native and circum-polar Indigenous political organizations.

"In the World of the Real People": celebrating Alaska Native cultures

In 1987, the Siberian delegation arrived in Alaska to sign the official documentation forming the Program. In commemoration of this his-toric moment, they delivered a speech at the UAA titled "In the World of the Real People" describing their travels in Alaska in advance of their arrival in Anchorage, which had included visits to Barrow, Kotzebue, Buckland, Nome, Little Diomede, and Fairbanks.[54] The choice of title for the talk was not accidental—"the Real People" is the English trans-lation of the words Iñupiat and Yup'ik. It was and is a term that many Alaskan Native peoples, Iñupiat and Yup'ik in particular, use as a form of self-designation.[55] For the Russian delegation to describe their visit as entering "the World of the Real People" was a gesture of political goodwill and a demonstration of their awareness of Indigenous Alaskan culture.

The speech included ambivalent reflections on their time spent in Anchorage and Fairbanks; although the Russian delegation described these visits as pleasant, they admitted that while in Alaska's major cities "our conscious and subconscious still clearly fixed the border between 'US' and 'THEM.'"[56] It was only in Iñupiat communities, "the land of the Real People," where they truly "forgot that we were located in American Alaska, and not in our native Taymira or Chukotka.... Everything happening was so similar to the personal contact with Soviet Chukchi, Yakuts, Eskimos, Nagacans ... that we were accustomed to."[57] One such interaction took place at a hospital in Kotzebue, where Program researchers expressed an interest in studying Alaskan Native peoples' "incredible tolerance to harsh climate conditions," but reserved a "smaller degree" of interest for "the traditional medicine of the Eskimos."[58]

In May 1988, the *Tundra Times*, the most prominent Alaskan Native newspaper, published two articles written by the Siberian visitors, and edited by Mala, about the time they had spent in Alaska. Statements about their fascination with Alaskan Native peoples' "incredible" acclimatization and their relative lack of interest in Alaskan Native medicine were left out.[59] Considering that, by the mid–1980s, groups like the ICC and the North Slope Borough were calling upon the Alaskan health research community to begin incorporating Indigenous healing traditions into western biomedicine, this may have been a tactical omission.[60] While the Program was promoted as an effort that would benefit Alaskan Native peoples, the participants stopped short of positioning Indigenous Peoples as legitimate producers of medical knowledge.[61]

The "recognition" era of multicultural liberalism was taking shape; in this context, observing and appreciating displays of authentic Indigenous culture signaled an enlightened moral sensibility.[62] Those involved in the Program were able to both take on this sensibility and continue to engage in research that situated Indigenous bodies as scientific objects. For instance, when the delegation arrived at the UAA to sign the agreement, Mala organized a series of public events to publicize the Program. These events, which were a mixture of cultural celebrations and performances and academic lectures, perfectly exemplified the ways that the Program incorporated these two seemingly oppositional views. The celebration began with a Potlatch, which was, according to advertisements, an opportunity for the public to "sample Native foods, singing, dancing, storytelling and sports."[63] It was followed by an open lecture on "Siberian Approaches to Human Adaptation to the North," delivered by Yuri Nikitin, an academician with RAMS and the Program's co-founder.[64] The celebrations also featured a benefit performance for the Program by Chuna McIntyre of the Nunamta Yup'ik Eskimo Dancers.[65] During this series of events, the audience might have learned what valuable scientific information Indigenous bodies had to offer, and simultaneously been reassured of the Program's moral acceptability by the presence of Indigenous performers.

Within "recognition" politics, the performance of "authentic" or "traditional" Indigenous cultures in the public sphere could easily be accommodated. And, as scholars have pointed out, performing traditional culture has also been a form of political action for Indigenous Peoples and a way to maintain and express their identities.[66] These performances likely had multiple meanings for both the performers and their audiences. For the performers, they might have been opportunities to demonstrate the resilience of their culture, or to show support for a project that claimed it would benefit Indigenous Peoples both medically—by improving their health—and politically—by helping to end the Cold War.[67] We might also think of the

Program's founders as "performing" by organizing an event that showcased their appreciation for Alaskan Native cultures and their recognition that the Far North was an Indigenous space.

"We in the North are the experts": improving Indigenous health

In *Breaking the Ice*, the narrator explained that one of the central concerns of the Program was "the delivery of health care to small rural communities." Physicians interviewed for the film explained that Alaska and Siberia struggled with the same logistical problems and prohibitive costs when trying to implement rural healthcare programs. It followed, they insisted, that the two regions could benefit from working collaboratively to develop solutions.[68] In making this argument, Program researchers were appealing directly to Alaskan Native people, who stood to benefit most from improved rural healthcare. In fact, the Alaska Federation of Natives (AFN) had been working steadily to improve rural healthcare, and had been advocating for better training and increased pay for Health Aides since the early 1970s.[69] Alaskan Native corporations, too, were involved in the CHAP program and in rural care; the Yukon-Kuskokwim Health Corporation had received a grant from the Office of Economic Opportunity in 1969 to "increase the quality, scope, and quantity of primary care services available to village residents ... by upgrading village health aides ..."[70]

One of the most emotional scenes in *Breaking the Ice* involved footage of an Alaska Native woman who, in her role as a Community Health Aide (CHA), was responsible for the health of the 260 people living in her village. She and her fellow Health Aides were trained in a diverse range of tasks, like basic dentistry, midwifery, the distribution of pharmaceuticals, and the acute care of injuries. By the 1980s, they had become the cornerstone of Alaska's rural primary care system—nearly 300 villages relied upon their services.[71] However, the CHA who appeared in the *Breaking the Ice* (her name is never given) was under "tremendous pressure" and found herself dealing with deaths she could not explain and devastated families she could not easily console.[72] According to the Program's initial design, developing ways to make the job of a CHA easier was one of its central aims.[73]

However, a focus on improving healthcare delivery was also a way to demonstrate that the Program's work was in line with the WHO's primary care agenda.[74] In fact, the Alaskan CHAP and the WHO's approach to primary care drew their inspiration from similar sources. Although the Alaskan system had its roots in an experimental tuberculosis chemotherapy aide program, health officials in Alaska also viewed the CHAP program as analogous to the Chinese Barefoot Doctors.[75] It was the success of the Barefoot Doctors system, and of other countries that relied upon "village health workers," which had convinced the WHO to develop models of primary care "based on the needs and participation of local communities."[76] Alaskan researchers felt they had something to offer this conversation, and saw the potential to position themselves as international experts in the delivery of healthcare to remote Indigenous communities.

In 1989, an article by Mala appeared in a WHO journal called the *World Health Forum*, which had been created to report on initiatives that reflected the WHO's post-Alma-Ata focus on primary care.[77] Mala's article discussed the challenges of rural healthcare in Alaska: "Many of the villages are inaccessible by road and few have sewerage systems, running water or reliable communication with urban centers." He explained that the Program hoped to help overcome these challenges by training more Indigenous residents of rural communities to work as health professionals.[78] *World Health Forum* also regularly featured articles that dealt with "traditional healing," providing optimistic coverage of rural health aide programs from around the world and explaining how traditional healers could be enlisted as health aides or encouraged to work collaboratively with them.[79] Combining traditional Chinese medicine and western biomedicine had, after all, been one of the hallmarks of the Barefoot Doctors system.[80] Although Mala did mention the need to make healthcare "culturally appropriate for the diverse target groups," his article, like the rest of the material relating to the Program, made no specific mention of Indigenous healing traditions.[81] Researchers involved in the Program were more interested in emphasizing their own expertise than in exploring the potential of Alaskan Native medical expertise.

Mala expressed an explicit desire to use the Program to help the WHO achieve its mandate for the future of circumpolar health research.[82] In 1981, a WHO working group on Polar Biomedical Research published a report which insisted that Arctic medical research was crucially important on a global scale, because the challenges of healthcare delivery in the Arctic—finding ways to serve small communities scattered across great distances—were shared by many other parts of the world. Research from the circumpolar regions, the report concluded, could be used to generate a "north-south dialogue."[83]

Such ideas were well-received by both Program researchers and the broader circumpolar health research community. In *Breaking the Ice*, one Alaskan physician, Tom Nightswander, mentioned that his work with CHAs was "applicable to much of the third world where we're using Indigenous health workers to provide services."[84] Mala was fond of this comparison and regularly emphasized the similarities between Alaska, Siberia, and the "Third World."[85] In this context, however, such comparisons were not lamentations over the challenges of circumpolar healthcare delivery. Instead, they were expressions of what researchers in the Far North had to offer the rest of the world. In 1984, J.D. Martin, the Director of Canada's Indian and Inuit Health Services, wrote to C. Earl Albrecht, the founder of the Alaskan circumpolar health movement: "We in the North are the experts in

Indigenous health care delivery and have much to offer of a practical nature to underdeveloped countries."[86] For Alaskan health researchers, framing the Program as an innovative primary care initiative offered clear political and professional advantages.

Conclusions: from Indigenous bodies to Indigenous knowledge

Mala and his colleagues enthusiastically emphasized their interest in issues like healthcare delivery and the training of Indigenous healthcare workers. But, in the end, the Program failed to include any investigations into primary care initiatives. This was partly a consequence of Mala's approach to creating a research agenda, which might be best described as broad, haphazard, and optimistic. Mala was not a researcher himself, and was ultimately most interested in forming relationships, doing promotional work, and engaging in political negotiations. Once the political work to establish the Program was finished, its progress stalled. Members of the UAA's administration soon became concerned about the direction (or lack thereof) that the Program was moving in.[87]

In March 1988, Mala received a letter from the President of the UAA, Donald O'Dowd, warning him that the Program's budget was running low. Even more seriously, O'Dowd felt that it was not producing results: "I believe," he wrote, "it is essential that the program develop an explicit research focus in the immediate future." He then forcefully suggested that a senior scientist be brought on to expedite this process. O'Dowd had already contacted Sven O.E. Ebbesson, a Swedish-born evolutionary biologist, anatomist, neuroscientist, and professor at the University of Alaska-Fairbanks, about taking over the Program.[88] With his political goals achieved, Mala had already shifted his focus towards fundraising for the UAA's new Institute for Circumpolar Health Studies. He readily ceded control of the Program to Ebbesson.[89]

Under Ebbesson's direction, the Program sponsored a range of collaborative projects on subjects like the risk factors for diabetes among Siberian and Alaska Yup'ik Peoples, long-term "Framingham-type" studies of stroke and cardiovascular disease amongst Alaskan Native peoples, the genetics of alcoholism in Arctic Indigenous Peoples, and investigations (through DNA research) into the evolutionary relationship between Iñupiat, Aleut, and Athabaskan Peoples in Alaska.[90] The Program had certainly developed a more rigorous scientific agenda; its research teams would produce more than 70 scientific papers between the late 1980s and 2011.[91] Ebbesson was proud of this work and felt that it had helped "Alaska Natives with their research-neglected health problems."[92] Like Mala before him, Ebbesson framed the Program as an effort to improve the health of Alaskan Native peoples.

Regardless of the specific research focus of the Program, Indigenous health (as a goal) and Indigenous Peoples (as study populations) remained rhetorically central. Program researchers had instrumentalized indigeneity and Indigenous Peoples to build useful alliances, to garner support for their project, to bolster their expertise, and to contribute to an important diplomatic process. The Program's founders acknowledged the importance of Indigenous history and culture in Arctic regions, and remained adamant about the scientific usefulness of Indigenous bodies. But, however much Program researchers claimed their work was oriented towards improving Indigenous health, they also remained adamant about their own authority to determine which health problems were urgent and how they might best be addressed.

This approach would soon be held up as insufficient. Even as the Program was taking shape, members of the ICC were attending circumpolar health meetings and imploring researchers to take seriously Indigenous healing traditions. They also began to develop strategies to ensure that Alaskan Native peoples had a seat at the table when health research agendas were being set to ensure that their interests were better represented.[93] By the early 1990s, circumpolar health meetings included a full "Indigenous Program" that facilitated the involvement of health aides and other Indigenous healthcare workers and provided a space to focus on subjects like Indigenous self-determination in healthcare.[94] The Program's major accomplishment was improving the scientific dialogue between Alaska and Siberia, but it also helped to situate the circumpolar health community as one that claimed to be invested in Indigenous health. By the 1990s, Alaskan Native peoples would hold them accountable for these claims, and insist that caring about Indigenous health meant caring about Indigenous knowledge.

Notes

1. Linda A. Chamberlain, "Breaking the Ice links Alaska with Soviet Union," *Tundra Times,* 25 May 1987.
2. Ibid. The film was produced by Douglas Barry, an Assistant Professor of Journalism and Public Communications at the UAA and directed by Edward Guiragos, who was an Alaska-based freelance television and film producer.
3. Institute for Circumpolar Health Studies Records (ICHS), UAA-0086, Box 3, Series 5, Folder 4, University of Alaska-Anchorage, Archives and Special Collections (UAA-ASC).
4. ICHS, UAA-0086, Box 17, Series 1, Subgroup, 1 Folder 6, UAA-ASC.
5. Alaskan Performing Artists for Peace, 1985–1989, MS 0279, Scrapbook, Alaska State Library Historical Collections.
6. On the Red Power movement, see *Daniel M. Cobb, Native Activism in Cold War America: The Struggle for Sovereignty* (Lawrence: University of Kansas Press, 2008); on global pan-indigeneity, see Ronald Niezen, *The Origins of Indigenism: Human Rights and the Politics of Identity* (Berkeley: University of California Press, 2003).
7. On the movement that led to ANCSA, see: William L. Iggiagruk *Hensley, Fifty Miles From Tomorrow: A Memoir of Alaska and the Real People* (New York: Farrar, Straus & Giroux, 2009); Donald Mitchell, *Take My Land,*

Take My Life: The Story of Congress's Historic Settlement of Alaska Native Land Claims, 1960–1971 (Fairbanks: University of Alaska Press, 2001). This system of political organization generated strong but ambivalent feelings amongst Native Alaskan Peoples. Canadian judge Thomas Berger was appointed by the ICC in the early 1980s to assess the impact of ANCSA, see: Thomas Berger, *Village Journey: The Report of the Alaska Native Review Commission* (New York: Hill and Wang, 1985); for more contemporary perspectives on the limitations of ANCSA, see Thomas Michael Swensen, Of Subjection and Sovereignty: Alaska Native Corporations and Tribal Governments in the Twenty-First Century," *Wicazo Sa Review*, vol. 30, no 1 (Spring 2015): 100–117 and Eve Tuck, "ANCSA as X-Mark: Surface and Subsurface Claims of the Alaska Native Claims Settlement Act," in *Transforming the University: Alaska Native Studies in the 21st Century, proceedings from the Alaska Native Studies Conference 2013*, eds. Beth Ginondidoy Leonard, Jean Táaw xiwaa Breinig, Lenora Ac'aralek Carpluk, Sharon Chilux Lind, and Maria Shah Tláa Williams (Minneapolis: Two Harbors Press, 2014), 240–272.

8. Stephen Haycox, *Frigid Embrace: Politics, Economics and Environment in Alaska* (Corvalis, OR: Oregon State University press, 2002), 98.

9. Niezen, *Origins of Indigenism*.

10. Niezen, *Origins of Indigenism*. The politics of recognition, it should be noted, has provided few meaningful solutions for indigenous peoples, because it has little to say about questions of sovereignty. As a framework for multicultural liberalism, it also tends to create crises over the limits of tolerance and debates about what forms of difference should be declared untenable; see Elizabeth A. Povinelli, *The Cunning of Recognition: Indigenous Alterities and the Making of Australian Multiculturalism* (Durham: Duke University Press, 2002). On the limitations of recognition as a political strategy, see: Glen Sean Coulthard, *Red Skin, White Masks: Rejecting the Colonial Politics of Recognition* (Minneapolis, University of Minnesota Press, 2014).

11. For instance, see Matthew Farish, "The Lab and the Land: Overcoming the Arctic in Cold War Alaska," *Isis*, 2013, 104(1): 1–29. Matthew S. Wiseman identifies a contemporaneous trend in Canadian Arctic research, see Matthew S. Wiseman, "Unlocking the 'Eskimo Secret': Defense Science in the Cold War Canadian Arctic, 1947–1954," *Journal of the Canadian Historical Association*, 2015, 26(1): 191–223.

12. On the rise and fall of primary care, see Randall M. Packard, *A History of Global Health: Interventions into the Lives of Other Peoples* (Baltimore: Johns Hopkins University Press, 2016), 227–265; Marcos Cueto, "The ORIGINS of Primary Health Care and SELECTIVE primary Health Care," *American Journal of Public Health*, November 2004, 94(11): 1864–1874.

13. ICHS, UAA-0086, Box 17, Series 1, Subgroup, 1 Folder 6, UAA-ASC.

14. Charles Hillinger, "Documentary U.S. Doctors Pay a Soviet House Call," *Los Angeles Times*, 25 September 1990, pg. 3.

15. ICHS, UAA-0086, Box 17, Series 1, Subgroup, 1 Folder 6, UAA-ASC.

16. Lael Morgan, *Eskimo Star: From the Tundra to Tinseltown: The Ray Mala Story* (Epicenter Press, 2011).

17. Morgan, *Eskimo Star*.

18. ICHS, UAA-0086, Box 5, Series 2b, Folder 4, UAA-ASC.

19. ICHS, UAA-0086, Box 3, Series 4, Folder 1, UAA-ASC. Mala was involved with the center for Alcoholism and Addiction studies at the UAA, per Bernard Segal, interview by Tess Lanzarotta, August 2016, Anchorage, Alaska.

20. Theodore A. Mala, "The Alaska-Siberia Program: 24 Years in Retrospect," *Alaska Medicine*, April/May/June 2007, 49(2): 46.

21. Theodore A. Mala, interview by Tess Lanzarotta, August 2016, Anchorage, Alaska.
22. ICHS, UAA-0086, Box 5, Series 2b, Folder 4, UAA-ASC.
23. Bernard Lown, *Prescription for Survival: A Doctor's Journey to End Nuclear Madness* (San Francisco, CA: Berrett-Koehler Publishers, 2008).
24. ICHS, UAA-0086, Box 17, Series 1, Subgroup, 1 Folder 6, UAA-ASC.
25. ICHS, UAA-0086, Box 3, Series 5, Folder 1, UAA-ASC.
26. ICHS, UAA-0086, Box 3, Series 4, Folder 1, UAA-ASC.
27. ICHS, UAA-0086, Box 3, Series 4, Folder 1, UAA-ASC.
28. ICHS, UAA-0086, Box 3, Series 4, Folder 1, UAA-ASC.
29. Farish, "The Lab and the Land": 1-29. More generally, international bio-scientific projects throughout the second half of the twentieth century framed indigenous bodies and biospecimens as the keys to understanding human evolution and adaptation, for instance, see Jenny Reardon, *Race to the Finish: Identity and Governance in an Age of Genomics* (Princeton: Princeton University Press, 2005); Joanna Radin, *Life on Ice: A History of New Uses for Cold Blood* (Chicago: University of Chicago Press, 2017).
30. Matthew Farish, "Frontier Engineering: from the global to the body in the Cold War Arctic," *The Canadian Geographer*, 2006, 50(6): 179.
31. Farish, "The Lab and the Land," 5.
32. Farish, "The Lab and the Land," 5.
33. Cold War era scientists had framed indigenous biospecimens as evidence of "the uncontaminated normal standard by which the citizen of modernity could measure his own pollution by techno scientific society," and as tools to track the adaptations that humans had developed to particular environments, see Radin, *Life on Ice*, 6–7.
34. Dalee Sambo Dorough, interview by Tess Lanzarotta, August 2016, Anchorage, Alaska. Sambo Dorough was involved in the early stages of the founding of the ICC and by 1982 was the Director of the Alaska Office of the ICC, and Special Assistant to the ICC President, Mary Simon. She was speaking here on the ways that the members of the ICC understood the relationship between Arctic Indigenous Peoples across Alaska, Canada, Greenland, and the Russian Far East. See also Lisa Stevenson, "The Ethical Injunction to Remember," in *Critical Inuit Studies*, eds. Pamela Stern and Lisa Stevenson (Lincoln: University of Nebraska Press, 2006): 175.
35. ICHS, UAA-0086, Box 17, Series 1, Subgroup, 1 Folder 6, UAA-ASC; Geri DeHoog, Ted Mala, and Joseph Campanella, *Breaking the Ice: the Alaska-Siberia Medical Research Program*, directed by Guiragos (1987; Anchorage: University of Alaska-Anchorage), VHS; The first F-15's were brought to Alaska in 1982. They were state of the art technology, part of an effort to modernize the American military and specifically to better equip the Air Force to intercept enemy aircraft and engage in air combat. In *Breaking the Ice*, they may have served as a further reminder that Cold War tensions were actually increasing in Alaska in the early 1980s. See: *Cold War in Alaska: a resource guide for teachers and students* (Anchorage: U.S. National Park Service, Alaska Regional Office, National Historic Landmarks Program, 2014).
36. ICHS, UAA-0086, Box 17, Series 1, Subgroup, 1 Folder 6, UAA-ASC.
37. Philip Taubman, "Soviet Proposes Arctic Peace Zone," *New York Times*, 2 October 1987.
38. ICHS, UAA-0086, Box 17, Series 1, Subgroup, 1 Folder 6, UAA-ASC; Geri DeHoog, Ted Mala, and Joseph Campanella, *Breaking the Ice: the Alaska-Siberia Medical Research Program*, directed by Guiragos (1987; Anchorage: University of Alaska-Anchorage), VHS.

39. Andrew Stuhl, "The politics of the 'New North': putting history and geography at stake in Arctic futures," *The Polar Journal*, 2013, 3(1): 96.
40. Information about performances from ICHS, UAA-0086, Box 17, Series 1, Subgroup, 1 Folder 6, UAA-ASC. Identity of performers confirmed by Theodore A. Mala, interview by Tess Lanzarotta, August 2016, Anchorage, Alaska.
41. Richard Mauer, "'Ice Wall' falls to rock 'n' roll," *Juneau Empire*, 27 February 1989, pg. 2.
42. Hal Spencer, "At Last, Warm Glances Across the Bering Strait," *New York Times*, 14 July 1988; see also: Victor Fischer, with Charles Wohlforth, *To Russia with Love: An Alaskan's Journey* (Fairbanks: University of Alaska Press, 2012).
43. Winthrop Griffith, "Detente on the Rocks," *New York Times Magazine*, August 1975.
44. Ibid.
45. Ted Mala, "The Alaska-Siberia medical program: 24 years in retrospect," *Alaska Medicine* 49(2) (2007): 47.
46. Ted Mala, "A bond for health," *World Health Forum*, vol. 10 (1989): 62; quotation regarding Siberian population from ICHS, UAA-0086, Box 6, Series 8, Folder 4, UAA-ASC.
47. ICHS, UAA-0086, Box 3, Series 5, Folder 4, UAA-ASC.
48. Tuck, "ANCSA as X-Mark," 242.
49. Steveson, "The Ethical Injunction to Remember,"176; Dalee Sambo Dorough, interview by Tess Lanzarotta, August 2016, Anchorage, Alaska.
50. Steveson, "The Ethical Injunction to Remember," 176; Also see Marybelle Mitchell, *From Talking Chiefs to a Native Corporate Elite: The Birth of Class and Nationalism Among Canadian Inuit* (Montreal; Kingston: McGill-Queen's University Press, 1996).
51. Jessica Shadian, "In Search of an Identity Canada Looks North," *American Review of Canadian Studies*, 2007, 33(3): 340.
52. *Inuit Circumpolar Conference, June 1977* (Barrow, AK: North Slope Borough, 1977); on the early work of the ICC, see Dalee Sambo "Indigenous human rights: The role of the Inuit at the United Nations Working Group on Indigenous Peoples, *Études/Inuit/Studies*, 1992, 16(1/2): 27–32.
53. *Inuit Circumpolar Conference, Fourth General Assembly, July 28-August 3, 1986, Resolutions and Workshop Reports* (Kotzebue, AK; 1986).
54. ICHS, UAA-0086, Box 5, Series 2a, Folder 2, UAA-ASC; information on the Soviet delegation's itinerary from ICHS, UAA-0086, Box 6, Series 4, Folder 8, UAA-ASC.
55. See Hensley, *Fifty Miles From Tomorrow*; Anne Fienup-Riordan and Lawrence D. Kaplan, eds., *Words of the Real People: Alaska Native Literature in Translation* (Fairbanks: University of Alaska Press, 2007).
56. ICHS, UAA-0086, Box 5, Series 2a, Folder 2, UAA-ASC.
57. ICHS, UAA-0086, Box 5, Series 2a, Folder 2, UAA-ASC.
58. ICHS, UAA-0086, Box 5, Series 2a, Folder 2, UAA-ASC.
59. ICHS, UAA-0086, Box 6, Series 4, Folder 8, UAA-ASC.
60. C. Earl Albrecht Papers (CAB), HMC-0375, Box 17, Folder 1, UAA-ASC; Frank Pauls Papers (FPP), HMC-0489, Box 12, Series 11, Folder 7, UAA-ASC.
61. Sven O. E. Ebbesson, who took over the Program in 1988, later reflected that it was intended to generate "meaningful, Native-driven participatory research." He went on to describe the 70,000 home visits he made over 20 years to explain his research on diabetes, heart disease and nutrition to participants. These projects may well have been viewed as valuable by the communities that were involved. However, it is not clear what, if any, role Alaska Native peoples

had in developing the research agenda and study design, or played in performing the research. See Sven O.E. Ebbesson, "The Legacy of the Alaska Siberia Medical Research Program: A Historical Perspective," *International Journal of Circumpolar Health*, 2011, 70(5): 584–593.
62. On this see Povinelli, *The Cunning of Recognition*.
63. ICHS, UAA-0086, Box 17, Subgroup 3, Series 2, Folder 4, UAA-ASC.
64. Nikitin had a particular research interest in the different patterns of disease and mortality that appeared in populations of newcomers to the North compared to those found in indigenous populations, and in studying the genetic or "ethnic" variations that underpinned these patterns. See Yuri P. Nikitin, "Some Health Problems of Man in the Soviet Far North," in *Circumpolar Health 84: Proceedings of the Sixth International Symposium on Circumpolar Health*, ed. Robert Fortuine (Seattle: University of Washington Press, 1985), 8–10.
65. Nikitin, "Some Health Problems of Man."
66. Laura R. Graham and H. Glenn Penny, "Performing Indigeneity: Emergent Identity, Self-Determination, and Sovereignty," in *Performing Indigeneity: Global Histories and Contemporary Experiences*, eds. Laura R. Graham and H. Glenn Penny (Lincoln: University of Nebraska Press, 2014), 2–31.
67. James Clifford has elaborated on the ways that stating "We Exist" through performances and publications can become a "powerful political act" for indigenous communities who have been "long marginalized or made to disappear, physically and ideologically ..." See James Clifford, *Returns: Becoming Indigenous in the Twenty-First Century* (Cambridge: Harvard University Press, 2013), 224. Ann Fienup-Riordan has also referred to post-ANCSA heritage projects in Alaska as demonstrations of "conscious culture" that embody both Alaska Indigenous cultural resurgence and the entanglement of this resurgence in the structures of the colonial state. See Ann Fienup-Riordan, *Hunting Tradition in a Changing World: Yup'ik Lives in Alaska Today* (New Brunswick, NJ: Rutgers University Press, 2000) and Clifford, *Returns*.
68. ICHS, UAA-0086, Box 17, Series 1, Subgroup, 1 Folder 6, UAA-ASC.
69. On the AFN and their interest in improving rural healthcare, see for instance *First Annual Statewide Seminar on Rural Health, December 12, 13, 14, 1973* (Anchorage: The Federation, 1974) and Philip Nice and Walter Johnson, *The Alaska Health Aide Program: A Tradition of Helping Ourselves* (Anchorage: Institute for Circumpolar Health Studies, 1998), 31.
70. Nice and Johnson, *The Alaska Health Aide Program*, 32.
71. Nice and Johnson, *The Alaska Health Aide Program*, 18-19; ICHS, UAA-0086, Box 17, Series 1, Subgroup, 1 Folder 6, UAA-ASC.
72. Nice and Johnson, *The Alaska Health Aide Program*.
73. Nice and Johnson, *The Alaska Health Aide Program*.
74. In earlier decades, the Alaskan circumpolar health community had focused on infectious disease outbreaks, but by the 1980s had begun to think about health more holistically, in terms of inequality and social justice. This was in part due to the influence of the WHO's emphasis on primary care. I am grateful to John Middaugh, former Alaska State Epidemiologist for elaborating this point. John Middaugh in discussion with the author, May 2017.
75. Nice and Johnson, *The Alaska Health Aide Program; Mala*, "Alaska native 'grass roots' movement": 85.
76. Packard, *A History of Global Health*, 242–243.
77. "The new Bulletin," *Bulletin of the World Health Organization*, December 2008, 86 (12): 916.
78. Mala, "A bond for health," 62–64.

79. For instance, see Wilbur Hoff and D. Nhlavana Maseko, "Traditional Medi-cine: Nurses and traditional healers join hands," *World Health Forum*, 1986, 7: 412–416; Wilbur Hoff, "Traditional Healers and Community Health," *World Health Forum*, 1992, 13: 182–187.
80. Victor W. Sidel and Ruth Sidel, "Barefoot in China, the Bronx, and Beyond," in *Comrades in Health: U.S. Health Internationalists Abroad and at Home*, eds. Anne-Emmanuelle Born and Theodore M. Brown (New Brunswick, NJ: Rutgers University Press, 2013), 124.
81. Mala, "A bond for health," 64.
82. Rob Stapleton, "Mala Travels to Soviet Union," *Tundra Times*, 21 September 1987.
83. Stapleton, "Mala Travels to Soviet Union".
84. ICHS, UAA-0086, Box 17, Series 1, Subgroup, 1 Folder 6, UAA-ASC.
85. Mala also explained that healthcare delivery models developed elsewhere in the United States failed in Alaska, because they assumed the existence of a robust transportation and communications infrastructure that simply did not exist, see Mala, "A bond for health," 62; ICHS, UAA-0086, Box 3, Series 4, Folder 1, UAA-ASC.
86. CAB, HMC-0375, Box 16, Folder 2, UAA-ASC.
87. The Program's lack of focus on indigenous healthcare and Mala's lack of involvement in directing the research agenda were confirmed to me by Bernard Segal, who was then the Director of the UAA's Center for Alcohol-ism and Addiction Studies and a prolific Program participant: Bernard Segal, interview by Tess Lanzarotta, August 2016, Anchorage, Alaska. John Mid-daugh confirmed Mala's policy interests and lack of research experience: John Middaugh in discussion with the author, May 2017. Mala, too, confirmed that he considered political work and relationship-building a primary goal of the Program: Theodore A. Mala, interview by Tess Lanzarotta, August 2016, Anchorage, Alaska.
88. ICHS, UAA-0086, Box 1, Series 3, Correspondence March 1988, UAA-ASC.
89. ICHS, UAA-0086, Box 1, Series 3, Folder 29, UAA-ASC.
90. More details on these studies can be found in Ebbesson, "The Legacy of the Alaska Siberia Medical Research Program," 584–593.
91. Ebbesson, "Legacy," 584.
92. Ebbesson, "Legacy," 585.
93. Dalee Sambo Dorough, interview by Tess Lanzarotta, August 2016, Anchor-age, Alaska; the text of Sambo Dorough's speech at the 6th International Sym-posium on Circumpolar Health can be found in CAB, HMC-0375, Box 17, Folder 1, UAA-ASC. In the speech, Sambo Dorough uses the term "tradi-tional tribal medicine," but because of her current discomfort with this termi-nology, I have used the term "indigenous healing traditions." Documentation of the ICC's interest in ensuring that the interests of Indigenous Peoples were represented in circumpolar health research was provided to me by Dalee Sambo Dorough, in e-mail message to author, September 29, 2016.
94. In 1993, for instance, the 9th International Congress on Circumpolar Health, in Reykjavik, Iceland, included an entire day on the subject of "Transfer-ring Responsibility for Health and Health Services to Indigenous Peoples," see John Middaugh Papers, USUAFV6-562, Box 2, Alaska & Polar Regions Collections, University of Alaska-Fairbanks.

Part 5

Epilogue: global Cold War— the Antarctic and the Arctic

16 Antarctic science and the Cold War

Adrian Howkins

The chapters in this collection demonstrate that the Cold War exerted a profound influence on the development of Arctic science, even as this research shaped the conflict between capitalism and communism. This chapter turns the focus southwards and suggests that there was also a strong and mutually reinforcing relationship between scientific research and the Cold War at the opposite end of the planet. The history of Antarctica during this period shares much in common with the history of the Arctic. In both places, the need to understand polar environments became a geopolitical expediency and the extreme nature of these environments offered a stage for demonstrating an ideologically charged "mastery of nature." However, there were also significant differences between the histories of Cold War science in the two polar regions. In Antarctica science helped constrain superpower rivalries to a much greater extent than in the Arctic. In seeking to promote a "continent dedicated to peace and science," for example, the 1959 Antarctic Treaty created a genuinely international system for governing the southern continent in a way that never occurred in the Arctic.

In presenting an overview of the historical relationship between science and politics in Antarctica, this chapter seeks to encourage a comparative perspective on the relationship between science and the Cold War in the polar regions. Such a comparison is somewhat arbitrary, and any comparative analysis needs to remember that the two polar regions are very different places: that both are at high latitude does not inevitably imply other similarities in geographies and histories.[1] The Arctic is an ocean surrounded by continents, while Antarctic is a continent surrounded by oceans; the human history of the two regions is vastly different. But there is value in taking a comparative perspective since important historical connections have existed between the two polar regions, not least during the period of the Cold War. The International Geophysical Year (IGY) of 1957–58, for example, focused on both polar regions as well as outer space, and this helped to create connections between the Arctic and Antarctica.[2] Perhaps most famously, the polar ice core research pioneered at the US Station at Camp Century in Greenland reached an important milestone at the Soviet Union's Vostok Base in East Antarctica with the extraction of a 400,000 year old ice core that helped to demonstrate a correlation between temperature and atmospheric carbon dioxide going back at least four glacial cycles.[3]

The focus of this chapter will be on the activities and interests of the United States and the Soviet Union in Antarctica since the experiences of these two countries offer some of the most obvious connections between science and politics during the Cold War. The two superpowers were members of a fairly small group of countries, which also included Britain, Norway, and Japan that had fairly extensive interests in both polar regions in the second half of the twentieth century. More common is the experience of countries that have a major focus on one or the other of the polar regions. The history of Canada, for example, has been profoundly shaped by engagement with the Arctic. But although Canada is a non-consultative member of the Antarctic Treaty System and a number of Canadians have made important contributions to Antarctic science, the country cannot be said to be a major player in the history of the southern continent.[4]

The ability of the United States and the Soviet Union to draw upon experiences in one polar region to support their interests in the other helps to highlight the colonial character of science in both the Arctic and Antarctica. While the presence of Indigenous populations in the Arctic makes the colonial intentions of scientific research more explicit (and often a lot more problematic), science has often had a colonial function in Antarctica as well. The idea of Antarctic science as colonial science is certainly complicated by the fact that the continent has no Indigenous population, but a similar settler colonial mentality can often be identified in both polar regions.[5] In a strategy that might be labeled "environmental authority," the production of useful knowledge about Antarctica has been used to justify imperial claims to the continent.[6] In Antarctica the fact that states such as the United States and the Soviet Union could draw upon scientific research in the Arctic helped to support assertions of environmental authority by reinforcing the idea of scientific expertise. In the Arctic, the same thing happened in reverse, with the added consequence that any comparison with the empty lands of Antarctica also helped to reinforce the rhetorical depopulation of northern territories that was often at the heart of colonial efforts in the region.

As well as highlighting the colonial dimensions of polar science, a comparative perspective creates a powerful argument against an overly deterministic reading of the relation between Cold War politics and science. Cold War rivalries which could lead to hostility in certain places could lead to cooperation in others. These differences have various causes, including simple contingency and chance. But an obvious conclusion from looking at the history of Antarctica alongside the history of the Arctic is that location matters, that context matters, and that personality matters. In the face of such differences it becomes very difficult to assert that the Cold War exerted a singular influence on polar science, or that polar science had an easily definable impact on the Cold War. Adding a discussion of Antarctica into an understanding of the history of Cold War science in the Arctic helps to show that almost nothing was inevitable in these histories.

Perhaps even more so than the Arctic, our historical understanding of Antarctica is skewed towards the Western side of the Cold War, and significant gaps remain in our knowledge of Soviet Antarctic policy. Nevertheless, sources exist to develop a solid outline of the Cold War history of the southern continent.[7] The first section of this chapter provides some background to US and Soviet interests in Antarctica and examines the role played by science in superpower rivalry in the period immediately following the Second World War. The second and third sections look respectively at the IGY of 1957–1958 and the Antarctic Treaty of 1959. These events are often portrayed as important turning points in the history of Antarctica, although there were also significant continuities. The fourth section turns to the history of superpower relations in Antarctica under the auspices of the Antarctic Treaty System (ATS) and the fifth section examines the end of the Cold War in Antarctica and briefly considers how the history of science has changed since the collapse of the Soviet Union. The conclusion reflects on what can be learned from a comparative perspective.

Science and early superpower rivalry in Antarctica

As the global rivalry between the Soviet Union and the United States began to escalate in the years after the Second World War, the histories of the two superpowers' engagement with Antarctica were in some ways quite similar and in other ways quite different. Both states could draw upon involvement in the somewhat hazy early nineteenth century history of discovery in the Antarctic continent.[8] Neither country, however, had direct involvement in the so-called "heroic era" of Antarctic exploration in the early twentieth century, partly as a result of their active interests in the Arctic during this period. In the case of Russia and the newly created Soviet Union, this lack of direct interest in the southern continent would continue until after the Second World War. For the United States, in contrast, the interwar period saw a major expansion of Antarctic activities, as the country seemed to be making up for lost time in Antarctic exploration. US activities during this period focused on aerial exploration, including the high-profile flights of Richard E. Byrd and Lincoln Ellsworth to the South Pole and across the continent.[9] There were significant Arctic connections to many of these endeavors, with Byrd, for example, making the first flight to the South Pole after making a similar (although less convincing) claim to being the first to fly over the North Pole.

In the winter of 1934, Richard Byrd followed up his aerial exploits with an effort to spend a year living alone at a meteorological station on the Ross Ice Shelf, 200 miles south of the Little America Station.[10] While there was some scientific value to this enterprise in the form of the collection of weather data, it was largely driven by the quest to achieve another "Antarctic first." Byrd returned to Antarctica in the late 1930s to establish what he hoped would be permanent American stations at Little America on the Ross Sea (West Base) and Stonington Island in the Antarctic Peninsula region (East Base). Scientific research was very much part of these activities, and the aerial surveying of recent expeditions continued.[11] Unfortunately for Byrd's ambitions, however, the geopolitical imperatives of the Second World War necessitated the hasty evacuation of these two bases, and a permanent American presence on the Antarctic continent would have to wait for almost another twenty years.

In the aftermath of the Second World War, the US returned with the largest expedition ever to sail to the Antarctic continent, known as Operation Highjump.[12] With an eye to preparing for possible conflict with the Soviet Union in the Arctic, its primary purpose was to familiarize the US Navy with operations in polar waters. But alongside its strategic imperatives, Highjump also had a scientific agenda, largely focused on aerial exploration and mapping. The science of Operation Highjump was not unproblematic, and the scientists involved in the expedition felt themselves to be very much playing second fiddle to the naval officers.[13] The surveying, for example, was flawed by a lack of ground-truth points to make sense of the myriad of aerial photos that were taken during the expedition. This necessitated a follow up expedition the following season known as Operation Windmill that utilized helicopter transport to fill in some of the missing ground points. Writing about the expedition the British historian of Antarctic science G.E. Fogg noted somewhat caustically that "for an expedition which involved 4,700 men, 13 ships and nine aircraft, three scientific publications … was a poor yield."[14] But the simple fact that scientists were involved on what was essentially a naval expedition demonstrates the continued overlap between science and politics in the Antarctic context.

Operation Highjump and Operation Windmill seem to have provoked the Soviet Union to reengage with the Antarctic continent over 125 years after the Russian naval Captain Thaddeus Von Bellingshausen had made what would later be claimed to be the first recorded sighting of the Antarctic continent.[15] In the immediate postwar period, Soviet Antarctic activities were focused on whaling, making use of whaling vessels that had been acquired during and after the war.[16] Soviet whaling had connections to the Arctic, although by the mid-twentieth century the focus of the industry had shifted to the southern hemisphere. There was a scientific component to Soviet whaling activities, which sought to give what was essentially an economically motivated expedition a veneer of scientific legitimacy. While research into the whales themselves was the focus of this research, Soviet scientists also began to study broader questions in Antarctic oceanography. For the next several decades the Soviet Union would play a leading role in the study of the Southern Ocean.[17] This research had potential strategic value, and Soviet expertise in this area would cause significant anxiety among US intelligence officers and policy-makers.

Alongside their science, the Soviet Union also engaged in some traditional games of *realpolitik*. The focus of these trouble-making activities was the

Antarctic Peninsula region, where the sovereignty dispute among Britain, Argentina, and Chile offered a clear opportunity to destabilize an important part of the Western alliance, and put the United States in the unenviable position of mediating between Britain and its allies in South America.[18] In 1949, for example, S. Mikhailov published an article in a journal titled *The Problems of Economics,* which sought to explain British-American conflict in Antarctica through the communist theory of imperial competition:

> The present US struggle for domination on the Antarctic pursues above all the aim of ousting Britain from the positions which she has seized. The rivalry between the USA and Britain for the Antarctic is in addition a reflection of the struggle of these two imperialist robbers for South America.[19]

Although Anglo-American relations were not always particularly close in Antarctica during this period, the rivalry was nowhere near as fierce as Soviet propaganda implied. British officials did not seem overly concerned by such obvious attempts to create discord, although the article's mention of the "probable" existence of uranium and oil in Antarctica certainly resonated with Cold War sensitivities.[20]

In 1924 the United States had outlined its basic position on Antarctic sovereignty in what became known as the Hughes Doctrine.[21] In a statement, Secretary of State Charles Evans Hughes declared that the United States did not recognize any sovereignty claims in Antarctica, based on the fact that the same legal standards for demonstrating sovereignty in the rest of the world applied to the southern continent. Over time this statement developed into a fully-articulated reservation of US rights to the entire continent. In 1949, the Geographical Society of the Soviet Union issued a statement saying that the Soviet Union had a right to participate in Antarctic affairs and this also developed into an official position of reserving rights to the continent.[22] As a consequence of these positions, the two superpowers found themselves taking an almost identical stance on the question of Antarctic sovereignty. Unlike most other countries with active interests in Antarctica in the middle of the twentieth century, neither the United States nor the Soviet Union showed much interest in making a formal claim to the continent. Despite their many differences, their similar positions in respect to Antarctic law created a commonality that would play a role in shaping the subsequent history of the southern continent.

The International Geophysical Year

The International Geophysical Year (IGY) of 1957–58 played an important role in the mid-twentieth century histories of both the Arctic and Antarctica. Although there were a number of similarities in IGY research in both polar regions, there were also significant differences. Perhaps most importantly, the uncertain legal and political situation of Antarctica contrasted with the more settled sovereignty arrangements in the Arctic. As a consequence, most IGY science in the Arctic took place in clearly defined national territories while research in Antarctica occurred in a much less settled political context. This helped to give the Antarctic section of the IGY a genuinely international character that was often lacking in the Arctic, despite the overarching co-operative agenda. A number of scientists on both sides of the Cold War divide worked in both polar regions during the IGY, and equipment was used in both north and south. In 1955, for example, in an effort to learn more about cosmic rays, the US icebreaking vessel *Atka* took cosmic ray recording equipment on a voyage to Antarctica that had been used the previous year by the Canadian *Labrador* icebreaker in the Northwest Passage.[23]

The fact that the IGY has come to occupy a central place in the twentieth century political history of Antarctica tells us much about the continued and intensifying overlaps between science and politics in the southern continent. This international scientific endeavor had consequences that would endure well beyond its eighteenth months of official activity. The IGY offered both superpowers an ideal opportunity to engage fully and openly with the Antarctic continent through scientific research. With both the United States and the Soviet Union reserving their rights to the whole of Antarctica, neither country felt the need to focus their IGY activities on just the sectors they claimed, as the seven claimant countries did. As a consequence, the superpowers had the widest research scope and their contributions exceeded that of all ten other nations engaged in IGY Antarctic science combined.[24]

The Antarctic section of the IGY had a "Jekyll and Hyde" personality, characterized by rivalry and cooperation.[25] On the one hand, the opportunity to conduct research as part of a massive international research project offered the two superpowers an opportunity to demonstrate their scientific and technological capabilities to a global audience. This led to a barely disguised competition to do more interesting and more logistically difficult science than their rival. On the other hand, the international context of the IGY and its aim of furthering knowledge about the Earth's geophysical processes encouraged significant cooperation around the world. This was especially the case in Antarctica, where a shared sense of environmental hostility helped foster a spirit of collaborative endeavor. A Soviet scientist, for example, worked at "Weather Central" at the US Little America station, which received meteorological data from around the continent and used it to produce weather forecasts of practical utility for all participants involved in the Antarctic section of the IGY.[26]

The underlying political rivalries of the Cold War offered an unprecedented opportunity for Antarctic scientists to win funding for their research. While a top-down perspective on this history might suggest that scientists were just pawns in the geopolitical games of Cold War, a focus on the scientists themselves shows how skilled they often were at exploiting the superpower conflict to win funding and logistical support for their research. Near the top of the system on the US side, for example, Lloyd Berkner ceaselessly promoted the IGY to political decision-makers.[27] While the launch of the Soviet Union's *Sputnik* satellite in 1957 was not good news for the United States from a geopolitical point of view, US scientists realized that it would lead to more funding for their research.[28]

Policy-makers in Washington D.C. initially hoped that they could promote international science in Antarctica without the participation of the Soviet Union, but such efforts proved elusive. At a planning meeting in Paris in 1955 for the Antarctic section of the IGY, a Soviet delegation showed up to the meeting two days late, despite the fact that they had not been invited.[29] After apologizing for his tardiness, the head of the Soviet delegation, V.V. Beloussov, explained to the assembled scientists that his country would be building a scientific station at the Geographic South Pole. According to reports from the conference, this statement caused quite a stir, since the United States had already declared its intention to build a station at the world's most southerly point. Likely knowing this already, Beloussov quickly took the opportunity to make an important concession: if the United States wanted to go to the South Pole then the Soviet Union would instead focus its efforts on building a station at the so-called "Pole of Relative Inaccessibility," the point of the continent furthest from the coastline and therefore the most difficult place to reach. The Soviet Union had been successful in reaching the equivalent pole of relative inaccessibility in the Arctic, suggesting that the last-minute change of plan was less spontaneous than the conference theatrics might have led many to assume.[30]

Despite Soviet attempts to destabilize the relations among their Cold War rivals in the Antarctic Peninsula region, growing communist interest also created a degree of unity. At the Antarctic IGY planning meeting held in Paris in 1955, delegates from Argentina and Chile proposed a so-called "gentleman's agreement," through which the participating countries would agree that IGY science would have no bearing on the political situation of Antarctica.[31] This suggestion initially received a less than enthusiastic reception from other delegates at the meeting, but as soon as it became clear that the Soviet Union was going to participate in the IGY with a large Antarctic science program of its own, the South American proposal received approval. This rhetorical separation of science from politics—however artificial it was in practice—would prove crucial for the future political arrangements of Antarctica.

Much of the publicity surrounding involvement of the United States and the Soviet Union in the Antarctic section of the IGY focused on their respective efforts to construct stations at the Geographic South Pole and the Pole of Relative Inaccessibility.[32] Interestingly, the two Cold War superpowers went about these efforts in very different ways. The United States relied almost entirely on airplanes to transport materials and equipment to the South Pole to construct their station there. On a preliminary expedition, Admiral George Dufek became leader of only the third ever expedition (after Roald Amundsen and Robert Falcon Scott) to stand at the South Pole when his plane landed there on October 31, 1956.[33] Even if these efforts involved less physical discomfort than earlier expeditions, they highlighted American technological expertise and logistical capabilities as well as the courage of the pilots, crews, and passengers. In contrast, the Soviet Union utilized an overland tractor caravan to transport the necessary materials to build a station at the Southern Pole of Inaccessibility. Images of these hefty machines with hammer and sickle flags moving relentlessly across the ice fit neatly with stereotypes of dogged Soviet determination. But the decision to travel by land offers a curious comparison with the Arctic, where Soviet aviators flew to the Northern Pole of Inaccessibility, and suggests that perhaps a deliberate effort was being made to contrast with the Antarctic logistics of the United States.

At the same time as functioning as propaganda tools in the Cold War, efforts to get to the Southern Geographic Pole and the Pole of Relative Inaccessibility produced valuable data in various scientific fields.[34] Based on largely speculative comparisons with the Arctic, one important question going into the IGY was the depth of the Antarctic Ice sheet, with some glaciologists proposing a thick ice hypothesis and others a thin ice hypothesis.[35] Although many of the measurements of ice depth took place in other parts of the continent, US and Soviet efforts at the center of the continent contributed to demonstrating that Antarctica's ice sheet fitted into the "thick ice" model with an average depth of around 2 km (1.24 miles), and therefore contained much more ice than the Arctic. Important observations also took place in the field of atmospheric science, with the respective stations at the Geographic South Pole and Pole of Relative Inaccessibility allowing year-round weather data to be collected from the center of the continent for the first time. These data helped to complete a picture of a near-constant high-pressure system over the pole. Also connected to atmospheric science, although not immediately influential, measurements of atmospheric carbon dioxide at the South Pole during the IGY helped to show an upward trend of this important greenhouse gas.[36]

Beyond the prestige projects of building stations at the South Pole and the Pole of Relative Inaccessibility, much important science was done by US and Soviet researchers in other parts of the continent.[37] Of the two superpowers, the United States was more geographically spread across the whole continent, while the Soviet Union's activities were concentrated in the sector claimed by Australia (something that caused particular concern to this country[38]). However, the Soviet Union's ongoing work in the field of oceanography widened its geographic scope, and continued to cause concern to policy-makers in Washington D.C. The research conducted by both superpowers made major contributions to the rapid increase in scientific knowledge about the Antarctic continent. As well as furthering

knowledge about Antarctica for its own sake, these scientific developments would also have important implications for the political history of
the southern continent.

The Antarctic Treaty

The 1959 Antarctic Treaty created a "continent dedicated to peace and
science" and has provided the international governing structure for
Antarctica since its ratification in 1961. The success of the Antarctic Treaty
in keeping the peace, promoting science, and, more recently, protecting the
environment has made it something of a model international treaty (including somewhat speculatively for something similar in the Arctic), and it has
been well studied by political scientists and international lawyers, as well as
by historians.[39] Within Antarctic historiography, much of the debate focuses
on the precise nature of the relationship between the IGY of 1957–58 and the
Antarctic Treaty. Some scholars argue that the Antarctic Treaty emerged
from an overflowing of scientific idealism, offering a rare example of what
can happen when scientific cooperation is allowed to triumph over political
squabbles.[40] Other scholars suggest that the connection was a little more
complicated than this, pointing out, for example, the importance of the
results of IGY scientific research in creating the impression that Antarctica
was a continent with little immediate economic or strategic potential.[41] Few
Antarctic historians, however, would deny that the IGY was an important
precursor to the international treaty that has promoted peace and science in
the southern continent since its signing on 1 December 1959.

As well as being closely connected to the IGY, the Antarctic Treaty can also
be seen as a product of the Cold War, albeit a somewhat unusual product.
Certain US policy makers had initially hoped to exclude the Soviet Union
from the political future of the southern continent, but they quickly realized
that this was unfeasible and the Soviet Union was included in the invitations
sent out by the United States in 1958 to the other eleven nations that had participated in IGY Antarctic science.[42] All eleven nations accepted these invitations, and the negotiations that led to the Antarctic Treaty took place in
Washington D.C. over the course of 1958 and 1959. At no point during these
negotiations was it inevitable that a treaty would emerge that could satisfy
all twelve countries, and on several occasions a complete breakdown looked
possible. In addition to the overarching Cold War conflict, continued political tension in the Antarctic Peninsula among Britain, Argentina, and Chile
threatened to destabilize the talks. As it turned out, the participation of the
Soviet Union in the negotiations helped to create a "common enemy" that was
threatening enough to encourage a degree of compromise among the other
eleven participants, and this proved important for moving things forward.

The successful negotiation of the Antarctic Treaty owed much to contingency and favorable circumstances. But clever diplomacy did play a role.
At the heart of the treaty Article IV suspends all sovereignty claims and

reservations of rights (including those of the United States and the Soviet Union) for the duration of the treaty, neither recognizing nor denying them.[43] Without definitively solving the problem of sovereignty, this article provided a workable means of bringing disputing parties together. Other important provisions of the Antarctic Treaty include demilitarization, inspections, and provisions for regular meetings. Also notable in the context of the Cold War is Article V, which prohibits nuclear explosions on the continent and has been regarded by some as the world's first nuclear test ban treaty.[44]

Not only did scientific research play an important role in determining who was invited to participate in the Antarctic Treaty negotiations, but references to science can be found running through the text of the treaty itself. A central goal of the treaty is to promote science, and science in turn is used as a tool to promote political cooperation. Article IX of the treaty stipulates that a country must first be conducting "substantial scientific research" in Antarctica before it can become a full consultative member. In much the same way as the decision just to invite IGY participants to the treaty negotiations, this functions as an effective way of being selective about which countries can be involved without appearing overly exclusive. In terms of the Cold War, this clause led the Soviet Union to promote Antarctic science among its eastern bloc partners in an effort to get more communist nations into the Antarctic Treaty. In 1977, Poland would become the first non-original signatory to become a full consultative member of the Antarctic Treaty.

It has frequently been noted that the signing of the Antarctic Treaty took place at an opportune moment in the history of the Cold War. As a number of commentators have observed, had negotiations been delayed, the relative goodwill that permitted the treaty could have been dissipated by the U2 spy plane incident of 1960, or by the Cuban Missile Crisis two years later. However, it is also important to note that significant tensions existed in superpower relations in Antarctica, even at the time of the signing of the Antarctic Treaty. This international treaty was not so much a solution to these tensions, as an effective way of managing them. In managing conflict, the superpowers continued to turn to science, which was by now becoming a familiar tool in Antarctic diplomacy. In the years that followed the signing and ratification of the Antarctic Treaty, scientific research continued to play a central role in keeping the peace in the southern continent. Activities became much less threatening when they were associated with science, and the rhetoric of scientific internationalism functioned as a unifying force. The underlying rivalries did not go away, but they were effectively muffled by the frequently repeated mantra of "peace and science."

Cold War science in the Antarctic Treaty System

An important consequence of the agreement to channel geopolitical tensions into scientific research was that a lot of high quality science has taken place under the auspices of the Antarctic Treaty System. From the early 1960s until the late 1980s, much of the research was driven by the United States and the Soviet Union, as well as by other scientifically active countries such as Britain, Australia, and New Zealand. The creation of the Scientific Committee on Antarctic Research (SCAR, initially the "special" committee) in the late 1950s provided a useful organization for coordinating Antarctic science.[45] Formal collaboration between the superpowers was a rarity, and an inherent sense of competition and rivalry remained in place, but there was certainly a willingness to share data, participate in conferences, and exchange scientific personnel.

For the United States, Antarctica continued to be an important stage for the National Science Foundation's formative idea of science as an "endless frontier."[46] Connected to the national mythology of a "frontier spirit," scientific research in places like Antarctica offered a seemingly endless sphere for new discoveries and advances. In the aftermath of the IGY, a consolidation of stations took place, which ultimately led to US stations at McMurdo, the South Pole and Palmer Station, which was opened in 1968 on Anvers Island in the Antarctic Peninsula region. The naming of the new station after Nathaniel Palmer, the American sealer who claimed one of the first sightings demonstrates that the United States was continuing to draw upon its long history of involvement in the Antarctic. But at the same time, the United States was looking into the future. Construction of a nuclear power station at McMurdo Station, which opened in 1963, can be seen as an assertion of capitalist technological capability. Nukey Poo, as the power station became affectionately known, may not have been a violation of the letter of the Antarctic Treaty, but certainly seemed to contravene its anti-nuclear spirit until its closure in 1973 and dismantling in 1978.

For the Soviet Union, Antarctic science from the 1960s to 1980s functioned as a valuable complement to the research that was taking place in the Arctic. Portions of its polar fleet could be switched from the northern hemisphere to the southern hemisphere with the changing seasons, and provided continued opportunities for Antarctic oceanography. In terms of terrestrial research, the Soviet Union quickly closed its IGY station at the Pole of Relative Inaccessibility (where distance from the coast made logistics challenging), and focused much of its inland efforts on the Vostok station. Like the United States, the Soviet Union also constructed a station in the contested Antarctic Peninsula region in 1968, and like their Cold War rivals they also sought to draw upon history by naming it Von Bellingshausen Station in honor of their claim to having made a first sighting of the Antarctic continent. In a demonstration of continued Cold War tensions beyond the immediate US-Soviet rivalry, Chile responded to the opening of Bellingshausen Station by opening their own station on King George Island, only two hundred meters (650 feet) away.

Although not limited to the United States and the Soviet Union, the exchange of scientific personnel caught the Cold War imagination. Given the tensions of the period, the idea that an American scientist could live among communist scientists at a Soviet Station seemed quite incredible. Books such as Gilbert Dewart's *Antarctic Comrades: An American with the Russians in Antarctica* showed that with personal connections geopolitical rivalries could be set aside and laughed at.[47] The program of scientific exchanges was closely linked to Article VII of the Antarctic Treaty, which gave all member states the right to inspect all stations, installations, and equipment. Although sometimes overlooked in attempts to explain the effectiveness of the Antarctic Treaty System, the right of any member state to inspect any other member state at any time in any place gave both sides in the Cold War the confidence that their rival was not conducting activities in Antarctica that might contravene the Treaty. Exchange scientists played a valuable role in this process by conducting observations in a non-threatening manner.

Continued Cold War rivalry within the overarching framework of the Antarctic Treaty created favorable conditions for science to flourish. Progress was made in many of the fields that had been stimulated by the IGY. Work, for example, continued on the investigation of the depth of the Antarctic sheet and the topography of the underlying bedrock. Scientists from the United States and Britain took in a lead in this research, strengthening an important Cold War alliance in Antarctica that had not always been completely harmonious.[48] The Soviet Union continued to play an important role in Antarctic oceanography, although the United States continued to keep up. The maintenance of permanent stations around the Antarctic continent provided opportunities for the systematic collection of weather data, which greatly contributed to understanding the southern continent's atmosphere and climate.

Geology was perhaps the discipline that generated the most significant changes in perceptions of Antarctica in the two decades following the signing of the Antarctic Treaty. Geology had not officially been part of the IGY research agenda, in part as a result of its capacity to create suspicions and rivalries. At the time of the Antarctic Treaty negotiations, perceptions that the southern continent contained little exploitable mineral wealth were widespread and contributed to a willingness to compromise over the political future of Antarctica. But over the course of the 1960s and 1970s developments in the field of Antarctic geology began to create an upswing in analyses of the southern continent's economic value. In particular, broader acceptance of the theory of continental drift from the mid-1960s began to encourage speculation by association, as scientists and policy-makers looked at the mineral wealth of territories that had formerly been connected to Gondwanaland and suggested that this might indicate what lies under the Antarctic ice.[49] Interestingly, Soviet Antarctic scientists were slower than others to accept the theory of continental drift.[50] But this did not stop the Soviet Union from speculating about the potential mineral wealth of the Antarctic continent.

Drilling in the Ross Sea in the early 1970s suggested evidence of hydrocarbon resources.[51] These developments took place at a time when increased mineral resource activity in the Arctic seemed to be demonstrating the potential economic viability of the polar regions.[52] Although nothing conclusive was found, within the next several years the Antarctic Treaty Consultative Meetings began to discuss the question of mineral exploitation, and in the 1980s these discussions moved into formal negotiations of a minerals regime for the Antarctic Continent. Both the United States and the Soviet Union were broadly supportive of these developments, in part because they saw themselves as well placed to take advantage of the developing arrangements. There was a shared sense that a mineral regime should be put in place before the issue of resource extraction became an immediate concern. This attests to the desire from all Antarctic Treaty signatories—including the United States and the Soviet Union—to maintain and strengthen the existing legal regime in the southern continent.

In the event, however, the minerals regime never came into effect. Environmentalist pressure from organizations such as Greenpeace and the Antarctic and Southern Ocean Coalition (ASOC), combined with other economic and political pressure convinced France and Australia not to sign the recently negotiated agreement.[53] Especially influential in this policy shift was the *Exxon Valdez* oil spill in Alaska, which appeared to show the particular vulnerability of the polar regions to environmental disaster.[54] When New Zealand announced in 1990 that it would not ratify the minerals regime, the convention effectively became redundant. In place of a minerals regime, the consultative members of the Antarctic Treaty put in place an Environmental Protocol to the Antarctic Treaty. Signed in 1991 and ratified in 1997, the "Madrid Protocol," as it became popularly known, put in place a number of measures to protect the Antarctic environment, including a complete prohibition of economic mineral activities for as long as the protocol remains in effect (an unspecified period, but with an option for these provisions to be discussed again after fifty years). The United States demonstrated its lack of enthusiasm for the Madrid Protocol by being the last country to ratify it. The Soviet Union, meanwhile, was distracted by events closer to home as the entire communist system began to disintegrate.

The end of the Cold War

In an interesting chronological coincidence, the collapse of the minerals negotiations in Antarctica and its replacement with the Madrid Protocol coincided with the fall of communism in Eastern Europe and the breakup of the Soviet Union. While the failure of the minerals regime was mostly driven by the environmental, economic and political concerns of France and Australia rather than by either of the superpowers, it is interesting to speculate about whether a more stable and robust Soviet Union might have done more to try to keep options open for mineral activities in the southern continent. The Soviet Union had something to gain and little to lose from economic mineral activities, and even the potential for resource extraction to sow discord may have worked to its advantage given that increased tensions would likely affect its Cold War rivals.

From at least the early 1980s, the economic struggles of the Soviet Union were being reflected in the reduction of resources that could be dedicated to Antarctic science. Among the most important contributions that the Soviet Union made to Antarctic Science in its final years of existence was the drilling of an ice-core at Vostok station, which had been built during the IGY. Vostok was located on one of the more stable accumulation zones of the East Antarctic Ice sheet, meaning that the ice underneath it formed consistent layers. Using a technique developed in the Arctic by the Danish Glaciologist Willi Dansgaard, investigations on the trapped air in this ice helped to demonstrate a correlation between periods of high carbon dioxide concentration and high temperatures in the Earth's climatic history.[55] This in turn suggested that the upward trends in atmospheric CO_2, which had been recorded at the South Pole along with other locations, could raise temperatures as predicted by laboratory models of greenhouse gases. As the two locations in the world with the most ice, the Arctic and Antarctica came to be seen as vulnerable to warming temperatures, with increased fears that melting polar ice could raise sea levels around the world.

Largely as a result of the importance of the Vostok drilling site for climate change research, there was a willingness among the wider Antarctic science community to help the Soviet Union—and then Russia—continue its drilling work at Vostok. First France and then the United States stepped in to offer financial support and technical assistance with the Vostok drilling program. This international collaboration suggests that science had taken on something of a life of its own, highlighted by a photo showing Russian, French and US flags at Vostok station (Figure 16.1). But taking a longer view of the history of connections between science and politics in the Antarctic continent suggests that efforts to support ice core research at Vostok Station were not purely scientific. Political power continued to underlie scientific research, and France and the United States were almost certainly well aware of the propaganda value of being seen to help out their beleaguered rival. At the end of the Cold War, just as at the beginning, Antarctic science and politics continued to be inextricably connected.

The reduction in Communist bloc activities brought about by the end of the Cold War was neither total nor permanent. Russia, in particular, maintained stations across the Antarctic continent in a continued demonstration of its political and scientific interests. The construction of a Russian Orthodox Church at Bellingshausen station in February 2004 represented a change in the form but not the substance of Russian Antarctic ambitions.[56] The newly independent nation of Ukraine retained an interest in Antarctic science through the purchase of Britain's Faraday Station in the Antarctic Peninsula. In another demonstration of the rhetorical value of science in Antarctica, Faraday was renamed as Vernadsky station in honor of the famous Ukrainian ecologist. Other former communist bloc

Figure 16.1 Ice core researchers from the Soviet Union, United States and France
at Vostok Station, Antarctica, c.1990.

Source: Public Domain (See https://commons.wikimedia.org/wiki/File:GISP2_team_photo_core37.
jpeg#/media/File:GISP2_team_photo_core37.jpeg).

countries such as Poland and Bulgaria continued to forge their own way
in Antarctica.

US interests in Antarctica have continued since the end of the Cold War,
but there has been no significant expansion of their activities. While the science itself has continued to advance in response to new scientific questions
and new techniques, the logistical capacity of the US Antarctic program has
stagnated somewhat, as demonstrated by recurrent problems with maintaining its ice-breaking vessels. The United States remains the preeminent
power in Antarctica, but its position is being challenged by the emergence of
so-called "New Polar Powers," including China and India.[57] China, in particular, has demonstrated an enthusiasm to extend its growing international
influence into the polar regions, and is increasingly coming to rival the United
States in terms of its Antarctic spending and infrastructure, if not yet its scientific output. It remains to be seen whether these new geopolitical challenges
will spur the United States to invest more in its Antarctic science program.

Conclusions

An interesting consequence of the end of the Cold War in Antarctica was
that it made available a fleet of ice-strengthened and ice-breaking vessels that
helped to stimulate the tourism industry in the southern continent. With the

post-Soviet state no longer able to afford their maintenance, the ships could be leased by tour companies to take tourists to the Antarctic continent. Attracted to the southern continent for a variety of reasons—not least of which its "last wilderness" mystique—Antarctic tourism has proved to be a lucrative industry, and tourist numbers have been going up year by year.[58] The fact that many of the same vessels work in the Arctic during the northern summer and the Antarctic during the southern summer creates another link between the two polar regions. Ships that had helped create scientific and geopolitical connections between north and south during the Cold War now create commercial connections. Direct links such as these provide an obvious justification for examining the history of Antarctica alongside the history of the Arctic.

In addition to the direct links between Antarctica and the Arctic there are a number of interesting similarities in their historical experiences. During the Cold War, both polar regions became objects of superpower attention and rivalry. The perceived hostility of polar environments created places where the superiority of the respective capitalist and communist systems could be tested and proved. This in turn connected to imperial visions of the polar regions as places to be conquered and colonized. Cold War tensions created fascinating overlaps between science and geopolitics in both polar regions. With science seen as offering strategic advantage, superpower tensions served to stimulate scientific research in Antarctica as well as the Arctic. Scientific investment undoubtedly produced geopolitical benefits, but these were almost always connected to advances in "pure science" as well. The study of climate change would be an obvious contemporary example, and a strong case could be made that we would have less understanding of a warming planet if it had not been for the stimulus that the Cold War provided for research in the polar regions.

Despite the historical links and similarities between the two polar regions there are also significant and ongoing differences. In Antarctica science functioned as much as a brake on Cold War tensions as it did an accelerator. The IGY created a framework for scientific cooperation across the southern continent, which was institutionalized by the 1959 Antarctic Treaty. It is easy to over-idealize the Antarctic Treaty, and to see it as a triumph of science over politics when in reality the politics never went away. But the Antarctic Treaty nevertheless represents a rare example of superpower cooperation, at a time when almost everywhere else in the world—including the Arctic—rivalry was the dominant trend. Despite the subsequent setting up of an Arctic Council to encourage cooperation in the far north, this system remains very different from the Antarctic Treaty. State sovereignty remains dominant in the north, whereas the political arrangements in the south might be described as a form of limited internationalization. In contemporary Antarctica, science is by far the predominant activity (despite the inroads being made by tourism), whereas in the Arctic science is just one activity among many. Despite the Cold War fantasies of populating Antarctica, even today nobody lives permanently on the southern continent, whereas the Arctic remains very much a lived-in region.

Putting a short summary of the Cold War history of science in Antarctica alongside a collection of chapters on the Cold War history of science in the Arctic reveals significant connections, similarities, and differences in the experiences of the two regions. The differences, in particular, are useful for raising new questions and offering new perspectives. Perhaps, for example, was there a little more cooperation in Cold War Arctic science than is sometimes assumed? Did rivalries and mutual suspicion continue to run more deeply in Antarctica than official narratives admit? The most valuable contribution of a comparative framing, however, is that it encourages us to identify and acknowledge both the similarities and the differences. By doing this, it quickly becomes apparent that there was no single trend in Cold War polar science: in some places science could exacerbate geopolitical tensions, elsewhere it could help to relax hostilities. Such observations discourage any sense of determinism or inevitability in the history, and instead encourage careful research and nuanced interpretation. The Cold War had a profound influence on the history of science in both polar regions, and science shaped the Cold War in both north and south. But these histories played out very differently in different places, and sweeping generalizations are difficult to make.

Notes

1. For a fuller discussion of these differences, see Adrian Howkins, *The Polar Regions: An Environmental History* (Cambridge, UK; Malden, MA: Polity, 2015).
2. An interesting contemporary account of the IGY is Walter Sullivan, *Assault on the Unknown; the International Geophysical Year*, (New York: McGraw-Hill, 1961).
3. Spencer R. Weart, *The Discovery of Global Warming*, Rev. and expanded ed. (Cambridge, Mass.: Harvard University Press, 2008).
4. For an interesting discussion of Canadian scientific work in Antarctica see Dean Beebe, *In a Crystal Land: Canadian Explorers in Antarctica*, (Toronto: University of Toronto Press, 1994).
5. Adrian Howkins, "Appropriating Space: Antarctic Imperialism and the Mentality of Settler Colonialism," in *Making Settler Colonial Space: Perspectives on Race, Place and Identity*, eds. Tracey Banivanua Mar and Penelope Edmonds (Houndmills, Basingstoke: Palgrave Macmillan, 2010), 29–52.
6. Adrian Howkins, *Frozen Empires: An Environmental History of the Antarctic Peninsula*, (New York, NY: Oxford University Press, 2017).
7. An ideologically influenced discussion of early Soviet activities in Antarctica is provided in Nikolai Andreevich Gvozdetskii, *Soviet Geographical Explorations and Discoveries*, (Moscow: Progress Publishers, 1974). More recent insights into Soviet Antarctic policy are provided in works such as Irina Gan, "'Will the Russians Abandon Mirny to the Penguins after 1959 ... Or Will They Stay?'," *Polar Record*, 2009, 45(2); "The Reluctant Hosts: Soviet Antarctic Expedition Ships Visit Australia and New Zealand in 1956," *Polar Record* 2009, 45(1); "Soviet Antarctic Plans after the International Geophysical Year: Changes in Policy," *Polar Record*, 2010, 46(3); "Russia, the Post-Soviet World

and Antarctica," in *The Emerging Politics of Antarctica* ed. Anne-Marie Brady (Abingdon: Routledge, 2013); V. V. Lukin, "Russia's Current Antarctic Policy," *The Polar Journal*, 2014, 4(1).

8. See, for example, Rip Bulkeley, "Bellingshausen's First Accounts of His Antarctic Voyage of 1819–1821," *Polar Record*, 2013, 49(1).

9. These expeditions are discussed in biographies such as Beekman H. Pool, *Polar Extremes: The World of Lincoln Ellsworth* (Fairbanks, Alaska: University of Alaska Press, 2002); Lisle Abbott Rose, *Explorer: The Life of Richard E. Byrd* (Columbia: University of Missouri Press, 2008).

10. Richard Evelyn Byrd, *Alone* (New York: G. P. Putnam's sons, 1938).

11. Discussed in Rose, *Explorer: The Life of Richard E. Byrd*.

12. G. E. Fogg, *A History of Antarctic Science, Studies in Polar Research* (Cambridge: Cambridge University Press, 1992), 166–67.

13. Fogg, *A History of Antarctic Science*, 166.

14. Fogg, *A History of Antarctic Science*, 167.

15. Thaddeus Bellingshausen and Frank Debenham, *The Voyage of Captain Bellingshausen to the Antarctic Seas 1819–1821* (London: Hakluyt Society, 1945).

16. See also J. N. Tønnessen and Arne Odd Johnsen, *The History of Modern Whaling* (Berkeley: University of California Press, 1982).

17. Fogg, *A History of Antarctic Science*, 214–15.

18. Discussions of the sovereignty dispute in the Antarctic Peninsula include Pablo Fontana, *La Pugna Antártica, El Conflicto Por El Sexto Continente:1939–1959* (Buenos Aires: Guazuvira Ediciones, 2014); Howkins, *Frozen Empires: An Environmental History of the Antarctic Peninsula*.

19. British Embassy Moscow (Chancery) to Northern Department FO, 17 December 1949. The National Archives, London, FO 371/74760.

20. British Embassy Moscow (Chancery) to Northern Department FO, 17 December 1949.

21. H.R. Hall, "The 'Open Door' into Antarctica: An Explanation of the Hughes Doctrine," *Polar Record*, 1989, 25(153).

22. Donald Rothwell, *The Polar Regions and the Development of International Law* (Cambridge: New York: Cambridge University Press, 1996), 65.

23. Walter Sullivan, *Quest for a Continent* (New York: McGraw-Hill, 1957), 322.

24. Sullivan, *Quest for a Continent*.

25. See, for example, Adrian Howkins, "Science, Environment, and Sovereignty: The International Geophysical Year in the Antarctic Peninsula Region," in *Globalizing Polar Science: Reconsidering the International Polar and Geophysical Years*, ed. Roger D. Launius, James Rodger Fleming, and David H. DeVorkin (New York: Palgrave Macmillan, 2010).

26. Fogg, *A History of Antarctic Science*, 303.

27. Allan A. Needell, *Science, Cold War and the American State: Lloyd V. Berkner and the Balance of Professional Ideals* (Amsterdam: Harwood Academic, 2000).

28. Roger D. Launius, "Sputnik and the Origins of the Space Age," *NASA*, https://history.nasa.gov/sputnik/sputorig.html.

29. An excellent discussion of this episode can be found in Oscar Pinochet de la Barra, *Medio Siglo De Recuerdos Antárticos: Memorias*, 1. ed. (Santiago de Chile: Editorial Universitaria, 1994).

30. See discussion in Howkins, *The Polar Regions:An Environmental History*.

31. Howkins, *Frozen Empires: An Environmental History of the Antarctic Peninsula*.

32. See, for example, Consuelo León Wöppke et al., *La Antártica Y El Año Geofísico Internacional. Percepciones Desde Fuentes Chilenas, 1954–58* (Valparaíso: Editorial Puntángeles Universidad de Playa Ancha, 2006).

33. George John Dufek, *Operation Deepfreeze*, 1 ed. (New York: Harcourt, 1957).
34. For a useful summary, see Richard S. Lewis, *A Continent for Science: the Antarctic Adventure* (New York: Viking Press, 1965).
35. Frank A. Simpson, *The Antarctic Today: A Mid-Century Survey by the New Zealand Antarctic Society* (Wellington: A.H. and A.W. Reed in conjunction with the NZ Antarctic Society, 1952).
36. Weart, *The Discovery of Global Warming*.
37. Sullivan, *Quest for a Continent*.
38. Gan, "The Reluctant Hosts."
39. For an interesting overview of the Antarctic Treaty, see Klaus Dodds, "50 Years On: Invited Reflections on the Antarctic Treaty," *Polar Record*, 2010, 46(236): 1–20. For an example of the Antarctic Treaty being used as a potential model for the Arctic see Timo Koivurova, "Environmental Protection in the Arctic and Antarctic: Can the Polar Regimes Learn from Each Other?," *International Journal of Legal Information*, 2005, 33.
40. An example of this approach is Paul Arthur Berkman, *Science Diplomacy: Antarctica, and the Governance of International Spaces* (Washington, D.C.: Smithsonian Institution Scholarly Press, 2011).
41. Howkins, *Frozen Empires: An Environmental History of the Antarctic Peninsula*.
42. Howkins, *Frozen Empires*.
43. For an overview of the Antarctic Treaty from a legal perspective, see, for example, Francisco Orrego Vicuña, *Derecho Internacional De La AntáRtida*, 1. ed. (Santiago: Dolmen Ediciones, 1994).
44. See, for example, Ryan Musto, "Cold Calculations: The United States and the Creation of Antarctica's Atom-Free Zone," *Diplomatic History*, 2017, 42(4).
45. D. W. H. Walton, Peter Clarkson, and C. P. Summerhayes, *Science in the Snow: Fifty Years of International Collaboration Through the Scientific Committee on Antarctic Research* (Cambridge: Scientific Committee on Antarctic Research, 2011).
46. A copy of this report can be found on the National Science Foundation's website: https://www.nsf.gov/od/lpa/nsf50/vbush1945.htm [accessed 8 May 2018].
47. Gilbert Dewart, *Antarctic Comrades: An American with the Russians in Antarctica* (Columbus: Ohio State University Press, 1989).
48. See, for example, Simone Turchetti et al., "On Thick Ice: Scientific Internationalism and Antarctic Affairs, 1957–1980," *History and Technology*, 2008, 24(4).
49. James D. Hansom and John E. Gordon, *Antarctic Environments and Resources: A Geographical Perspective* (New York: Longman, 1998), 32.
50. Fogg, *A History of Antarctic Science*, 260.
51. Hansom and Gordon, *Antarctic Environments and Resources,* 32.
52. See, for example, Peter Coates, *The Trans-Alaska Pipeline Controversy: Technology, Conservation, and the Frontier* (Bethlehem: Lehigh University Press, 1991).
53. Hansom and Gordon, *Antarctic Environments and Resources,* 284.
54. Howkins, *The Polar Regions: An Environmental History*.
55. Weart, *The Discovery of Global Warming*.
56. Lukin, "Russia's Current Antarctic Policy."
57. Anne-Marie Brady, *The Emerging Politics of Antarctica*, Routledge Advances in International Relations and Global Politics (London; New York: Routledge, 2013).
58. See the annual statistics available on the IAATO webpage: https://iaato.org/home [accessed 8 May 2018].

Index

Note: *Italicized* page numbers refer to figures, **bold** page numbers refer to tables